AF719826

Computational Methods and Experimental Measurements XIX

&

Earthquake Resistant Engineering Structures XII

WIT*PRESS*

WIT Press publishes leading books in Science and Technology.
Visit our website for new and current list of titles.
www.witpress.com

WIT*eLibrary*

Home of the Transactions of the Wessex Institute.
The WIT eLibrary provides the international scientific community with immediate and permanent access to individual papers presented at WIT conferences.
Visit the WIT eLibrary at www.witpress.com.

NINETEENTH INTERNATIONAL CONFERENCE ON
COMPUTATIONAL METHODS AND EXPERIMENTAL MEASUREMENTS
&
TWELFTH INTERNATIONAL CONFERENCE ON EARTHQUAKE RESISTANT
ENGINEERING STRUCTURES

CMEM XIX & ERES XII

CONFERENCE CHAIRMEN

FOR CMEM XIX

W. P. De Wilde
Free University of Brussels, Belgium
Member of WIT Board of Directors

G. M. Carlomagno
University of Naples "Federico II", Italy

FOR ERES XII

W. P. De Wilde
Free University of Brussels, Belgium
Member of WIT Board of Directors

INTERNATIONAL SCIENTIFIC ADVISORY COMMITTEE

FOR CMEM XIX

R. Cerny
K. El Khamlichi Drissi
J. Everett
M. Goto
M. Hadfield
H. Horii
S. Ivorra
Y. Jaluria
I. Kotchergenko
G. Manos
E. Menga
D. Poljak
H. Sakamoto
S. Syngellakis
Y. Villacampa
B. Wang
G. Zappala
V. Zhmud

FOR ERES XII

F. J. Baeza
M. Favvata
S. Grasso
Y. Ibrahim
S. Ivorra
Y. Kimura
J. U. Klugel
M. Phocas
M. C. Porcu
L. Romera
I. Roselli
S. Syngellakis
A. D. Tsonos

CMEM XIX ORGANISED BY

Wessex Institute, UK

University of Naples "Federico II", Italy

ERES XII ORGANISED BY

Wessex Institute, UK

CMEM XIX SPONSORED BY

WIT Transactions on Engineering Sciences

International Journal of Computational Methods and Experimental Measurements

ERES XII SPONSORED BY

WIT Transactions on the Built Environment

The International Journal of Safety and Security Engineering

WIT Transactions

Wessex Institute
Ashurst Lodge, Ashurst
Southampton SO40 7AA, UK

Senior Editors

H. Al-Kayiem
Universiti Teknologi PETRONAS, Malaysia

G. M. Carlomagno
University of Naples Federico II, Italy

A. H-D. Cheng
University of Mississippi, USA

J. J. Connor
Massachusetts Institute of Technology, USA

J. Th M. De Hosson
University of Groningen, Netherlands

P. De Wilde
Vrije Universiteit Brussel, Belgium

N. A. Dumont
PUC-Rio, Brazil

A. Galiano-Garrigos
University of Alicante, Spain

F. Garzia
University of Rome "La Sapienza", Italy

M. Hadfield
University of Bournemouth, UK

S. Hernández
University of A Coruña, Spain

J. T. Katsikadelis
National Technical University of Athens, Greece

J. W. S. Longhurst
University of the West of England, UK

E. Magaril
Ural Federal University, Russia

S. Mambretti
Politecnico di Milano, Italy

W. J. Mansur
Federal University of Rio de Janeiro, Brazil

J. L. Miralles i Garcia
Universitat Politècnica de València, Spain

G. Passerini
Università Politecnica delle Marche, Italy

F. D. Pineda
Complutense University, Spain

D. Poljak
University of Split, Croatia

F. Polonara
Università Politecnia delle Marche, Italy

D. Proverbs
Birmingham City University, UK

T. Rang
Tallinn Technical University, Estonia

G. Rzevski
The Open University, UK

P. Skerget
University of Maribor, Slovenia

B. Sundén
Lund University, Sweden

Y. Villacampa Esteve
Universidad de Alicante, Spain

P. Vorobieff
University of New Mexico, USA

S. S. Zubir
Universiti Teknologi Mara, Malaysia

Editorial Board

B. Abersek University of Maribor, Slovenia
Y. N. Abousleiman University of Oklahoma, USA
G. Alfaro Degan Università Roma Tre, Italy
K. S. Al Jabri Sultan Qaboos University, Oman
D. Almorza Gomar University of Cadiz, Spain
J. A. C. Ambrosio IDMEC, Portugal
A. M. Amer Cairo University, Egypt
S. A. Anagnostopoulos University of Patras, Greece
E. Angelino A.R.P.A. Lombardia, Italy
H. Antes Technische Universitat Braunschweig, Germany
M. A. Atherton South Bank University, UK
A. G. Atkins University of Reading, UK
D. Aubry Ecole Centrale de Paris, France
H. Azegami Toyohashi University of Technology, Japan
J. M. Baldasano Universitat Politecnica de Catalunya, Spain
J. Barnes University of the West of England, UK
J. G. Bartzis Institute of Nuclear Technology, Greece
S. Basbas Aristotle University of Thessaloniki, Greece
A. Bejan Duke University, USA
M. P. Bekakos Democritus University of Thrace, Greece
G. Belingardi Politecnico di Torino, Italy
R. Belmans Katholieke Universiteit Leuven, Belgium
D. E. Beskos University of Patras, Greece
S. K. Bhattacharyya Indian Institute of Technology, India
H. Bjornlund University of South Australia, Australia
E. Blums Latvian Academy of Sciences, Latvia
J. Boarder Cartref Consulting Systems, UK
B. Bobee Institut National de la Recherche Scientifique, Canada
H. Boileau ESIGEC, France
M. Bonnet Ecole Polytechnique, France
C. A. Borrego University of Aveiro, Portugal
A. R. Bretones University of Granada, Spain
F-G. Buchholz Universitat Gesanthochschule Paderborn, Germany
F. Butera Politecnico di Milano, Italy
W. Cantwell Liverpool University, UK
C. Capilla Universidad Politecnica de Valencia, Spain
D. J. Cartwright Bucknell University, USA
P. G. Carydis National Technical University of Athens, Greece
J. J. Casares Long Universidad de Santiago de Compostela, Spain
A. Chakrabarti Indian Institute of Science, India
F. Chejne National University, Colombia
J-T. Chen National Taiwan Ocean University, Taiwan
J. Chilton University of Lincoln, UK
C-L. Chiu University of Pittsburgh, USA
H. Choi Kangnung National University, Korea
A. Cieslak Technical University of Lodz, Poland
C. Clark Wessex Institute, UK
S. Clement Transport System Centre, Australia
M. C. Constantinou State University of New York at Buffalo, USA
M. da C Cunha University of Coimbra, Portugal
W. Czyczula Krakow University of Technology, Poland
L. D'Acierno Federico II University of Naples, Italy
M. Davis Temple University, USA
A. B. de Almeida Instituto Superior Tecnico, Portugal
L. De Biase University of Milan, Italy
R. de Borst Delft University of Technology, Netherlands
G. De Mey University of Ghent, Belgium
A. De Naeyer Universiteit Ghent, Belgium
N. De Temmerman Vrijie Universiteit Brussel, Belgium
D. De Wrachien State University of Milan, Italy
L. Debnath University of Texas-Pan American, USA
G. Degrande Katholieke Universiteit Leuven, Belgium
S. del Giudice University of Udine, Italy
M. Domaszewski Universite de Technologie de Belfort-Montbeliard, France

K. Dorow Pacific Northwest National Laboratory, USA
W. Dover University College London, UK
C. Dowlen South Bank University, UK
J. P. du Plessis University of Stellenbosch, South Africa
R. Duffell University of Hertfordshire, UK
A. Ebel University of Cologne, Germany
V. Echarri University of Alicante, Spain
K. M. Elawadly Alexandria University, Egypt
D. Elms University of Canterbury, New Zealand
M. E. M El-Sayed Kettering University, USA
D. M. Elsom Oxford Brookes University, UK
F. Erdogan Lehigh University, USA
J. W. Everett Rowan University, USA
M. Faghri University of Rhode Island, USA
R. A. Falconer Cardiff University, UK
M. N. Fardis University of Patras, Greece
A. Fayvisovich Admiral Ushakov Maritime State University, Russia
H. J. S. Fernando Arizona State University, USA
W. F. Florez-Escobar Universidad Pontifica Bolivariana, South America
E. M. M. Fonseca Instituto Politécnico do Porto, Instituto Superior de Engenharia do Porto, Portugal
D. M. Fraser University of Cape Town, South Africa
G. Gambolati Universita di Padova, Italy
C. J. Gantes National Technical University of Athens, Greece
L. Gaul Universitat Stuttgart, Germany
N. Georgantzis Universitat Jaume I, Spain
L. M. C. Godinho University of Coimbra, Portugal
F. Gomez Universidad Politecnica de Valencia, Spain
A. Gonzales Aviles University of Alicante, Spain
D. Goulias University of Maryland, USA
K. G. Goulias Pennsylvania State University, USA
W. E. Grant Texas A & M University, USA
S. Grilli University of Rhode Island, USA
R. H. J. Grimshaw Loughborough University, UK
D. Gross Technische Hochschule Darmstadt, Germany
R. Grundmann Technische Universitat Dresden, Germany
O. T. Gudmestad University of Stavanger, Norway
R. C. Gupta National University of Singapore, Singapore
J. M. Hale University of Newcastle, UK
K. Hameyer Katholieke Universiteit Leuven, Belgium
C. Hanke Danish Technical University, Denmark
Y. Hayashi Nagoya University, Japan
L. Haydock Newage International Limited, UK
A. H. Hendrickx Free University of Brussels, Belgium
C. Herman John Hopkins University, USA
I. Hideaki Nagoya University, Japan
W. F. Huebner Southwest Research Institute, USA
M. Y. Hussaini Florida State University, USA
W. Hutchinson Edith Cowan University, Australia
T. H. Hyde University of Nottingham, UK
M. Iguchi Science University of Tokyo, Japan
L. Int Panis VITO Expertisecentrum IMS, Belgium
N. Ishikawa National Defence Academy, Japan
H. Itoh University of Nagoya, Japan
W. Jager Technical University of Dresden, Germany
Y. Jaluria Rutgers University, USA
D. R. H. Jones University of Cambridge, UK
N. Jones University of Liverpool, UK
D. Kaliampakos National Technical University of Athens, Greece
D. L. Karabalis University of Patras, Greece
A. Karageorghis University of Cyprus
T. Katayama Doshisha University, Japan
K. L. Katsifarakis Aristotle University of Thessaloniki, Greece
E. Kausel Massachusetts Institute of Technology, USA
H. Kawashima The University of Tokyo, Japan
B. A. Kazimee Washington State University, USA
F. Khoshnaw Koya University, Iraq
S. Kim University of Wisconsin-Madison, USA
D. Kirkland Nicholas Grimshaw & Partners Ltd, UK
E. Kita Nagoya University, Japan
A. S. Kobayashi University of Washington, USA
D. Koga Saga University, Japan
S. Kotake University of Tokyo, Japan

A. N. Kounadis National Technical University of Athens, Greece

W. B. Kratzig Ruhr Universitat Bochum, Germany

T. Krauthammer Penn State University, USA

R. Laing Robert Gordon University, UK

M. Langseth Norwegian University of Science and Technology, Norway

B. S. Larsen Technical University of Denmark, Denmark

F. Lattarulo Politecnico di Bari, Italy

A. Lebedev Moscow State University, Russia

D. Lesnic University of Leeds, UK

D. Lewis Mississippi State University, USA

K-C. Lin University of New Brunswick, Canada

A. A. Liolios Democritus University of Thrace, Greece

D. Lippiello Università degli Studi Roma Tre, Italy

S. Lomov Katholieke Universiteit Leuven, Belgium

J. E. Luco University of California at San Diego, USA

L. Lundqvist Division of Transport and Location Analysis, Sweden

T. Lyons Murdoch University, Australia

L. Mahdjoubi University of the West of England, UK

Y-W. Mai University of Sydney, Australia

M. Majowiecki University of Bologna, Italy

G. Manara University of Pisa, Italy

B. N. Mandal Indian Statistical Institute, India

Ü. Mander University of Tartu, Estonia

H. A. Mang Technische Universitat Wien, Austria

G. D. Manolis Aristotle University of Thessaloniki, Greece

N. Marchettini University of Siena, Italy

J. D. M. Marsh Griffith University, Australia

J. F. Martin-Duque Universidad Complutense, Spain

T. Matsui Nagoya University, Japan

G. Mattrisch DaimlerChrysler AG, Germany

F. M. Mazzolani University of Naples "Federico II", Italy

K. McManis University of New Orleans, USA

A. C. Mendes Universidade de Beira Interior, Portugal

J. Mera Polytechnic University of Madrid, Spain

J. Mikielewicz Polish Academy of Sciences, Poland

R. A. W. Mines University of Liverpool, UK

C. A. Mitchell University of Sydney, Australia

K. Miura Kajima Corporation, Japan

A. Miyamoto Yamaguchi University, Japan

T. Miyoshi Kobe University, Japan

G. Molinari University of Genoa, Italy

F. Mondragon Antioquin University, Colombia

T. B. Moodie University of Alberta, Canada

D. B. Murray Trinity College Dublin, Ireland

M. B. Neace Mercer University, USA

D. Necsulescu University of Ottawa, Canada

B. Ning Beijing Jiatong University, China

S-I. Nishida Saga University, Japan

H. Nisitani Kyushu Sangyo University, Japan

B. Notaros University of Massachusetts, USA

P. O'Donoghue University College Dublin, Ireland

R. O. O'Neill Oak Ridge National Laboratory, USA

M. Ohkusu Kyushu University, Japan

G. Oliveto Universitá di Catania, Italy

R. Olsen Camp Dresser & McKee Inc., USA

E. Oñate Universitat Politecnica de Catalunya, Spain

K. Onishi Ibaraki University, Japan

P. H. Oosthuizen Queens University, Canada

E. Outa Waseda University, Japan

O. Ozcevik Istanbul Technical University, Turkey

A. S. Papageorgiou Rensselaer Polytechnic Institute, USA

J. Park Seoul National University, Korea

F. Patania Universitá di Catania, Italy

B. C. Patten University of Georgia, USA

G. Pelosi University of Florence, Italy

G. G. Penelis Aristotle University of Thessaloniki, Greece

W. Perrie Bedford Institute of Oceanography, Canada

M. F. Platzer Naval Postgraduate School, USA

D. Prandle Proudman Oceanographic Laboratory, UK

R. Pulselli University of Siena, Italy

I. S. Putra Institute of Technology Bandung, Indonesia

Y. A. Pykh Russian Academy of Sciences, Russia

A. Rabasa University Miguel Hernandez, Spain

F. Rachidi EMC Group, Switzerland

K. R. Rajagopal Texas A & M University, USA

J. Ravnik University of Maribor, Slovenia

A. M. Reinhorn State University of New York at Buffalo, USA

G. Reniers Universiteit Antwerpen, Belgium

A. D. Rey McGill University, Canada

D. N. Riahi University of Illinois at Urbana-Champaign, USA

B. Ribas Spanish National Centre for Environmental Health, Spain

K. Richter Graz University of Technology, Austria

S. Rinaldi Politecnico di Milano, Italy

F. Robuste Universitat Politecnica de Catalunya, Spain

A. C. Rodrigues Universidade Nova de Lisboa, Portugal

G. R. Rodríguez Universidad de Las Palmas de Gran Canaria, Spain

C. W. Roeder University of Washington, USA

J. M. Roesset Texas A & M University, USA

W. Roetzel Universitaet der Bundeswehr Hamburg, Germany

V. Roje University of Split, Croatia

R. Rosset Laboratoire d'Aerologie, France

J. L. Rubio Centro de Investigaciones sobre Desertificacion, Spain

T. J. Rudolphi Iowa State University, USA

S. Russenchuck Magnet Group, Switzerland

H. Ryssel Fraunhofer Institut Integrierte Schaltungen, Germany

S. G. Saad American University in Cairo, Egypt

M. Saiidi University of Nevada-Reno, USA

R. San Jose Technical University of Madrid, Spain

F. J. Sanchez-Sesma Instituto Mexicano del Petroleo, Mexico

B. Sarler Nova Gorica Polytechnic, Slovenia

S. A. Savidis Technische Universitat Berlin, Germany

A. Savini Universita de Pavia, Italy

G. Schleyer University of Liverpool, UK

R. Schmidt RWTH Aachen, Germany

B. Scholtes Universitaet of Kassel, Germany

A. P. S. Selvadurai McGill University, Canada

J. J. Sendra University of Seville, Spain

S. M. Şener Istanbul Technical University, Turkey

J. J. Sharp Memorial University of Newfoundland, Canada

Q. Shen Massachusetts Institute of Technology, USA

G. C. Sih Lehigh University, USA

L. C. Simoes University of Coimbra, Portugal

A. C. Singhal Arizona State University, USA

J. Sladek Slovak Academy of Sciences, Slovakia

V Sladek Slovak Academy of Sciences, Slovakia

A. C. M. Sousa University of New Brunswick, Canada

H. Sozer Illinois Institute of Technology, USA

P. D. Spanos Rice University, USA

T. Speck Albert-Ludwigs-Universitaet Freiburg, Germany

C. C. Spyrakos National Technical University of Athens, Greece

G. E. Swaters University of Alberta, Canada

S. Syngellakis Wessex Institute, UK

J. Szmyd University of Mining and Metallurgy, Poland

H. Takemiya Okayama University, Japan

I. Takewaki Kyoto University, Japan

C-L. Tan Carleton University, Canada

E. Taniguchi Kyoto University, Japan

S. Tanimura Aichi University of Technology, Japan

J. L. Tassoulas University of Texas at Austin, USA

M. A. P. Taylor University of South Australia, Australia

A. Terranova Politecnico di Milano, Italy

T. Tirabassi National Research Council, Italy

S. Tkachenko Otto-von-Guericke-University, Germany

N. Tomii Chiba Institute of Technology, Japan

T. Tran-Cong University of Southern Queensland, Australia

R. Tremblay Ecole Polytechnique, Canada

I. Tsukrov University of New Hampshire, USA

R. Turra CINECA Interuniversity Computing Centre, Italy

S. G. Tushinski Moscow State University, Russia

R. van der Heijden Radboud University, Netherlands

R. van Duin Delft University of Technology, Netherlands

P. Vas University of Aberdeen, UK

R. Verhoeven Ghent University, Belgium

A. Viguri Universitat Jaume I, Spain

S. P. Walker Imperial College, UK

G. Walters University of Exeter, UK

B. Weiss University of Vienna, Austria

T. W. Wu University of Kentucky, USA

S. Yanniotis Agricultural University of Athens, Greece

A. Yeh University of Hong Kong, China

B. W. Yeigh University of Washington, USA

K. Yoshizato Hiroshima University, Japan

T. X. Yu Hong Kong University of Science & Technology, Hong Kong

M. Zador Technical University of Budapest, Hungary

R. Zainal Abidin Infrastructure University Kuala Lumpur, Malaysia

K. Zakrzewski Politechnika Lodzka, Poland

M. Zamir University of Western Ontario, Canada

G. Zappalà National Research Council, Italy

R. Zarnic University of Ljubljana, Slovenia

Computational Methods and Experimental Measurements XIX

&

Earthquake Resistant Engineering Structures XII

EDITORS

W. P. De Wilde
Free University of Brussels, Belguim

G.M. Carlomagno
University of Naples "Federico II", Italy

WIT*PRESS* Southampton, Boston

Editors:

W. P. De Wilde
Free University of Brussels, Belguim

G.M. Carlomagno
University of Naples "Federico II", Italy

Published by

WIT Press
Ashurst Lodge, Ashurst, Southampton, SO40 7AA, UK
Tel: 44 (0) 238 029 3223; Fax: 44 (0) 238 029 2853
E-Mail: witpress@witpress.com
http://www.witpress.com

For USA, Canada and Mexico

Computational Mechanics International Inc
25 Bridge Street, Billerica, MA 01821, USA
Tel: 978 667 5841; Fax: 978 667 7582
E-Mail: info@compmech.com
http://www.witpress.com

British Library Cataloguing-in-Publication Data

A Catalogue record for this book is available
from the British Library

ISBN: 978-1-78466-333-9
eISBN: 978-1-78466-334-6

CMEM XIX ISSN: 1746-4471 (print)
CMEM XIX ISSN: 1743-3533 (on-line)
ERES XII ISSN: 1746-4498 (print)
ERES XII ISSN: 1743-3509 (on-line)

The texts of the papers in this volume were set individually by the authors or under their supervision. Only minor corrections to the text may have been carried out by the publisher.

No responsibility is assumed by the Publisher, the Editors and Authors for any injury and/or damage to persons or property as a matter of products liability, negligence or otherwise, or from any use or operation of any methods, products, instructions or ideas contained in the material herein. The Publisher does not necessarily endorse the ideas held, or views expressed by the Editors or Authors of the material contained in its publications.

© WIT Press 2019

Open Access: All of the papers published in this journal are freely available, without charge, for users to read, download, copy, distribute, print, search, link to the full text, or use for any other lawful purpose, without asking prior permission from the publisher or the author as long as the author/copyright holder is attributed. This is in accordance with the BOAI definition of open access.

Creative Commons content: The CC BY 4.0 licence allows users to copy, distribute and transmit an article, and adapt the article as long as the author is attributed. The CC BY licence permits commercial and non-commercial reuse.

Contents

Part I – Computational Methods and Experimental Measurements XIX

Preface

Part I

Computational Methods and Experimental Measurements XIX

WIT*eLibrary*
Home of the Transactions of the Wessex Institute.
Papers published in this volume are archived in the WIT eLibrary in volume 125 of WIT Transactions on Engineering Sciences (ISSN 1743-3533).
The WIT electronic-library provides the international scientific community with immediate and permanent access to individual papers presented at WIT conferences.
Visit the WIT eLibrary at www.witpress.com.

Preface

This book contains the majority of the papers presented at the 19th International Conference on Computational Methods and Experimental Measurements (CMEM19), held in Seville (Spain) in 2019 and organised by the University of Naples 'Federico II' and the Wessex Institute of Technology. The series of these events is unique in this particular field and it deserves its long tradition. Indeed, the first conference took place in Washington DC in 1981 followed by a meeting on board the Queen Elizabeth II Ocean Liner (1984); Porto Carras, Greece (1986); Capri (1988); Montreal (1991); Siena (1993); Capri (1995); Rhodes (1997); Sorrento (1999); Alicante (2001); Sani Beach, Greece (2003); Malta (2005); Prague (2007); the Algarve (2009); the New Forest, UK (2011); A Coruña, Spain (2013), Opatija, Croatia (2015) and Alicante (2017), with uninterrupted success. A key objective of the conference is to offer to the scientific and technical community an international atmosphere to analyse the interaction between computational methods and experimental measurements with their corresponding characteristics, the main concern and significance being devoted to their mutual and beneficial merging.

It is well known that the advances in computers efficiency and numerical techniques are creating a steady growth of computational simulations which currently impact on both an ever-widening range of engineering problems, as well as on our everyday behaviours. Nevertheless, even if these simulations are continuously expanding and improving, there still exists the need of their validation, especially for the most recent ones, which can be only achieved by performing dedicated experimental tests. Furthermore, it has to be observed that, because of their continuous development, experimental techniques are becoming more and more multifaceted and sophisticated so that, often, both their execution as well as data collection can only be performed by means of computers. Finally, it must be stressed that, for a great number of measurements, the obtained data must be compulsory processed by means of numerical methods, a typical example being found in most of the many optically based techniques.

This volume contains a substantial number of excellent scientific papers, which present several advanced approaches to modern research questions. The papers contained in this book, as well as those from previous conferences since 1993, have been archived in the eLibrary of the Wessex Institute (https://www.witpress.com/elibrary) where they are permanently accessible to the international community. The Editors are very grateful to all the authors for their valuable contributions and to the Members of the International Scientific Advisory Committee, as well as other colleagues, for their help in reviewing the papers collected in the book.

The Editors
Seville, 2019

BEHAVIOUR OF SMALL FATIGUE CRACKS IN CU-5.5NI-1.28SI ALLOY ROUND-BAR SPECIMENS

MASAHIRO GOTO[1], TAKAEI YAMAMOTO[1], SEUNG ZEON HAN[2], JEE HYUK AHN[2], JUNICHI KITAMURA[1], RYOUTAROU TAKANAMI[1], TERUTOSHI YAKUSHIJI[3] & JEHYUN LEE[4]
[1]Department of Mechanical Engineering, Oita University, Japan
[2]Korea Institute of Materials Science, Republic of Korea
[3]National Institute of Technology, Oita College, Japan
[4]Changwon National University, Republic of Korea

ABSTRACT

Cu-5.5Ni-1.28Si alloy processed by air-cooling after solution heat treatment was conducted to prevent the generation of discontinuous precipitates which bring a detrimental effect on mechanical properties, and stress-controlled fatigue tests were carried out at room temperature. The change in the surface damage during stressing was monitored by means of both direct observation and plastic replication technique, showing the crack initiation at grain boundaries, crystallographic slip planes and twin boundaries. The growth rate of a small crack can be determined by a term $\sigma_a{}^n l$ when $n = 5.4$, not by the stress intensity factor range.

Keywords: fatigue, copper alloy, crack propagation, microstructure.

1 INTRODUCTION

Cu–Ni–Si alloys used for lead frame, connector applications are commercially manufactured through conventional thermomechanical processes, including aging, which induces the continuous precipitation of δ-Ni_2Si in the Cu matrix [1]–[3]. Previous studies have focused on Cu–Ni–Si alloys with low solute concentrations (e.g. < 3Ni) [3]–[6] to develop alloys with enhanced tensile properties and electrical conductivity. However, it is well known that an enhancement in the tensile strength of precipitation-hardened Cu alloys is inevitably accompanied by a reduction in electrical conductivity. For Cu–Ni–Si alloys with high solute concentrations, on the other hand, phase transformation studies have been conducted mainly to control the microstructure of the alloy [7], [8]. In addition, discontinuous precipitates (DPs) that were fibre-shaped, stable δ-Ni_2Si intermetallic compounds were formed in the Cu matrix [9], [10]. The formation of DPs in Cu–Ni–Si alloys is detrimental to their mechanical properties. It was shown that air cooling after solution heat treatment (SHT) encourages formation of disc-shaped δ-Ni_2Si precipitates instead of discontinuous precipices [11].

For the envisaged structural applications, a fatigue mechanism, such as crack initiation and propagation, should be clarified. The fracture surfaces of Cu-Ni-Si alloys analysed by a scanning electron microscope (SEM) have showed intergranular facets, leading to a hypothesis that grain boundaries (GBs) were an origin of fatigue crack [12], [13] and fatigue cracks may be propagated along GBs. Up to now, however, there was a small number of clear evidences for crack nucleation and propagation mechanisms [14]. In the evaluation of fatigue life of smooth members, the growth behaviour of a small crack should be clarified, because that the crack growth life from an initial size (grain order size) to 1 mm accounted for about 70% of the fatigue life of plain specimens of many metals and alloys. However, the growth rate of small cracks cannot be determined by a linear elastic fracture mechanics [15], [16].

In the present study, fatigue tests of Cu-5.5Ni-1.28Si alloy processed by air cooling after SHT were carried out in the room temperature. The objective of this study is to investigate the microstructure produced by air cooling at SHT and the behaviour of initiation and propagation of a surface crack.

WIT Transactions on Engineering Sciences, Vol 125, © 2019 WIT Press
www.witpress.com, ISSN 1743-3533 (on-line)
doi:10.2495/CMEM190011

2 EXPERIMENTAL PROCEDURES

Cu and Si 99.99% pure and 99.9% pure Ni were used as alloying elements for the fabrication of the Cu–5.5wt%Ni–1.28wt%Si alloy ingots by induction melting. The ingots were cold-rolled with 80% reduction in thickness, followed by SHT at 980°C for 1 h. After the SHT, the samples were cooled by water quenching or air cooling, followed by aging at 500°C (Fig. 1). Fig. 2 shows the microstructure with aging times 3 and 6 h of (a) water quenched and (b) air-cooled alloys. Dark/tarnished phases in water-quenched microstructure after 3 h aging were DPs. After 6 h aging, the microstructure was almost completely transformed into DP phases. The formation of DPs in Cu–Ni–Si alloys is detrimental to their mechanical properties. To suppress the formation of DPs, air cooling, instead of water quenching, was applied after SHT. Fig. 2(b) shows the microstructure of air-cooled alloy, showing no generation of DPs even after 6 h aging. Accordingly, the microstructure, tensile properties and fatigue characteristics were studied on solution treated, air cooled and 6 h aged Cu-5.5Ni-1.2Si alloy. The aging duration of 6 h at 500°C yielded near-peak hardness.

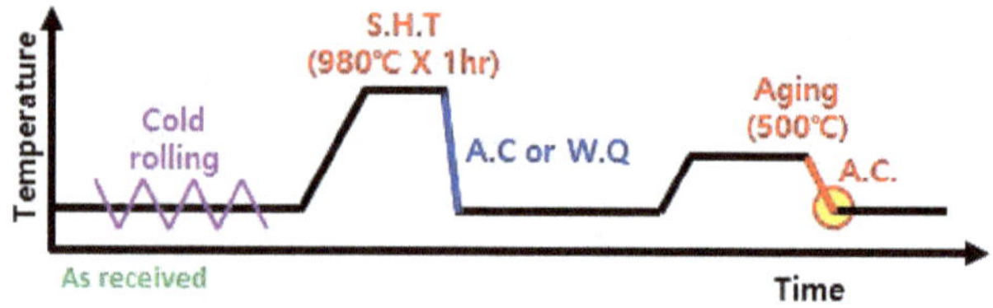

Figure 1: Schematic illustration of thermomechanical process.

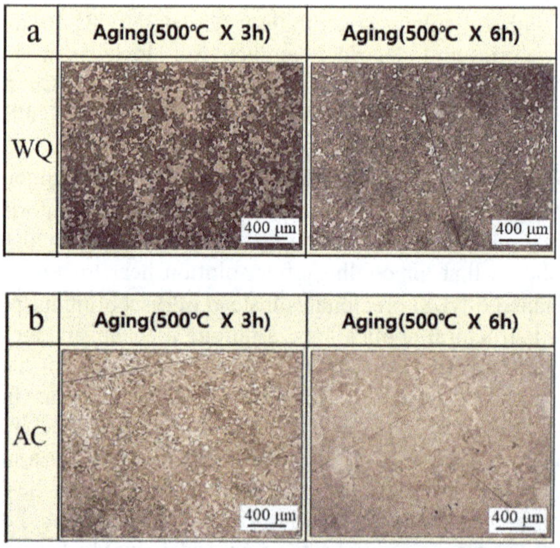

Figure 2: Microstructure after 3 and 6 h aging. (a) Water-quenched; and (b) Air-cooled alloy.

The microstructure was observed using an optical microscope (OM, GX51. Olympus) and a SEM (JSM-6610LV, JEOL). The characterization of precipitates was carried out using a 200 kV field-emission transmission electron microscope (TEM, JEOL-2100F. JEOL) equipped with an energy-dispersive X-ray spectroscopy (EDS) detector along with a scanning TEM. The 3-mm-diameter, 100-μm-thick disk was prepared for TEM observation by mechanical polishing with a digitally enhanced precision specimen grinder (DEPS-101, Total Solution) and dimpling by a dimple grinder (656 dimple grinder, Gatan).

Tensile tests were conducted on universal testing machine (model: Instron 4206) using 4 mm diameter round tensile specimens at a strain rate of 0.017 s^{-1} at room temperature. The microhardness was measured using a Vickers hardness tester with an applied load of 1 N. Round bar fatigue specimens (Fig. 3) with 5 mm in diameter were machined from the plate of aged alloy. In some specimens, a shallow partial notch (Fig. 3(b); 2.5 mm radius and 0.1 mm depth) was machined for monitoring of crack initiation behavior. The fatigue strength reduction factor for the partial notch was close to 1, meaning that notched specimens could be considered plain. All fatigue specimens were electrolytically polished (approximately 25 μm from the surface layer) prior to mechanical testing to remove any preparation-affected surface layer. Polishing was carried out at 25°C using an electrolyte consisting of 600 mL of phosphoric acid, 300 mL of distilled water, and 100 mL of sulfuric acid. All fatigue tests were carried out at room temperature using a rotating bending fatigue machine (constant bending-moment type) operating at 50 Hz. The fatigue damage on the specimen surface and on the fracture surface was observed by using OM and SEM. The crack length, l, was measured along the circumferential direction of the surface using a plastic replication technique. The stress value referred to is that of the nominal stress amplitude, σ_a, at the minimum cross-section (5 mm diameter).

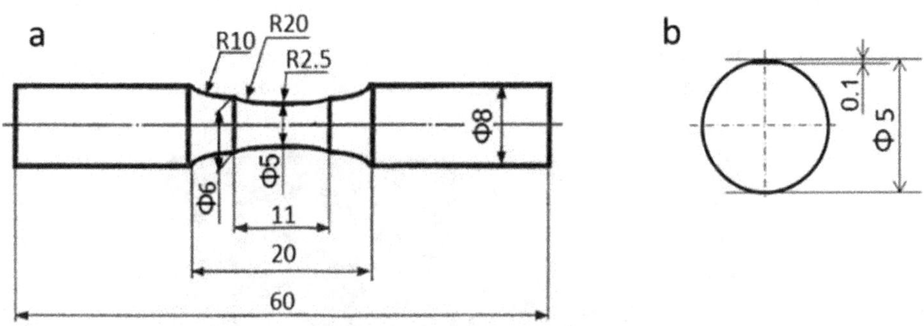

Figure 3: Schematic illustration of fatigue specimen (in mm). (a) Shape and dimensions; and (b) The details of a partial notch.

3 EXPERIMENTAL RESULTS AND DISCUSSION

Fig. 4(a) shows an optical micrograph of the Cu–5.5Ni–1.28Si alloy. Fig. 4(b) and (c) shows TEM images of the matrix. Disc-shaped precipitates were observed, which were identified as δ-Ni_2Si intermetallic compounds by an optical diffractogram (Fig. 4(d)). Fig. 5 shows EDS mapping, showing Ni_2Si precipitations in Cu matrix.

The stress–strain curves of the air-cooled and water-quenched specimens are shown in Fig. 6. The mechanical properties of the specimens are summarized in Table 1. Indeed, the tensile strength of the water quenched alloy was lower than that of the water cooled one, because of the generation of discontinuous precipitates (Fig. 1(a)).

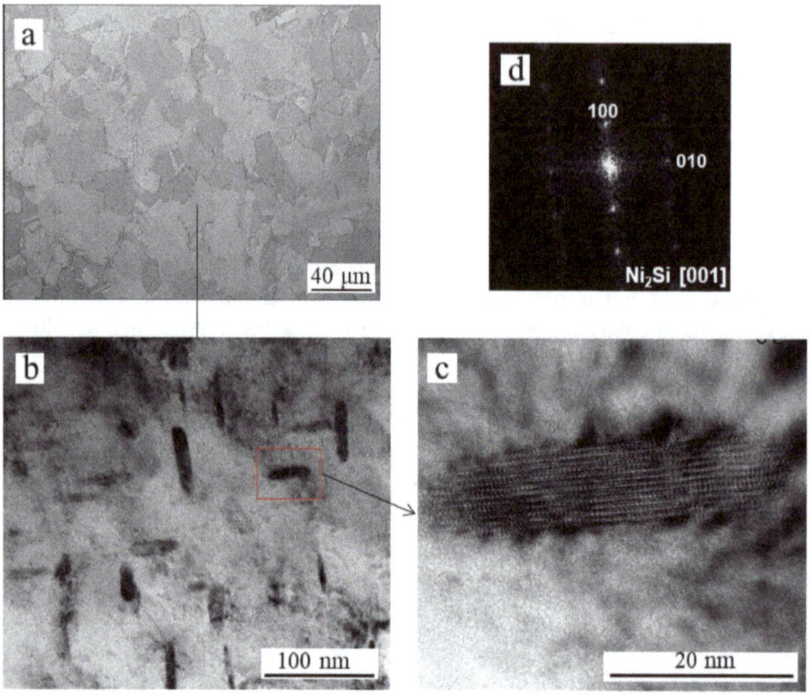

Figure 4: Microstructure of 6 h aged alloy with air cooling after SHT. (a) OM micrograph; (b), (c) TEM images; and (d) selected area deflection pattern.

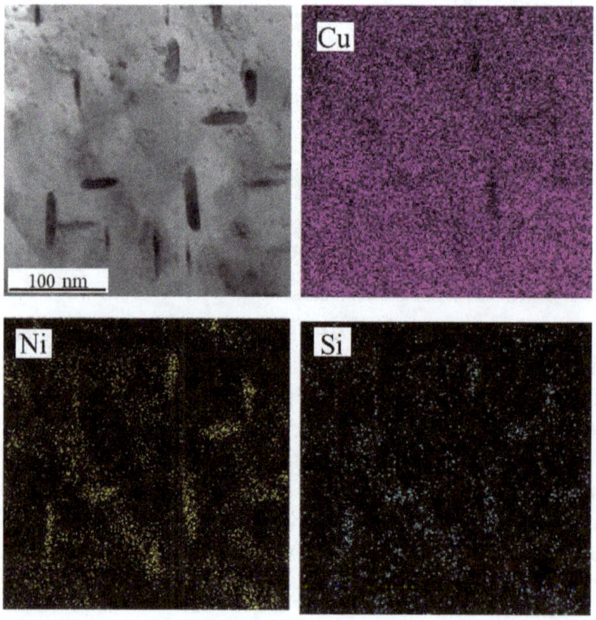

Figure 5: EDS mapping of Cu-5.5Ni-1.28Si alloy.

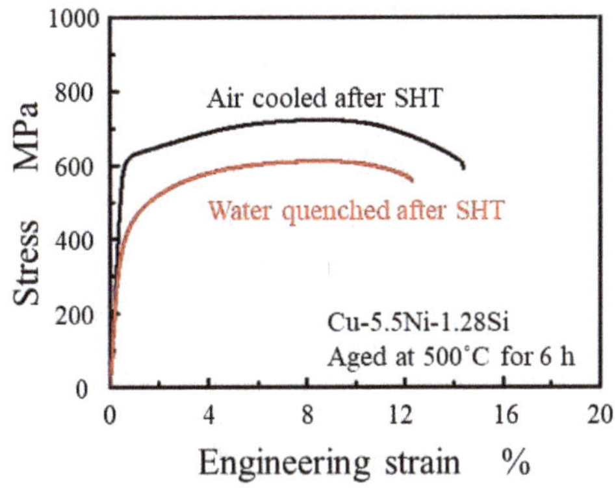

Figure 6: Stress–strain curves of the air-cooled and water-quenched specimens.

Table 1: Mechanical properties of Cu-5.5Ni-1.28Si alloys.

Materials	σ_u [MPa]	Φ [%]	H_V
Air cooling	725	14.4	233
Water quenching	616	12.3	187

σ_u: Ultimate tensile strength,
Φ : Tensile elongation, *Hv*: Vickers hardness

The stress vs. number of cycles to failure curves of the air-cooled specimens are shown in Fig. 7. A solid and an open circle shows the S–N plots for without and with notch, respectively. There was no significant difference in S–N plots between without and with notch, meaning the results obtained from the notched specimens can represent that of the plain specimens. A solid triangle shows the S–N plot for an annealed copper (99.99% Cu). Fig. 7 is replotted in Fig. 8 in terms of stress amplitude normalized by tensile strength, σ_a/σ_u. The ratios of the fatigue limit stress at $N = 10^7$ cycles, σ_w, to the tensile strength, σ_u, was 0.29 which was lower than that of annealed Cu, $\sigma_a/\sigma_u = 0.37$.

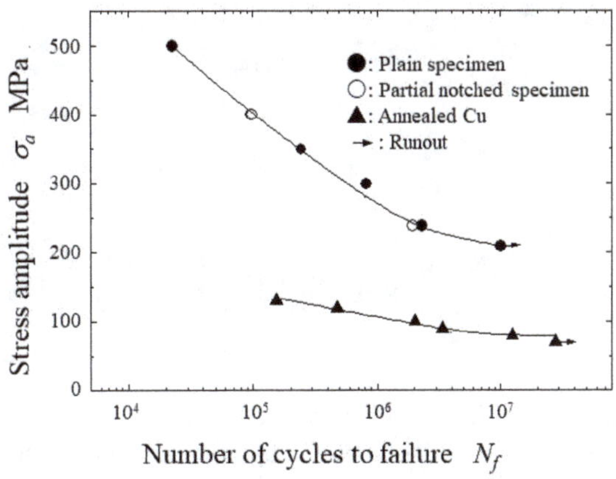

Figure 7: S–N curves of the air-cooled Cu-5.5Ni-1.28Si and annealed Cu specimens.

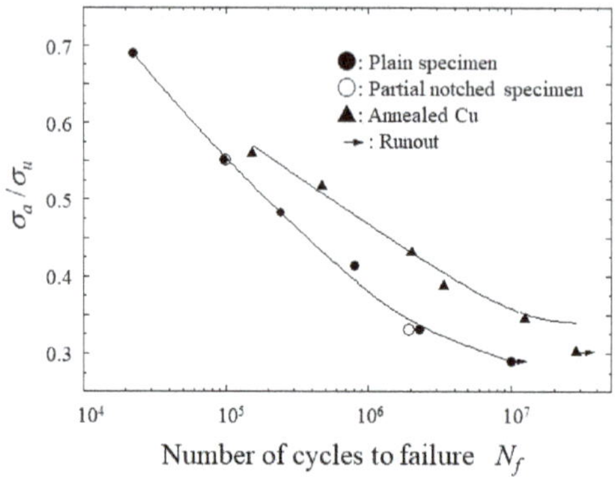

Figure 8: σ_a/σ_u vs. N_f relationship of the air-cooled Cu-5.5Ni-1.28Si and annealed Cu specimens.

The changes in the surface morphologies of the air-cooled specimens during cyclic stressing are presented in Fig. 9. To clarify the particular microstructural feature where a crack was initiated, the specimen surface was etched prior to the fatigue test. After the OM analysis of the etched surface, a layer of a few micrometers was polished by buffing to remove the surface area damaged by etching, which would affect the crack initiation behaviour during stressing. The fatigue tests were then conducted. After a fatigue crack was initiated to grain-size length, the fatigue test was interrupted to etch for microstructural observation and then resumed with the etched specimen. Two cracks labelled with I and II were formed at $N = 8.5 \times 10^5$ on the buffed surface fatigued at $\sigma_a = 240$ MPa (Fig. 9). The length of crack-I at $N = 8.5 \times 10^5$ was $l = 70$ μm. A microstructural comparison between the as-received and as-fatigued specimens suggested that the cracks I and II were initiated at twin boundaries and grain boundaries (GBs), respectively, as indicated by the red arrow in Fig. 9. In this alloy, cracks were initiated at crystallographic slip planes, GBs, and twin boundaries.

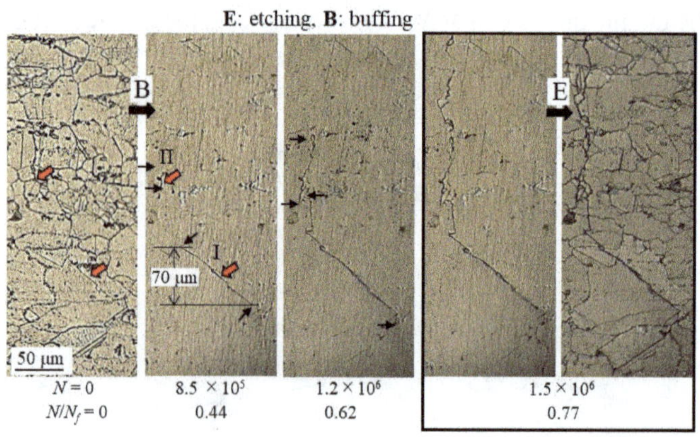

Figure 9: The change in surface states at $\sigma_a = 240$ MPa, ($N_f = 1.93 \times 10^6$): To prevent the crack initiation from GBs damaged by etching, etched specimens were fatigued after polishing off a layer of a few micrometres from the surface. After a grain-sized crack was initiated, the surface was etched to reveal the microstructure.

Fig. 10 shows the crack growth curves (ln$l$ vs. N) of the air-cooled specimens. Fig. 10 is replotted in Fig. 11 in terms of number of cycles normalized by fatigue life, N/N_f. Except for both initial (l < 0.1 mm) and final (l > 2.0 mm) stages of fatigue, the relation can be approximated by a linear relation, whereas the relation for σ_a = 240 MPa deviated from the linear relation. In the ln$l$ vs. N, the linearity means that the crack growth rate, dl/dN, is proportional to the crack length [17], [18].

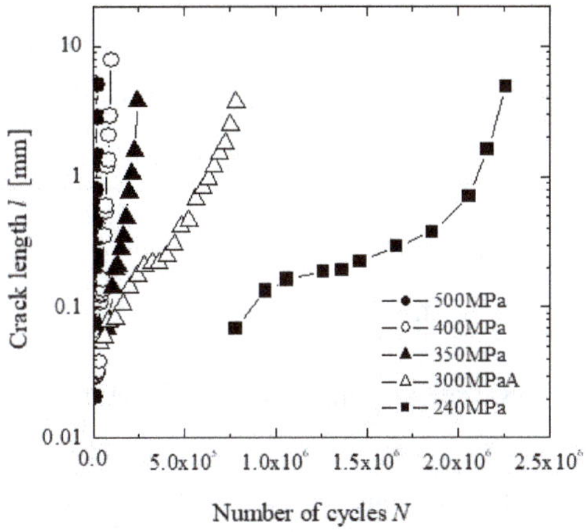

Figure 10: Crack growth curves (ln$l$ vs. N correlation).

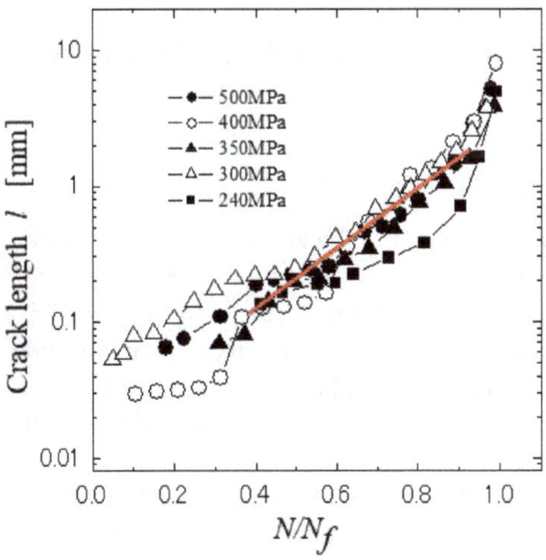

Figure 11: Crack growth curves (ln$l$ vs. N/N_f correlation).

Fig. 12 shows the crack growth rate as the function crack length: (a) σ_a = 300 MPa and (b) σ_a = 350 MPa. The crack growth rate in ranging from 0.1 to 2.0 mm is nearly proportional to the crack length excepting for the threshold region ($dl/dN < 10^{-6}$ mm/c).

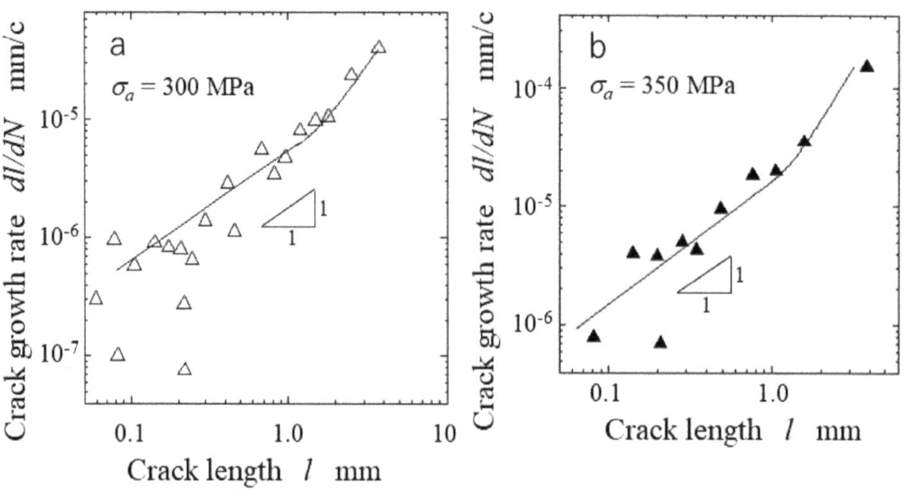

Figure 12: The dl/dN vs. l relationship: (a) $\sigma_a = 300$ MPa; and (b) $\sigma_a = 350$ MPa.

It has been generally accepted that the growth rate of a large/long crack can be unified in terms of stress intensity factor range (SIFR, ΔK) that is an effective parameter for estimation of crack growth rate (CGR) when the condition of small scale yielding at a crack tip is satisfied. However, the growth rate of a small surface-crack, e.g. crack length of less than 1 mm, cannot be evaluated by SIFR [15], [16] unlike a long crack propagating under a low stress amplitude.

Nisitani et al. [17] and Goto and Nisitani [18] proposed eqn (1) for determining the CGR at high stress amplitudes:

$$\frac{dl}{dN} = C_1 \sigma_a^n l \quad (1)$$

where C_1 and n are material constants. The growth data of small cracks in various metals and alloys indicated that dl/dN of a mechanically small crack is uniquely determined by the term $\sigma_a^n l$ and not by ΔK [15]–[18]. The expression $\sigma_a^n l$ ($n = 3$) was first proposed by Frost and Dugdale [19], who applied it to comparatively large cracks in which the condition of small-scale yielding nearly held. Now $\sigma_a^3 l$ can be considered as an approximation for ΔK, whereas the $\sigma_a^n l$ term proposed by Nisitani and Goto is a parameter for crack propagation under large-scale yielding.

Fig. 13 shows the growth data of a small surface-crack with the dl/dN vs. $\sigma_a^n l$ correlation. The values of n in eqn (1) were 5.4. Excepting a lower stress amplitude ($\sigma_a < 300$ MPa), the CGR can be uniquely determined by $\sigma_a^n l$. Accordingly, the CGR of small cracks propagating at $\sigma_a \gtrsim 300$ MPa can be approximately estimated by eqn (1). It has been shown that the CGR of small cracks in annealed copper is determined by eqn (1) in the range of $\sigma_a \gtrsim 80$ MPa [20].

SEM fractographs of the specimen fatigued at $\sigma_a = 400$ MPa and documented at 0.03 and 1.0 mm beneath the specimen surface are shown in Fig. 14. A notable difference in fracture morphology was observed with depth. The fracture morphology of the subsurface area

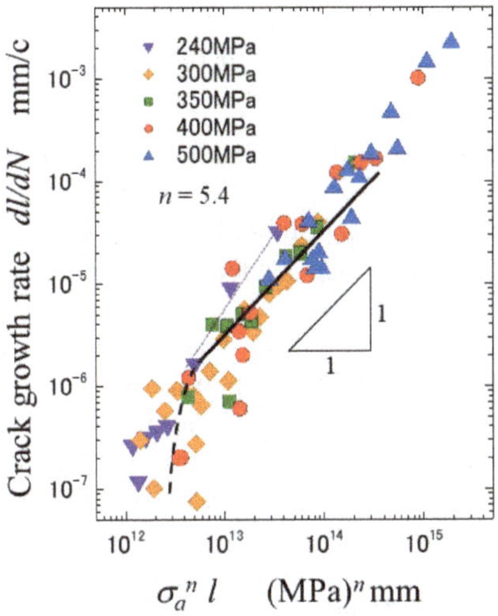

Figure 13: Growth data of small surface-cracks (dl/dN vs. $\sigma_a{}^n l$ correlation: $n = 5.4$).

formed by initial crack propagation ($dl/dN < 10^{-6}$ mm/c) appeared to be associated with the remaining microstructural inhomogeneity. For example, uneven surfaces composed of facets were observed; these were probably formed by crack propagation in slip planes within grains and along GBs. In general, a microstructure-sensitive fracture can occur rather easily in the near-threshold and initial stages of crack growth, where the driving force for crack propagation is relatively weak. It has been demonstrated that the propagation of an initial small crack is affected by the presence of microstructural inhomogeneities, such as GBs, phase boundaries, and nonmetallic inclusions [21], [22]. By contrast, the fracture surface far from the specimen surface ($dl/dN > 10^{-6}$ mm/c) was relatively flat with a smaller occurrence of morphological features associated with microstructural inhomogeneity. Striations (Fig. 14(b)) were frequently observed on the fracture surface, suggesting that crack propagation was largely controlled by mechanical fatigue.

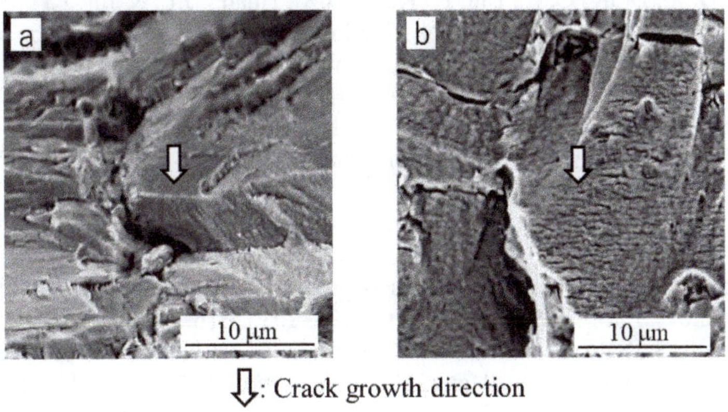

Figure 14: SEM fractographs of the specimen fatigued at σ_a = 400 MPa. (a) 0.03 mm beneath the surface; and (b) 1.0 mm beneath the surface.

4 CONCLUSIONS

The main findings of this study on the behavior of fatigue cracks in Cu-5.5Ni-1.28Si alloy can be summarized as follows:

1) To prevent the generation of discontinuous precipitates which brings a detrimental effect on mechanical properties and to encourage formation of continuous precipitates, air-cooling was applied after SHT, followed by aging at 500°C. The matrix was strengthened by disc-shaped δ-Ni_2Si intrametric compounds with a few tens of nanometers diameter.
2) The fatigue cracks in the alloy were found to originate from the GBs, crystallographic slip plane and twin boundaries.
3) Except for threshold range ($dl/dN < 10^{-6}$ mm/c), the growth rate of a small-surface crack propagating at high stress amplitudes ($\sigma_a \gtrsim 300$ MPa), can be uniquely determined by a term $\sigma_a{}^n l$. n was material constant and its value was 5.4.

ACKNOWLEDGEMENTS

This study was supported by a Grant-in-Aid for Scientific Research (Kiban-B) (KAKENHI: No. 18H01340) and for Encouragement of Scientists (No. 18H00244) from the Japan Society for the Promotion of Science, as well as the National Research Foundation of Korea (NRF) grant funded by the Korea government (MSIP) (2018R1A5A6075959). The authors are very grateful to the members of the Strength of Materials Laboratory of Oita University for their excellent experimental assistance. Thanks are also extended to the members of the Korea Institute of Materials Science, for the fabrication of the Cu–Ni–Si system alloys.

REFERENCES

[1] Corson, M.G., Electrical conductor alloys. *Electr. World*, **89**, pp. 137–139, 1927.

[2] Fujiwara, H., Sato, T. & Kamio, A., Effect of alloy composition on precipitation behavior in Cu-Ni-Si alloys. *J. Jpn. Inst. Metals*, **62**, pp. 301–309, 1998.

[3] Hu, T., Chen, J.H., Liu, J.Z., Liu, Z.R. & Wu, C.L., The crystallographic and morphological evolution of the strengthening precipitates in Cu–Ni–Si alloys. *Acta Mater.*, **61**, pp. 1210–1219, 2013.

[4] Locker, S.A. & Noble, F.W., Precipitate microstructure in a Cu-Ni-Si alloy. *J. Mater. Sci.*, **29**, pp. 218–226, 1994.

[5] Zhao, D.M., Dong, Q.M., Liu, P., Kang, B.X., Huang, J.L. & Jin, Z.H., Structure and strength of the age hardened Cu–Ni–Si alloy. *Mater. Chem. Phys.*, **79**, pp. 81–86, 2003.

[6] Srivastava, V.C., Schneider, A., Uhlenwinkel, V. & Bauckhage, K., Effect of thermomechanical treatment on spray formed Cu–Ni–Si alloy. *Mater. Sci. Technol.*, **20**, pp. 839–848, 2004.

[7] Lei, Q. et al., Phase transformation behavior in a Cu-8.0Ni-1.8Si alloy. *J. Alloys Compds.*, **509**, pp. 3617–3622, 2011.

[8] Shen, L., Li, Z., Zhao, Y., Wang, Y., Dong, Q. & Wang, M., Phase transformation behavior in a Cu-10Ni-3Al-0.8Si alloy. *Mater. Chem. Phys.*, **173**, pp. 421–428, 2016.

[9] Han, S.Z. et al., Increasing toughness by promoting discontinuous precipitation in Cu–Ni–Si alloys. *Philos. Mag. Lett.*, **96**, pp. 196–203, 2016.

[10] Semboshi, S., Sato, S., Iwase, A. & Takasugi, T.. Discontinuous precipitates in age-hardening Cu–Ni–Si alloy. *Mater. Characterization*, **115**, pp. 39–45, 2016.

[11] Han, S.Z., Lee, J., Lim, S.H., Ahn, J.H., Kim, K. & Kim, S., Optimization of conductivity and strength in Cu-Ni-Si alloys by suppressing discontinuous precipitation. *Met. Mater. Inter.*, **22**, pp. 1049–1054, 2016.

[12] Lockyer, S.A. & Noble, F.W., Fatigue of precipitate strengthened Cu-Ni-Si alloy. *Mater. Sci. Technol.*, **15**, pp. 1147–1153, 1999.

[13] Sun, Z., Laitem, C. & Vincent, A., Dynamic embrittlement during fatigue of a Cu-Ni-Si alloy. *Mater. Sci. Eng. A*, **528**, pp. 6334–6337, 2011.

[14] Goto, M. et al., Role of microstructure on initiation and propagation of fatigue cracks in precipitate strengthened Cu-Ni-Si alloy. *Inter. J. Fatigue*, **87**, pp. 15–21, 2016.

[15] Miller, K.J. & de los Rios, E.R. (eds), *The Behaviour of Short Fatigue Cracks*, EGF-Pub 1, MEP, 1986.

[16] Ravichandran, K.S., Ritchie, R.O. & Murakami, Y. (eds), *Small Fatigue Cracks: Mechanics, Mechanisms and Applications*, Elsevier, 1999.

[17] Nisitani, H., Goto, M. & Kawagoishi, N., A small-crack growth law and its related phenomena. *Eng. Fract. Mech.*, **41**, pp. 499–513, 1992.

[18] Goto, M. & Nisitani, H., Fatigue life prediction of heat-treated carbon steels and low alloy steels based on a small-crack growth law. *Fatigue Fract. Eng. Mater. Struct.*, **17**, pp. 171–185, 1994.

[19] Frost, N.E. & Dugdale, D.S., The propagation of fatigue cracks in sheet specimen. *J. Mech. Phys. Solids*, **6**, pp. 92–110, 1958.

[20] Goto, M., Kamil, K., Han, S.Z., Euh, K., Kim, S.S. & Yokoho, Y., Effects of grain refinement due to severe plastic deformation on the growth behavior of small cracks in copper. *Inter. J. Fatigue*, **50**, pp. 63–71, 2013.

[21] Goto, M., Statistical investigation of the behaviour of small-cracks and fatigue life in carbon steels with different ferrite grain sizes. *Fatigue Fract. Eng. Mater. Struct.*, **17**, pp. 635–649, 1994.

[22] Goto, M. & Knowles, D.M., Initiation and propagation behavior of microcracks in Ni-based superalloy Udimet 720 LI. *Eng. Fract. Mech.*, **60**, pp. 1–18, 1998.

NUMERICAL MODELING OF HIGH-VELOCITY, PROJECTILE PENETRATING CONCRETE BLOCKS REINFORCED BY TEFLON SHEETS

AYA E. ELHOZAYEN[1], MOHAMED Y. LASSI[2] & WALID A. ATTIA[3]
[1]The British University in Egypt, Egypt
[2]University of Prince Mugrin, Saudi Arabia
[3]Cairo University, Egypt

ABSTRACT

Recently, the concern about protecting people and structures has increased due to the increasing number of terrorist attacks. This paper presents a numerical simulation of plain concrete blocks with an unconfined compressive strength of 35 MPA, reinforced by Teflon sheets and subjected to ballistic impact by a high-velocity rigid projectile (960 m/s). The reinforcement sheets are modeled with different thicknesses and located at different depths from the face of the concrete target. The Teflon material was chosen due to its high impact strength over a wide range of temperatures. A validation model was conducted and results showed a good agreement with previous experimental results. The investigation presents the development of a finite element accurate model using AUTODYN 3D. The Lagrangian formulation numerical technique is used to model both the projectile and the concrete target. The main findings showed an enhancement in the penetration resistance of concrete target when reinforced by Teflon sheets compared to the concrete resistance without reinforcement, where the projectile depth of penetration was reduced by 64.8% and the full damage depth of the concrete target was also reduced by 58%, which demonstrates the great performance of the chosen reinforcement in the shock wave propagation.
Keywords: ballistic, impact, concrete, Teflon, projectile, numerical modeling.

1 INTRODUCTION

As one of the most important and effective construction materials, concrete may be subjected to extreme dynamic loads such as blast and impact loading causing penetration and perforation and leading to failure of the structure; this impact may be caused by terrorist attacks or high-velocity fragments resulted from accidental explosions.

In the last decade, many empirical and numerical models and experimental studies were conducted to analyze the behavior of normal reinforced concrete subjected to impact loading with different range of velocity.

Experimental studies of plain and reinforced concrete targets subjected to impact loading were of great interest to researches such as Rajput et al. [1] who conducted penetration tests on different thicknesses of plain and reinforced concrete plates to study the influence of impact of hard steel projectile on their ballistic performance; while Iqbal et al. [2] compared the ballistic limit and the damaged area of plain and reinforced concrete when subjected to normal impact of hardened steel projectiles.

On the other hand, many numerical models were conducted to study and analyze the penetration and perforation of different elements of normal reinforced concrete, such as Smith and Cusatis [3] who studied the impact of a projectile on regular strength concrete (RSC) and high-strength concrete (HSC) using the Lattice Discrete Particle Model (LDPM). Jiricek and Foglar [4] simulated the impact of regular and sub-caliber projectile on military bunkers (concrete blocks) using Autodyne 3D. Another numerical simulation was conducted using Autodyne by Tawadrous et al. [5]; they modeled numerically the penetration of a high velocity projectile to concrete blocks reinforced by ceramic.

WIT Transactions on Engineering Sciences, Vol 125, © 2019 WIT Press
www.witpress.com, ISSN 1743-3533 (on-line)
doi:10.2495/CMEM190021

Some researches combined simulations and experiments to compare the results and validate the proposed numerical models; such as Almusallam et al. [6] who compared the effect of CFRP strengthening on the response of reinforced concrete slabs when modeled experimentally and numerically using LS-DYNA; the analysis was conducted for the impact of hemispherical steel projectile with varying velocity and the slabs. Pavlovic and Fragassa [7] evaluated the resistance level of reinforced concrete barriers subjected to ballistic impact; the study was done with explicit Finite Element (FE) simulations and the results were compared to experiments on similar RC barriers. Kamal and Eltehewy [8] investigated the penetration resistance of concrete blocks reinforced by different number of layers of woven wire steel mesh experimentally and numerically using a steel blunt-nose projectile with a diameter of 23 mm with striking velocity about 980 m/s; they found an overall reduction in penetration depth and damage of target when using wire meshes as a reinforcement.

This paper aims to enhance the penetration resistance of plain concrete blocks to high velocity projectile impact by using tetrafluoroethylene polymer sheets (PTFE) known as Teflon as reinforcement for plain concrete. Teflon was chosen due to its excellent impact strength over a wide range of temperatures according to the Teflon handbook published by DuPont [9].

2 METHOD AND MATERIALS

Impact is defined as a high force or shock applied over a very short time from a body to another one and inducing significant local contact deformation. From the perspective of conservation of energy, the kinetic energy of the impacting body will be partially converted to strain energy in the target and partly dissipated through friction and local plastic deformation, and the strain energy is radiated away as stress waves.

The mechanism of penetration differs from one material to another; this section will discuss the penetration mechanism in concrete, metals, ceramics, polymers and composites. During ballistic impact, a pressure wave is induced and expands radially through the concrete target, leading to interfering stress wave that results in rapid change in multiaxial stress states and strain rates of the target. Material local failure can occur in four main stages [10] as shown in Fig. 1:

1. Penetration: the projectile tunneling into the target by a length called the penetration depth, formation of a cone-like crack under the projectile.
2. Spalling and scabbing: the target material ejecting from the front and rare faces.
3. Perforation: the projectile passing completely through the target with or without a residual velocity.
4. Recovery phase: it is the period during which the crater recovers or contracts slightly.

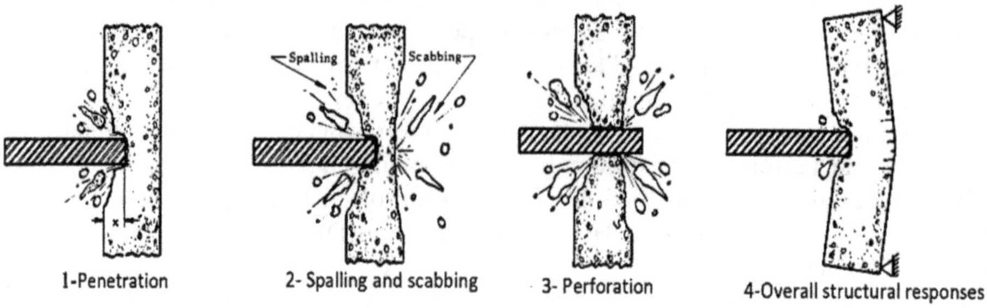

Figure 1: Projectile impact main stages [10].

The penetration of polymer plates occurs by elastic dishing, petalling, cone cracking, and plugging, the projectile, pushing a cylinder of material ahead, reaches the rear surface and perforation occurs. The mechanism of failure can be explained by tensile failure, increase and cohesion of planar cracks, formation of voids, and shear instabilities [11].

2.1 Numerical analysis

With the increasing power and speed of numerical FE codes and explicit dynamic solvers, it is possible to use numerical analysis and computer simulation to model any ballistic impact case. The hydrocode Autodyne is one of the most efficient codes used to simulate ballistic impact problems and it is used in this paper to model the depth of penetration and the damage in concrete targets. The Lagrangian formulation was chosen to simulate both the projectile and the concrete target.

In Lagrangian formulation, each grid point has different material coordinates and elements are created by connecting the grid points. The element or the mesh deforms with the body, i.e. both the grid nodes and the material points change position as the body deforms but the position of the material points relative to the grid nodes remains fixed as shown in Fig. 2. The grid points are forced to move with the element when the body is subjected to deformation.

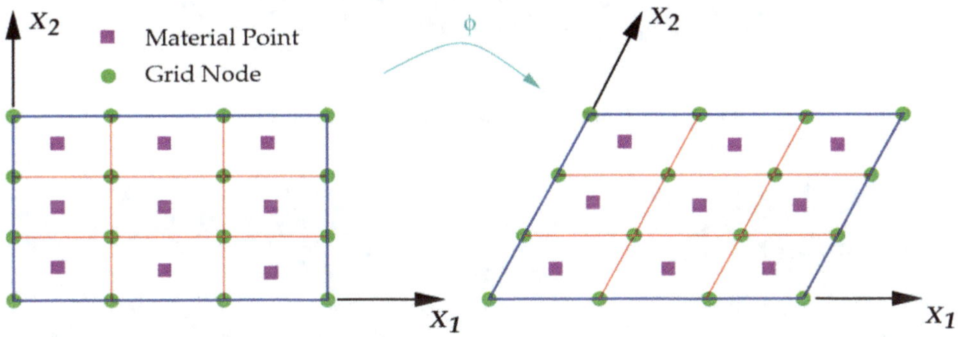

Figure 2: Lagrangian mesh before and after body deformation.

The main difficulty of Lagrangian mesh is the severe distortion that usually occurs causing stability problems and breaking the discretization. Deformations also cause the mesh to become ill-conditioned and result in decreasing stable time steps and mesh entanglement. This problem can be solved by defining an erosion mechanism that can remove the distorted elements to allow the calculation to continue. The main advantages of this numerical formulation are the fast solution due to fewer computations per cycle; the high accuracy in describing the material interfaces; the automatic satisfaction of the mass conservation; the ability to apply various boundary conditions; and the availability to handle damage and plasticity.

2.2 Geometry and mesh description

The dimensions of the simulated concrete block target are 500 mm × 500 mm × 400 mm, the concrete block is symmetric with respect to X and Y axis (X = 0 and Y = 0); to take advantage of the symmetry condition and to reduce the computational domain, the block is represented by its quarter (250 mm × 250 mm × 400 mm) as shown in Fig. 3. The concrete region local to the projectile (100 mm × 100 mm) which experiences large deformations is defined by a fine mesh of (2.5 mm × 2.5 mm × 3 mm) in *i*, *j* and *k* directions and a grade zoning is defined for the region far from the projectile in *i* and *j* directions, where little or no deformations occur, to reduce the computational time. The projectile geometry has a diameter of 23 mm and a length of 64 mm; it is modeled as a ¼ cylinder; the geometry and mesh description of projectile are shown in Fig. 4. The model boundary condition is fixed from the side and the projectile has a velocity of 960 m/s in Z-direction.

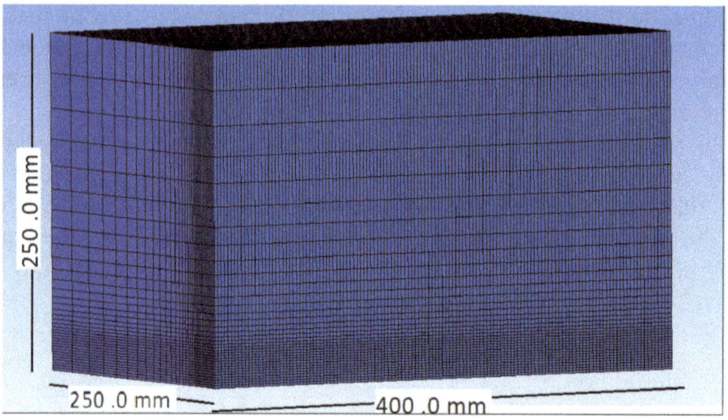

Figure 3: Concrete block geometry and meshing.

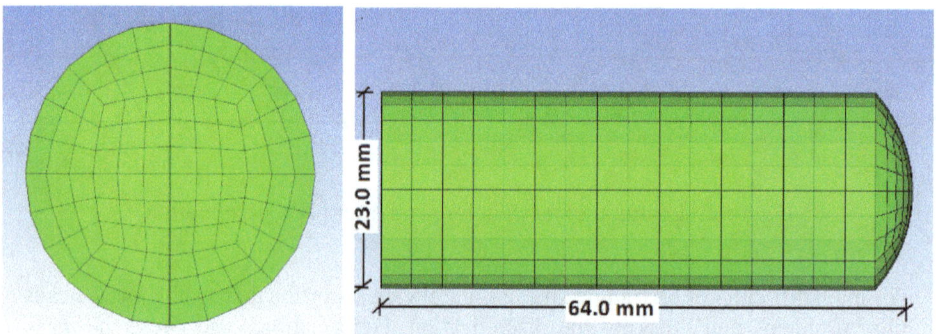

Figure 4: Projectile geometry and meshing.

2.3 Material description

The material models are chosen from AUTODYN library to simulate the used materials as following: STEEL4340 is used to simulate the projectile, while CONC-35MPA is used to simulate the plain concrete of unconfined compressive strength 35 MPA; and TEFLON for the polymer PTFE (known as Teflon). The details of the materials description are shown in Tables 1 to 3 where the Equation of State, the strength model and the failure model for each material are presented as well as the erosion model if it existed.

Table 1: Material constants for steel (STEEL4340).

Parameter	Value	Parameter	Value
Equation of state	Johnson Cook	Hardening constant (KPa)	1.79E+07
Reference density (gm/cm^3)	7.83	Hardening exponent	0.26
Bulk modulus (KPa)	1.67E+08	Strain rate constant	0.014
Reference temperature (K)	300	Thermal-softening exponent	1.03
Specific heat (C.V.) (j/kg K)	477	Melting temperature (k)	1793
Strength model	Johnson Cook	Ref. strain rate (/s)	1.00
Shear modulus (KPa)	7.98E+07	Erosion model	Geometric strain
Yield Stress (KPa)	1.73E+06	erosion strain	3.50

Table 2: Material constants for concrete (CONC-35MPA).

Parameter	Value	Parameter	Value
Equation of state	P-alpha	Failure surface parameter A	2
Reference density (gm/cm^3)	2.7	Failure surface exponent N	0.7
Porous density (gm/cm^3)	2.350	Tens./comp. meridian ratio	0.6805
Porous sound speed (m/s)	2.92E+03	Brittle to ductile transit	0.0105
Initial compaction pressure (KPa)	2.33E+04	G (elas.)/G (elas.-plas.)	2
Solid compaction pressure (KPa)	6.00E+06	Compaction curve	Standard
Compaction exponent n	3.00	Elastic strength/ft	0.7
Solid EOS	Polynomial	Elastic strength/fc	0.53
Bulk modulus A1 (KPa)	3.53E+07	Use cap on elastic surface	1
Parameter A2 (KPa)	3.96E+07	Fractured strength constant	1.50
Parameter A3 (KPa)	9.04E+06	Fractured strength exponent	0.61
Parameter B0	1.22	Compressive strain rate	0.032
Parameter B1	1.22	Tensile strain rate exponent	0.025
Parameter T1 (KPa)	3.53E+07	Max. fracture strength ratio	1.00E+20
Parameter T2 (KPa)	0.00	Failure model	RHT concrete
Reference temperature (K)	300.00	Damage constant D1	0.04
Specific Heat (C.V.) (j/kg K)	654.00	Damage exponent D2	1.00
Strength Model	RHT concrete	Min. strain to failure	0.01
Shear modulus (KPa)	1.67E+07	Residual shear modulus frac.	0.13
Compressive strength fc (KPa)	3.50E+04	Tensile failure model	Hydro tens.
Tensile strength ft/fc	0.088	Erosion model	Geometric strain
Shear strength fs/fc	0.18	Erosion strain	1.045

Table 3: Material constants for teflon (TEFLON).

Parameter	Value	Parameter	Value
Equation of state	Shock	Strength model	von Mises
Reference density (gm/cm^3)	2.16	Shear modulus (KPa)	2.33E+06
Gruneisen coefficient	0.9	Yield Stress (KPa)	5.00E+04
Parameter C1 (m/s)	1.34E+03	Failure model	Hydro (Pmin)
Parameter S1	1.93	Hydro tensile limit	-1.00E+06
Reference temperature (K)	0	Reheal	yes
Specific heat (j/kg K)	0	Crack softening	no

3 NUMERICAL AND VALIDATION MODELS

An experimental study was conducted by Mohamed et al. [12] on a concrete block with the same dimensions (500 mm x 500 mm x 400 mm) having a compressive strength of 35 MPa and a density of 2,350 mg/cm^3; the block was subjected to a similar projectile of diameter 23 mm and having a striking velocity about 970 m/s as shown in Fig. 5. The concrete target was supported by the steel frame along its perimeter to prevent movement in both directions, then it was fired by the projectile. Fig. 6 shows the concrete block before and after impact. The recorded depth of penetration was 400 mm and the front and rear faces were fully damaged.

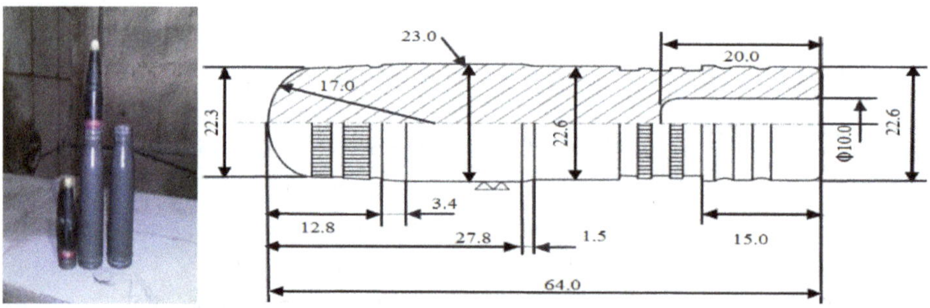

Figure 5: Dimensions of the projectile.

Figure 6: Concrete block before and after impact.

The first model in this paper is used for numerical validation. The model that was simulated using the dimensions and properties of the experimental model, recorded a depth of penetration of 390 mm when it was subjected to a 23 mm projectile with a velocity of 960 m/s as shown in Fig. 7; this shows a good agreement with the experiment result of the plain concrete block subjected to the same projectile. Fig. 8 shows the predicted damage contour of concrete target, it is obvious that the block was fully damaged.

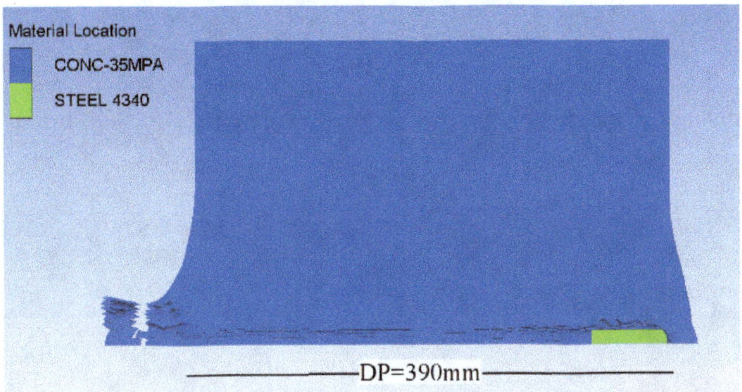

Figure 7: Depth of penetration for target with no reinforcement = 390 mm.

Figure 8: Damage contour for concrete without reinforcement: side section, front face and rear face.

The concrete target models simulated in this paper are reinforced by the chosen material (Teflon) with different thickness and at different locations. Five different thinness and locations are suggested to enhance the impact behavior of concrete block and increase its resistance to penetration of high velocity projectile. The first two models consist of two layers of 9 mm each and the other three models consist of one layer of 18 mm. Fig. 9(a) to 9(e) show the layers location and thickness of models 1 to 5 respectively.

4 RESULTS AND DISCUSSION

The results of the simulation are captured and presented in this section in terms of the projectile depth of penetration as well as the degree of concrete target damage.

The interaction between the concrete target models with Teflon layers reinforcements and the projectile was captured and represented in this section. The results show that the

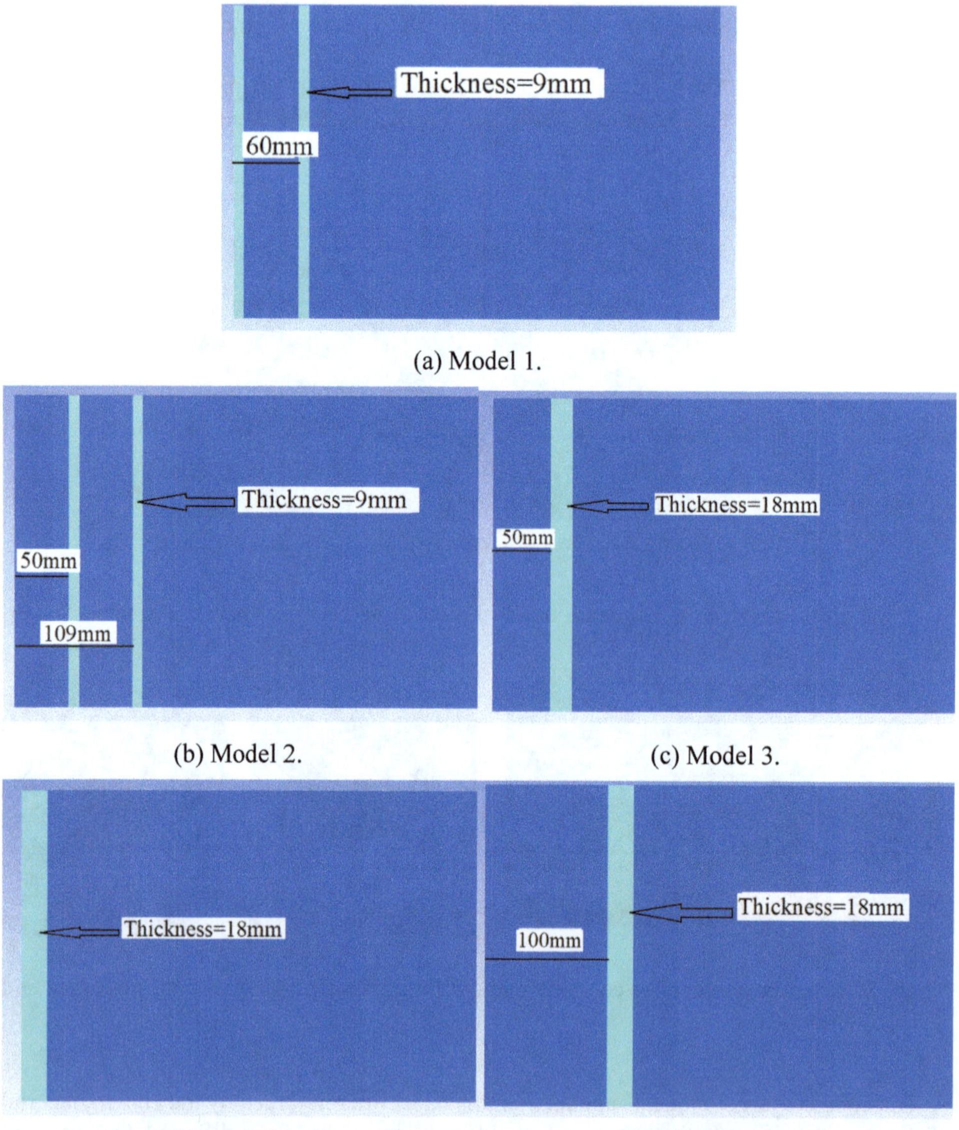

Figure 9: Teflon sheets location and thickness.

minimum depth of DP occurred in model 1, and also the minimum depth of TD. The Teflon reinforcement in model 1 is divided into two layers situated at distances 0 and 60 mm from the face of concrete target; a DP of 137 mm was recorded which means that the resistance of concrete is increased by about 65% compared to its resistance without the Teflon reinforcement. For model 2, the Teflon is also divided into two layers but situated deeper at distances 50 and 109 mm; the DP and TD recorded for this model show also an enhancement in concrete resistance but they are higher than model 1, which means that, when the reinforcement is located at lower depth, the resistance to ballistic impact is greater than when it is located at greater depth. For the models 3, 4 and 5, only one Teflon layer of thickness 18 mm is used as a reinforcement at distances 0, 50 and 100 mm respectively. The lowest DP and TD are recorded for model 3, while model 4 recorded higher DP and TD than model 3, and model 5 recorded the highest values. It can be concluded from these results also that the deeper the Teflon sheets are located in the concrete target, the higher values for DP and TD are obtained. Comparing the results of model 1 and 3, it is obvious that projectile DP in model 1 is lower; which means that when dividing the Teflon reinforcement into two layers of the same total thickness, we decrease the DP by about 14% but the total damage TD remains the same.

Table 4 summarizes the results of the five models concerning the depth of penetration of the projectile into the concrete target (DP) and the percentage of decrease of the DP with respect to the concrete model without reinforcement as well as the depth of the area of concrete subjected to total damage (TD). Contour plots of deformation and material damage are shown in Fig. 10(a) to 10(j).

Table 4: The results of simulated models.

Model #	Projectile depth of penetration (DP)	% of reduction in DP	Depth of concrete total damage (TD)	% of reduction in TD
Model 1	137 mm	64.8%	168 mm	58.0%
Model 2	164 mm	57.9%	224 mm	44.0%
Model 3	156 mm	60.0%	168 mm	58.0%
Model 4	164 mm	57.9%	189 mm	52.7%
Model 5	183 mm	53.1%	210 mm	47.5%

5 CONCLUSION

This paper presents an investigation of the ballistic performance of concrete blocks reinforced by Teflon sheets at different locations and with different thicknesses and subjected to high-velocity rigid projectile (960 m/s). The finite element models were numerically simulated using the hydrocode Autodyne 3D, and some results were compared to previous experimental work found in literature. The effect of the used material on the penetration depth of projectile and the resulted damage of concrete are presented and explained: the projectile depth of penetration can be decreased from 390 mm to 137 mm with an improvement of 64.8%, and the full damage depth is decreased from 400 mm to 168 mm with an improvement of 58%, which means a very good enhancement in penetration resistance and the degree of damage of concrete target.

From the previous analysis, we can conclude that the usage of tetrafluoroethylene polymer sheets, known as Teflon, as reinforcement for plain concrete blocks can be considered as a successful anti-penetration composite structure.

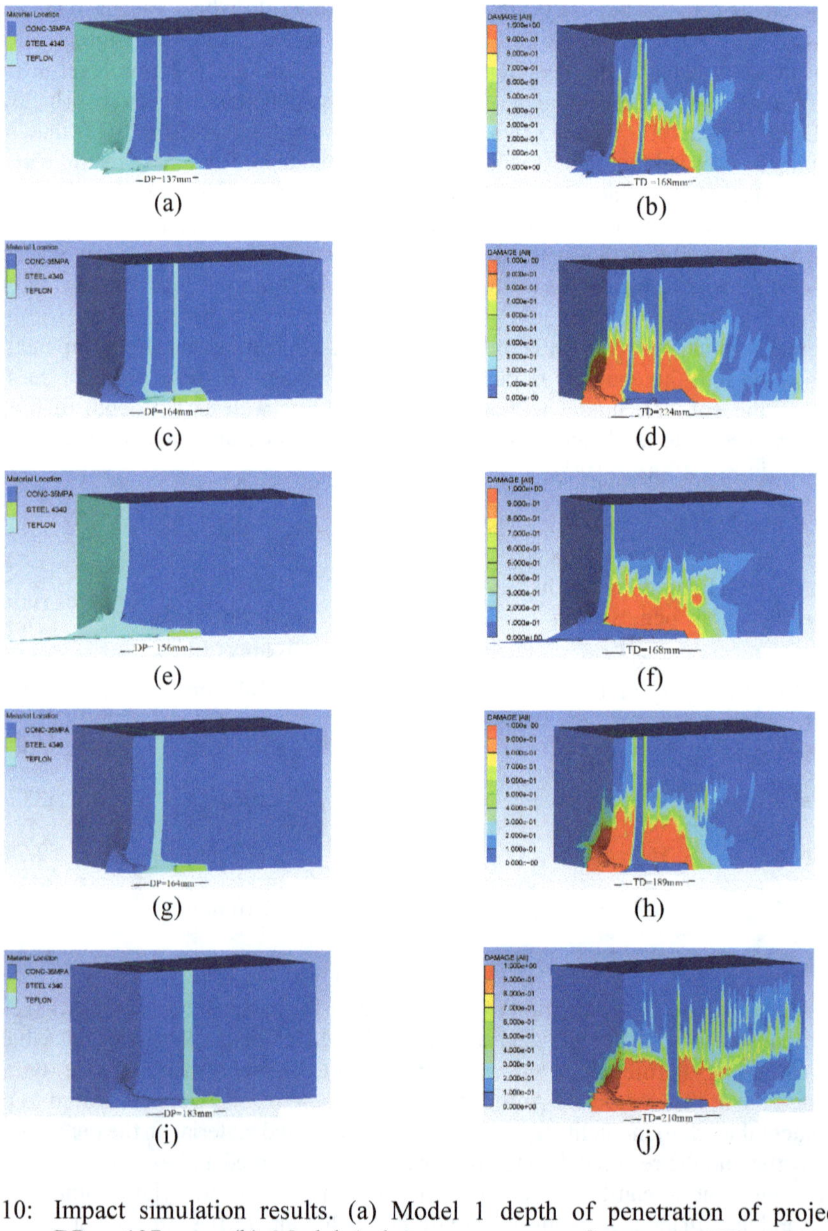

Figure 10: Impact simulation results. (a) Model 1 depth of penetration of projectile, DP = 137 mm; (b) Model 1 damage contours of concrete, TD = 168 mm; (c) Model 2 depth of penetration of projectile, DP = 164 mm; (d) Model 2 damage contours of concrete, TD = 224 mm; (e) Model 3 depth of penetration of projectile, DP = 156 mm; (f) Model 3 damage contours of concrete, TD = 168 mm; (g) Model 4 depth of penetration of projectile, DP = 164 mm; (h) Model 4 damage contours of concrete, TD = 189 mm; (i) Model 5 depth of penetration of projectile, DP = 183 mm; and (j) Model 5 damage contours of concrete, TD = 210 mm.

REFERENCES

[1] Rajput, A., Jakka, R., Pruthvik, B.M. & Iqbal, M.A., Experimental study of plain and reinforced concrete targets subjected to impact loading. *Procedia Engineering,* **144**, pp. 124–131, 2016.

[2] Iqbal, M.A., Rajput, A. & Bhargava, P., Plain and reinforced concrete targets subjected to projectile impact. *Procedia Engineering*, **173**, pp. 138–144, 2017.

[3] Smith, J. & Cusatis, G., Numerical analysis of projectile penetration and perforation of plain and fiber reinforced concrete slabs. *International Journal for Numerical and Analytical Methods in Geomechanics*, **41**, pp. 315–337, 2017.

[4] Jiricek, P. & Foglar, M., Numerical simulation of the effect of regular and sub-caliber projectiles on military bunkers. *EPJ Web of Conferences*, **94**, 04049, 2015.

[5] Tawadrous, R.I., Attia, W.A. & Laissy, M.Y., Using ceramic plates as shielding for concrete blocks against projectile penetration. *HBRC Journal*, **12**(3), pp. 263–271, 2016.

[6] Almusallam, T., Al-Salloum, Y., Alsayed, S., Iqbal, R. & Abbas, H., Effect of CFRP strengthening on the response of RC slabs to hard projectile impact. *Nuclear Engineering and Design*, **286**, pp. 211–226, 2015.

[7] Pavlovic, A. & Fragassa, C., Investigating the resistance of reinforced barriers to high velocity projectiles. *Engineering Structures*, **174**, pp. 384–395, 2018.

[8] Kamal, I.M. & Eltehewy, E.M., Projectile penetration of reinforced concrete blocks: Test and analysis. *Theoretical and Applied Fracture Mechanics*, **60**(1), pp. 31–37, 2012.

[9] DuPont, Ptfe T. Teflon PTFE®.

[10] Kennedy, R.P., A review of procedures for the analysis and design of concrete structures to resist missile impact effects. *Nuclear Engineering and Design*, **37**(2), pp. 183–203, 1976.

[11] Wright, S.C., Fleck, N.A. & Stronge, W.J., Ballistic impact of polycarbonate – An experimental investigation. *International Journal of Impact Engineering*, **13**(1), pp. 1–20, 1993.

[12] Mohamed, M.E., Eltehawy, E.M., Kamal, I.M. & Aggour, A.A., Experimental Analysis of Reinforced Concrete Panels Penetration Resistance. pp. 1–10, 2009.

CROWD-FLOW SIMULATION USING THE POTENTIAL PATHLINE METHOD

SEONGWON NAM
Korea Railroad Research Institute, South Korea

ABSTRACT
A new computational method for crowd-flow dynamics was developed to simulate pedestrian flow. It is a simulation method that uses the concept of potential pathline. The physical domain to be analysed is divided into small numerical cells, and potential values are gradually assigned starting from the exit cell. Based on the potential values, the unit direction vector for each cell can be obtained and the pathline from origin to destination drawn. These pathlines are the basic routes for individual pedestrians to move between the origin and the destination. Then, a pedestrian might change the route according to the cell conditions of each location and moves to the destination. Among the many ergonomic factors, the following four were used as the main parameters to determine the route: walking speed, visual angle, straightness probability, and right-hand traffic. It can also be applied to a three-dimensional interlayer movement model. Using the developed analysis program, the author simulated the pedestrian flow for various sample models.
Keywords: computational method, pedestrian flow, rule-based system, way-finding, potential pathline.

1 INTRODUCTION

When a pedestrian looks for a path in a certain space, the person finds a path by various environmental factors, physical characteristics and experience. Environmental factors include brightness, colour, landmark, path size, vertical and horizontal patterns, signage, etc., and are mainly external factors [1], [2]. Among the many physical characteristics of pedestrians who determine the route and move, representative factors include walking speed, visual angle, straightness, and walking on the right [3], [4]. Crowd flow is common in subway transit stations, terminals, large complex buildings, and stadiums. Because metropolitan subway transit stations are places where passenger flow is overcrowded, predicting crowd flows is important not only for comfort but also for safety [5]. Passengers in subway transit stations during commute times tend to move according to their experience and physical characteristics of the individual rather than external environmental factors. In this study, the author developed a computational method to simulate the crowd flow using the concept of potential pathline of computational fluid dynamics.

2 POTENTIAL PATHLINE METHOD

Crowd-flow dynamics (CFD) is a subject that is often studied in the field of transportation and architectural design / planning, but it is physically similar to particle flow, so there are many similarities between the two phenomena. If individual pedestrians are considered as particles, the movement of particles is the movement of individuals, the volume fraction of particles is the density of population, and the trajectory of particle is similar to the pedestrian's movement between origin and destination. Table 1 compares the similarities between the two phenomena. Many methods have been proposed to analyse the crowd flow, which can be divided into discrete models and continuous models.

Representative discrete models are the discrete element model [6], [7], cellular automata model [8], [9], lattice gas discrete model [10], and agent-based model [11]. On the contrary, continuous models are the social force model [12], [13] and forces model [14]. In the discrete model, individual pedestrians move in units of cells. That is, one cell has an area for one

WIT Transactions on Engineering Sciences, Vol 125, © 2019 WIT Press
www.witpress.com, ISSN 1743-3533 (on-line)
doi:10.2495/CMEM190031

Table 1: Similarity between particle flow and pedestrian flow.

Item	Particle flow	Pedestrian flow
Carrier	Fluid flow	Unit direction vector
Mover	Particle	Pedestrian
Density	Volume fraction	Population density
Velocity	Particle speed	Walking speed
Governing equation	The equation of particle motion	The equation of pedestrian position
Decision	*None*	*Thinking*

person, and the movement of the pedestrian moves in each cell unit. On the other hand, in the continuous model, the movement coordinates at each time are calculated by velocity and time. The crowd-flow simulation method developed in this study can trace the individual trajectory in a bilateral passenger flow as well as a multi-layer flow by the Lagrangian approach. When each pedestrian finds a path, the pathline that follows the potential value of each computational grid is the basic path, and the alternative path is searched and moved according to the real-time situation of each cell. When looking for alternative routes, the main parameters of physical characteristics are walking speed, visual angle, straightness, and walking on the right.

In Fig. 1, the factors written in blue letters are environmental external ones and the factors written in black letters are physical characteristic factors. Because external factors depend on how the environment or building is designed, it is difficult to assess quantitatively. The crowd-flow issues are mainly due to individual physical factors that are colliding or dense. In this study, physical characteristics were quantitatively modelled on the assumption that environmental factors are the same. Particle flow has many similarities with pedestrian flow. FVM (Finite Volume Method) and FEM (Finite Element Method) are typical computational fluid dynamics methods for particle flow. These computational fluid dynamics methods have the following characteristics. The space to be analysed is divided into small calculation grids, and the physical property values in each calculation grid are taken as representative average values. Once the flow properties in each computational grid are determined, the particles in the grid are moved by the equation of motion of the particle.

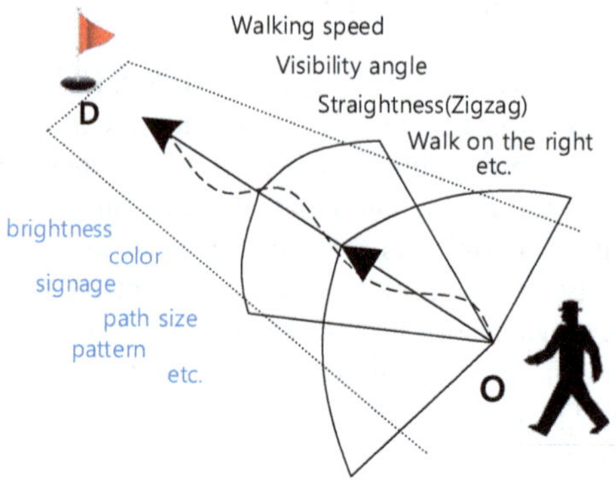

Figure 1: Factors to determine way-finding.

To illustrate the potential pathline Method developed in this study, the author set up a physical model as shown in Fig. 2. It is a building space with two entrances and one exit, with a wall in the middle. The computational grid and coordinates for the crowd-flow analysis for this model are shown in Fig. 3. The computation domain is divided by the square grid of length *dL*, and the I and J addresses are given in the x and y directions, respectively. It is assumed that the grid size can be adjusted according to the problem, and that only one person can occupy each grid. The average walking speed of people is assumed to be 4 km/h, and the horizontal and vertical lengths of the grid are 0.5 m each. The next step is to calculate the potential value for each computational grid. In the potential pathline method, this process is very simple. As a boundary condition, it is sufficient to assign a potential value of 1 to the exit grid. Grids adjacent to the exit grid have a potential value of 2. That is, if the potential value of grid (I, J) is M, the potential values of the eight surrounding grids have M + 1 values. In this way, a potential value map for the entire region can be constructed by sequentially calculating the potential values for the remaining grids. Fig. 4 shows the calculated potential value map.

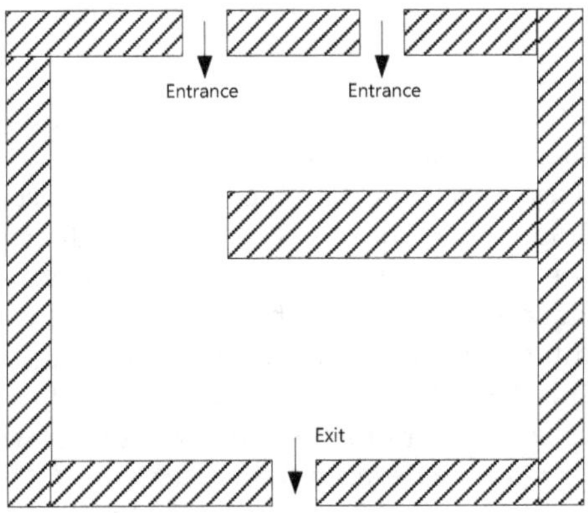

Figure 2: Physical model.

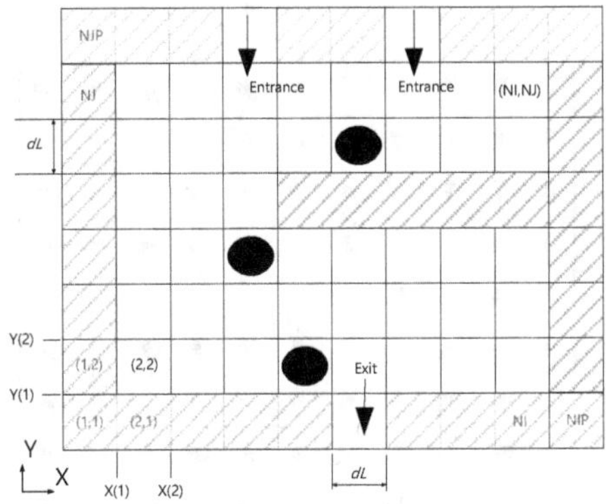

Figure 3: Computational grids and coordinates.

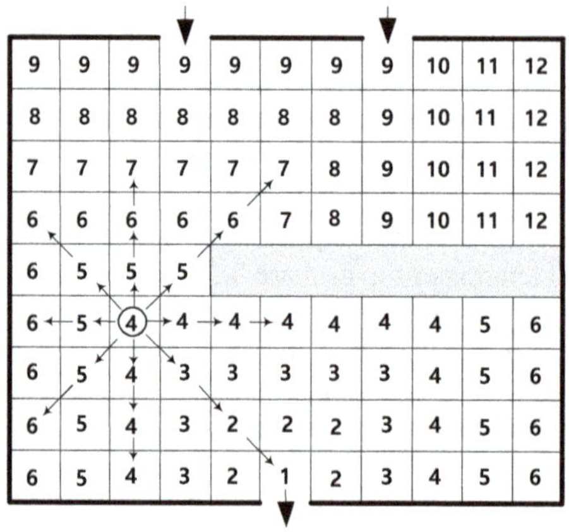

Figure 4: Potential value map.

The next procedure is to calculate the unit direction vector of each grid using a potential value map. In Fig. 4, there are eight grids around the grid with a potential value of 4. By comparing potential values in each direction, a unit direction vector can be set in the direction having the smallest average value. Potential values in each direction are 4-4-4 in the east, 3-2-1 in the southeast, 4-4-4 in the south, 5-6 in the southwest, 5-6 in the west, 5-6 in the northwest, 5-6-7 in the north, and 5-6-7 in the northeast. Therefore, it can be seen that the unit direction vector in this grid is southeast. Fig. 5 shows the direction vector set for all the grids and pathline. It is the pathline that connects the lines that follow the direction vector in each grid. If there are no other obstructions, the pedestrians will move along the pathline towards the exit. This can also be deduced from automobile GPS navigation system. When pedestrian set the origin and destination, the route is displayed. Normally, pedestrian will follow the path, but if an obstruction occurs on the path, it will lead to another alternate path. Potential pathline method also finds alternate grid in a similar way. In Fig. 6, let us assume that a person is in grid (I, J) and the direction to move is grid (I, J + 2).

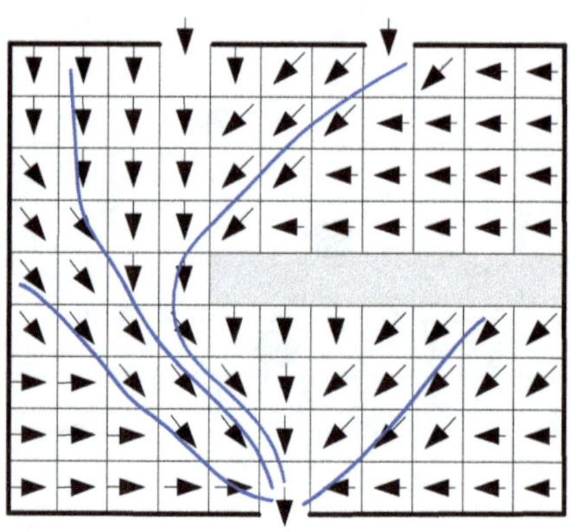

Figure 5: Direction vector and pathline.

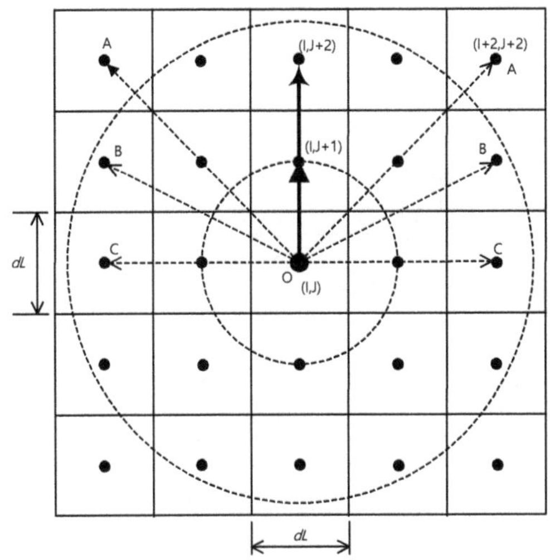

Figure 6: Alternative grid finding.

If the pedestrian is a fast person, it will go to (I, J + 2) in the big circle, and if it is slow, it will move to (I, J + 1) in the small circle. If there are other pedestrians or obstacles already in the grid in the direction he wants to proceed, he will find an alternate grid among the adjacent grids. If he is a person with narrow visibility angle, he will find an alternate grid within the range of angle AOA. If he is a person with wide visibility angle, he will find an alternate grid within the range of angle COC. Generally, pedestrians do not walk straight on a given route. According to ergonomics, pedestrians often walk in zigzags. In this case, he moves to a grid that deviates slightly from the specified direction. In addition, whether he is walking on right side when he encounters an opposite pedestrian or not also affects the route selection. The equation of the pedestrian's position considering the above-mentioned factors is as follows:

$$\bar{\boldsymbol{S}}_{ij}^{t+dt} = \left\{x_{ij}^t + \alpha V_0 \cos\left(\theta_1 \pm \theta_2 \pm \frac{\theta_3}{\beta}\right)\Delta t\right\}\bar{\imath}$$

$$+\left\{y_{ij}^t + \alpha V_0 \sin\left(\theta_1 \pm \theta_2 \pm \frac{\theta_3}{\beta}\right)\Delta t\right\}\bar{\jmath}, \qquad (1)$$

where $\bar{\boldsymbol{S}}_{ij}^{t+dt}$ is a new position vector at time $t + dt$, x_{ij}^t and y_{ij}^t are positions in the x and y directions at time t, V_0 is walking speed, θ_1 is the tangential angle of the pathline at (i, j), θ_2 is the angle deviating from the straight line according to the straightness probability, θ_3 is the visibility angle, α is the pedestrian group number (specifying a slow pedestrian or a fast pedestrian), β is the coefficient of right walking. If the pedestrian is straight ahead, θ_2 is 0, and θ_3 is 0 if there are no other pedestrians or obstacles in the direction to move. And, if pedestrians are walking on the right side of the road, this coefficient β is 2. Fig. 7 is a flow chart of the algorithm for finding a new grid to move in unit time. The calculation procedure is as follows.

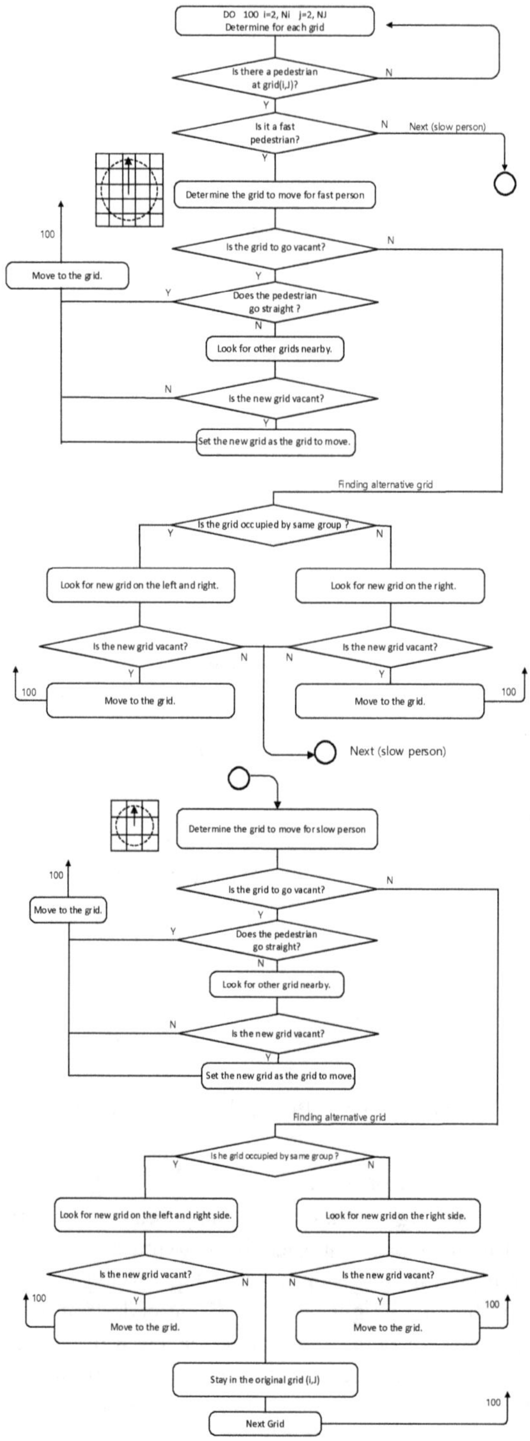

Figure 7: Flow chart to determine the grid to move.

Walking speed, visibility angle, straightness, and right-hand traffic are assigned to each pedestrian as input data. First, it is decided whether there is a pedestrian in the grid (I, J). If so, decide whether the pedestrian is a fast pedestrian or a slow pedestrian. If he is a fast pedestrian, look for the grid to move to in the big circle, and if it is a slow pedestrian, look for the grid in the small circle. If the new grid on the direction vector side is empty, move to that grid. If the newly found grid is occupied by the pedestrian of the same group, it looks for a new grid to move among the left and right grids, and if it is occupied by the pedestrians of another group, it searches for a new grid to move among grids in the rightward direction. If the newly found grids in the previous routine are not empty, follow the slow pedestrian procedures to find the new grid to move. If the pedestrian can't find the grid to move in these procedures, he will stay in his original position.

3 RESULTS AND DISCUSSION

In order to compare the validity and performance of potential pathline method developed in this study, the results using this method were compared with those of other programmes (Fig. 8). The target model is that 20 pedestrians pass through an L-shaped passage of 2.0 m in width and 20 m in length. The results of the Exodus show that the pedestrian distribution after 10 seconds is relatively long. The distribution of pedestrians at 10 and 16 seconds is similar, and differs from that at 4 seconds. The results of Pynamics show that, after 10 seconds, when turning left at the curved path, there is a tendency to outward. This is somewhat unrealistic, as pedestrians generally tend to choose the shortest distance. In contrast, these phenomena did not appear in the results using the potential pathline method.

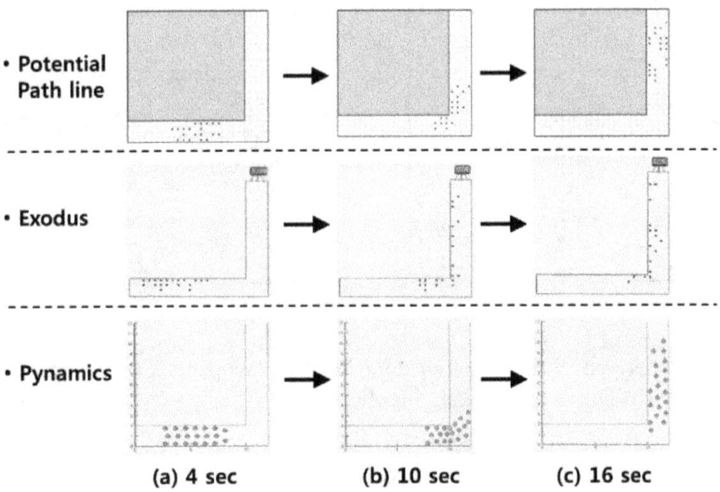

Figure 8: Comparison with other programmes results.

For the next comparison model, the results of Helbing et al.'s study are given as an example [13]. 200 people in a room of 15 m in width and 15 m in height move at a speed of 1 m/s through 1 m wide exit on the right, and Fig. 9 shows a snapshot. The crowd-flow characteristics in this model are arching and clogging around the exit. Both analytical results are qualitatively similar. But, the results of Helbing et al.'s research show that the arcs are almost the same radius from the exit, which is somewhat excessive. In contrast, the results of the potential pathline method are much more natural.

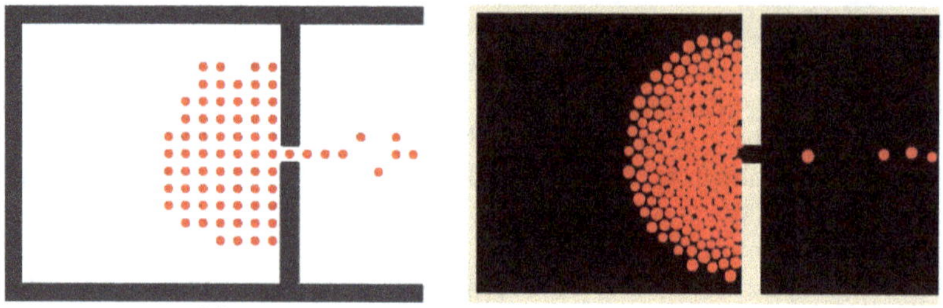

(a) (b)

Figure 9: Comparison with Helbing et al.'s result (snapshot). (a) Potential pathline method; (b) Helbing et al.'s result [13].

Fig. 10 shows a comparison of snapshot of the effect of door location. The author simulated the change of the paths of the audience leaving the theatre after the play. The author compared the cases where the doors are only on the rear side of the theatre and those in each direction. In the case of Fig. 9(a), since the doors are at the rear, the path of the audience in front of the stage is lengthened, and it can be seen that the exit is stagnated. Comparing the figures after 10 seconds, it seems that the audience in the case of Fig. 9(b) exits much more efficiently.

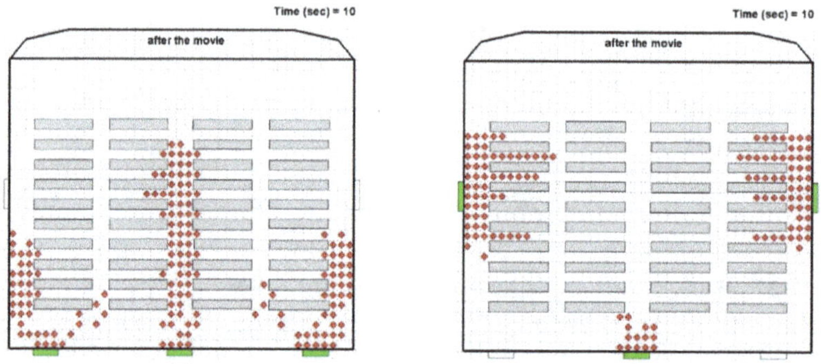

(a) (b)

Figure 10: Comparison of the effect of door location (snapshot). (a) Doors are in the back side; (b) Doors are in each direction.

4 CONCLUSIONS

Crowd-flow simulation is often used as a design tool for the structural design of a railway station, terminal, complex multiple building, and stadium. In this study, a new numerical analysis method was developed to simulate pedestrian flow. It is a computational analysis method using the concept of potential pathline. The domain to be analysed is divided into small calculation cells, and a potential value is gradually given starting from the exit cell. Based on the potential values, the direction vector for each cell can be obtained and the pathline following it drawn. These pathlines are the basic routes for individual pedestrians to move between the origin and the destination. In this simulation method, walking speed, visibility angle, straightness, and right-hand traffic were used as main parameters. Using the developed simulation programme for various pedestrian flow cases, the pedestrian flow at

L-shaped route, evacuation from a room, pedestrian count flow in passage, and exit patterns for the doors' location were simulated.

ACKNOWLEDGEMENT

This research was supported by a grant (PK1901C4) from the R&D programme of Korea Railroad Research Institute (KRRI), Republic of Korea.

REFERENCES

[1] Hughes, R.L., The flow of human crowds. *Annual Review of Fluid Mechanics*, **35**, pp. 169–182, 2003.

[2] Moussaid, M., Helbing, D. & Theraulaz, G., How simple rules determine Pedestrian Behavior and Crowd Disasters. *PNAS*, **108**(17), pp. 6884–6888, 2011.

[3] Polus, A., Schofer, J.L., & Ushpiz, A., Pedestrian flow and level of service. *J. Transp. Eng. Proc. ASCE*, **109**, pp. 46–57. 1983.

[4] Nam, S., Development of potential pathline method for railway passenger flow analysis. *Journal of the Korean Society for Railway*, **21**(8), pp. 830–837, 2018.

[5] Daamen, W., *Modelling Passenger Flows in Public Transport Facilities*. PhD thesis, Delft University of Technology, 2004.

[6] Nam, S., Development of algorithm for passenger flow analysis based on DEM. *Journal of the Korean Society for Railway*, **8**(4), pp. 337–341, 2005.

[7] Nam, S., Analysis of pedestrian flow characteristics in subway station. *Journal of the Korean Society for Railway*, **9**(3), pp. 271–276, 2006.

[8] Nagal, K. & Schreckenber, M., A cellular automation model for freeway traffic. *Journal of Physics*, **2**(12), pp. 2221–2229, 1992.

[9] Burstedde, C., Klauck, K. & Schadschneider, A., Simulation of pedestrian dynamics using a two-dimensional cellular automaton. *Physica A*, **295**, pp. 507–525, 2001.

[10] Guo, R. & Huang, H., A mobile lattice gas model for simulating pedestrian evacuation. *Physica A*, **387**, pp. 580–586, 2008.

[11] Heliovaara, S., Korhonen, T., Hostikka, S. & Ehtamo, H., Counterflow model for agebt-based simulation of crowd dynamics. *Building and Environment*, **48**, pp. 89–100, 2012.

[12] Helbing, D. & Molnar, P., Social force model for pedestrian dynamics. *Physical Review E*, **51**, pp. 4282–4286, 1995.

[13] Helbing, D., Farkas, I. & Vicsek, T., Simulating dynamical features of escape panic. *Nature*, **407**(28), pp. 487–490, 2000.

[14] Lohner, R., On the modelling of pedestrian motion. *Applied Mathematical Modelling*, **34**, pp. 366–382, 2010.

CFD ANALYSIS OF LOW FREQUENCY OSCILLATIONS IN NEWTONIAN AND NON-NEWTONIAN FLUIDS IN A VERTICAL PIPE

MADURANGA AMARATUNGA, HERIMONJA A. RABENJAFIMANANTSOA & RUNE W. TIME
Department of Energy and Petroleum Engineering, University of Stavanger, Norway

ABSTRACT

The effect of vertically imposed, low frequency oscillations on the rheology of shear-thinning non-Newtonian fluids are studied numerically by computational fluid dynamics (CFD). Both Newtonian (water) and inelastic time-independent fluids of power-law (Poly-Anionic Cellulose (PAC)) are used as test fluids. Unsteady state simulations were performed using ANSYS Fluent version 18.0 for a vertical 2-D pipe geometry (ID = 50 mm and H = 850 mm) and sinusoidal, vertical oscillations to the liquid body itself were imposed with a user-defined function (UDF). The multiphase volume of fluid (VOF) method with realizable k-ε method was used to impose the turbulence nature of the flow for the cases with water while the cases with non-Newtonian fluids were simulated under the laminar condition. Oscillations of different low frequency values (0.25–5 Hz) and different velocity amplitudes (0.1–0.3 m/s) were tested numerically. The dynamic variation of velocity and shear rate within the oscillated, bulk liquid medium is demonstrated. The flow inside the vertical pipe acts plug like at higher frequencies for both Newtonian and non-Newtonian fluids. The air–liquid interface becomes unstable with small disruptive peaks for the cases with water at higher velocity amplitudes while that is very calm for the laminar cases with non-Newtonian fluids. The achieved velocity gradients and the resultant shear rate variation are low with the increased PAC concentration due to the viscous resistance. However, the instantaneous velocity profiles display a progressively more complex structure with increased frequency and velocity amplitude, revealing the presence of alternating upward/downward motion. These alternating velocity profiles confirm the varying shear field present within a drilling pipe at different frequencies and velocity amplitudes while the variation of the shear field is more dependent on the velocity amplitude.

Keywords: frequency, velocity amplitude, displacement amplitude, shear rate, dynamic velocity, Computational Fluid Dynamics.

1 INTRODUCTION

Drilling fluids may experience different frequencies and oscillations while they are circulating through the well. For an instance, in some cases, gas kicks may arise at underbalanced drilling conditions, where there is a possibility of severe oscillations and even extreme accelerations might take place until pressure equilibrium is reached by friction losses.

The shear-thinning behaviour of typical water-based drilling fluids contributes for a reduction in flow friction against the drill pipe walls. The apparent viscosity of such fluids decreases with the shear rate, which will eventually alter the dynamical behaviour of the drilling fluids. That is the main reason why the oscillatory motion should be considered so importantly when shear-thinning non-Newtonian drilling fluids are circulating through the well. Time and Rabenjafimanantsoa [1] describe this phenomenon further that, the oscillation frequency of such system is mainly related to mass inertia and to the gas compressibility, while the amplitude decay can be attributed to the liquid viscosity via friction. Since the shear rate and shear stress of drilling fluids may vary substantially over time under such scenarios, it is vital to design them accordingly to achieve desired solids carrying capacities and solids separation capacities.

WIT Transactions on Engineering Sciences, Vol 125, © 2019 WIT Press
www.witpress.com, ISSN 1743-3533 (on-line)
doi:10.2495/CMEM190041

Understanding of dynamic velocity variation and thus the change of rate of shear within a vertically oscillating liquid body is hard due to complex rheology of the non-Newtonian fluid and its behaviour at different oscillating conditions. Computational Fluid Dynamics (CFD) modelling is an advanced tool to study the various aspects of dynamic velocity variations within such complex systems. In literature, studies can be found related to oscillatory flow in pipelines or simply in tubes [2]–[6]. However, almost all of them have investigated the flow enhancement or retardation when the liquid containing pipe is oscillated by different means and confirm that pulsatile flow of shear-thinning non-Newtonian fluids enhances the flow within pipelines [2]. Experimental and numerical study performed by Mena et al. [4] to examine the flow enhancement of viscoelastic fluids through pipes vibrated longitudinally reveals that the maximum enhancement achieved at maximum frequency and amplitude. But, the dominating factor in the oscillating flow of viscoelastic liquids in pipes is the shear dependent viscosity. According to the studies performed by Torralba et al. [7], it has been identified that, the remarkable enhancement in the dynamic response of the viscoelastic behaviour in the presence of a pulsating environment could be attributed to a resonant effect due to the elastic behaviour of the fluid and the geometry of the container as well.

However, the studies that address the effect of oscillatory motion on the rheology of non-Newtonian fluids are limited. Amaratunga et al. [8] have studied the impact of horizontal oscillation on non-Newtonian rheology in a square-cross sectional liquid column. Their efforts in quantification of the change of shear field close to the walls of the tank prove the fact that, the oscillatory motion which is experienced by shear-thinning non-Newtonian fluids will result in increased settling due to the reduction of flow resistance.

The objective of this study is to numerically investigate the effect of oscillation frequency and velocity amplitude on the dynamic velocity field within the liquid medium and to interpret its consequences on non-Newtonian rheology and settling of particles. The work is a part of the study, which aims to investigate the effect of vibration/oscillation on particle settling in shear-thinning non-Newtonian drilling fluids.

2 COMPUTATIONAL MODEL AND METHODOLOGY

Two-dimensional, two-phase CFD simulations were performed to investigate the effect of frequency and amplitude of oscillatory motion on the velocity field of a liquid column.

2.1 The geometry, boundary conditions and material properties

ANSYS DesignModeler 18.0 and ANSYS Meshing 18.0 were used for the geometry and mesh generation, respectively. A vertical pipe with an internal diameter of 50 mm and a length of 850 mm was used as shown in Fig. 1. A piston arrangement was made to provide the sinusoidal oscillatory motion to the liquid body itself and an air layer was considered on top of the liquid surface to study the interface behaviour during oscillations. Both Newtonian (water) and non-Newtonian fluids were simulated. For the non-Newtonian case 2 g/L and 4 g/L Poly-Anionic Cellulose (PAC) in water were used. The viscosity of the PAC solution was defined as a non-Newtonian power-law fluid in ANSYS Fluent.

A grid-independence study was carried out by performing number of simulations with different grid sizes to ensure satisfactory results without any influence on the grid size used. In this way, a grid size of 1 mm was selected as the maximum element size and the geometry was meshed with triangular cells. The outlet boundary was set as an "outflow" and the sidewalls were defined as deforming boundaries. The physical and rheological properties of the test fluids are listed in Table 1. The values for water and air are from Fluent data base and for PAC are from laboratory experiments.

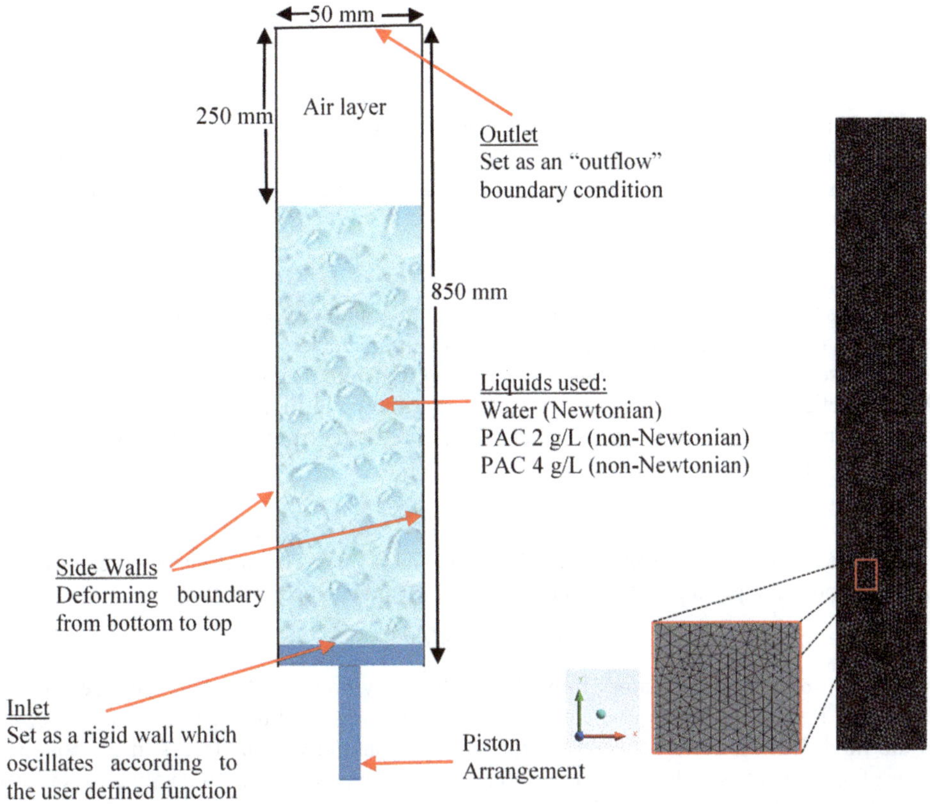

Figure 1: The geometry and the grid.

Table 1: Properties of the test materials.

Material	Density (kg/m^3)	Newtonian viscosity (Pa.s)	Non-Newtonian power-law parameters			
			Consistency index (K) ($Pa.s^n$)	Behaviour index (n) (-)	Minimum viscosity limit (Pa.s)	Maximum viscosity limit (Pa.s)
Water	998.2	0.0010	–	–	–	–
Air	1.225	1.79e-5	–	–	–	–
PAC 2 g/L	1000	–	0.057	0.78	0.013	0.036
PAC 4 g/L	1000	–	0.194	0.71	0.024	0.124

2.2 CFD model

The governing equations, which form the basis of the flow for a generalized incompressible Newtonian fluid in a pipe, can be written as [9]:

$$\text{continuity} \qquad \nabla . \boldsymbol{U} = 0 \tag{1}$$

motion $$\rho \frac{DU}{Dt} = -\nabla p + [\nabla . \eta \dot{\gamma}] + \rho \boldsymbol{g}, \tag{2}$$

where, $\boldsymbol{U}$ is the velocity vector, t is the time, $\boldsymbol{g}$ is the gravitational acceleration. The term on the left-hand side of eqn (2) represents the mass per unit volume times acceleration, while the right-hand side represents the pressure force, viscous force and the gravity force per unit volume respectively. The non-Newtonian fluids considered in the simulations are of power-law type, i.e.:

$$\eta = k\dot{\gamma}^{n-1}, \tag{3}$$

where, η is the apparent viscosity, k is the consistency index, $\dot{\gamma}$ is the shear rate and n is the behaviour index of the non-Newtonian fluid. The transverse sinusoidal oscillatory motion was imposed on the liquid body itself mimicking a piston arrangement from the bottom in the y-direction normal to the pipe cross section as shown in Fig. 1. In this case of a simple harmonic motion, the displacement of the liquid body y(t) from its middle position can be written by eqn (4):

$$y(t) = a \sin(\omega t), \tag{4}$$

where, a is the piston displacement amplitude and $\omega = 2\pi f$. The time varying velocity $V(t)$ of the piston is simply the differentiation of eqn (4) and can be written as,

$$V(t) = A \cos(\omega t), \tag{5}$$

where $A = a\omega$. For an oscillatory flow, the oscillatory/vibrational Reynolds number (Re_v) is defined as $Re_v = (a\omega D/\upsilon)$ [6], where D is the pipe diameter and υ is the kinematic viscosity of the fluid. It is the ratio between the forces of inertia and viscous forces in the boundary layer of the flow caused due to the oscillatory effect. For all the cases, the values of Re_v were in the laminar regime (< 455) for both PAC fluids and in turbulent regime (> 9000) for water. The test conditions for the different simulation cases with different oscillation properties are given in Table 2.

Table 2: Conditions for the different simulation cases.

Case #	Frequency (f) (Hz)	Velocity amplitude (A) (m/s)
Case 1	0.25	0.1
Case 2	0.50	0.1
Case 3	0.75	0.1
Case 4	1.00	0.1
Case 5	5.00	0.1
Case 6	1.00	0.2
Case 7	1.00	0.3

The inlet was defined as a "rigid body" in order to facilitate the provision of different properties (frequency and velocity amplitude) of the oscillatory motion using a user-defined function (UDF). The user-defined function was important to define the geometry and motion with time, since the zone adjacent to the inlet wall is deforming and moving at the same time.

The multiphase volume of fluid (VOF) method was employed to specify the air and fluid phases. The pressure based, realizable k-ε turbulent model was used for all the cases with water and the test cases for non-Newtonian fluids were simulated under laminar conditions.

Under the dynamic mesh settings, diffusion-based smoothing method with a diffusion parameter of 2 was utilized to update the volume mesh in the deforming regions.

All the simulation cases were solved over five oscillation periods in order to acquire more generalized result. It did not affect the accuracy of the result, but increased the simulation time in proportion. The time step size and the number of time steps were determined after a considerable amount of computational experimentation based on the frequency of the particular simulation case and solution convergence. The simulated liquid velocities and shear rates were extracted along a radial line located 200 mm below the initial liquid surface and displayed in Figs 3–7.

3 RESULTS AND DISCUSSION

The results of the 2-D transient simulations performed using ANSYS Fluent to investigate the effect of oscillatory motion on dynamic velocity and shear field are reported in this section. The displacement and the corresponding velocity of the piston motion at different positions within a complete phase cycle are illustrated in Fig. 2. In the following, the positions A, B, C and D refer to the different positions within a complete phase cycle as shown in Fig. 2.

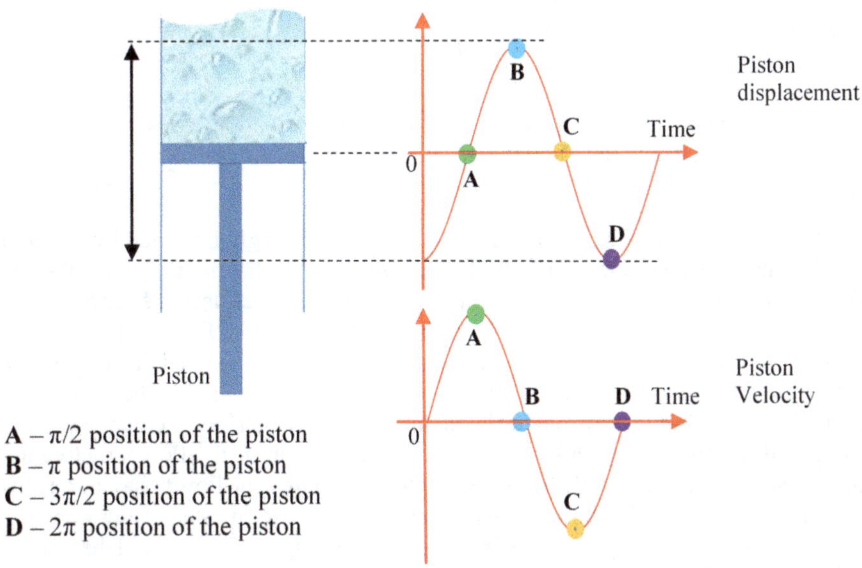

Figure 2: Displacement and velocity profiles of piston arrangement at different phase positions.

3.1 Effect of oscillatory motion on air–liquid interface behaviour

It is quite important to study the behaviour of the interface between the two fluids, when the oscillations take place. Fig. 3 shows the air–liquid interface patterns at different positions within a complete phase cycle only for water (representing the turbulent scenario) and for PAC 2 g/L (representing a laminar flow situation) for brevity. At t = 0, the surface of the fluid is subjected to zero parallel shear stresses and when the vertical oscillation starts to act, it induces surface waves on the air–liquid interface and thus enhances undesired dynamic mixing behaviour. Ito et al. [10] describe that, the pressure perturbations due to the time dependent fluid acceleration and the viscous/capillary forces due to the fluid–wall relative motion are the two main factors affecting the stability of the oscillating interface.

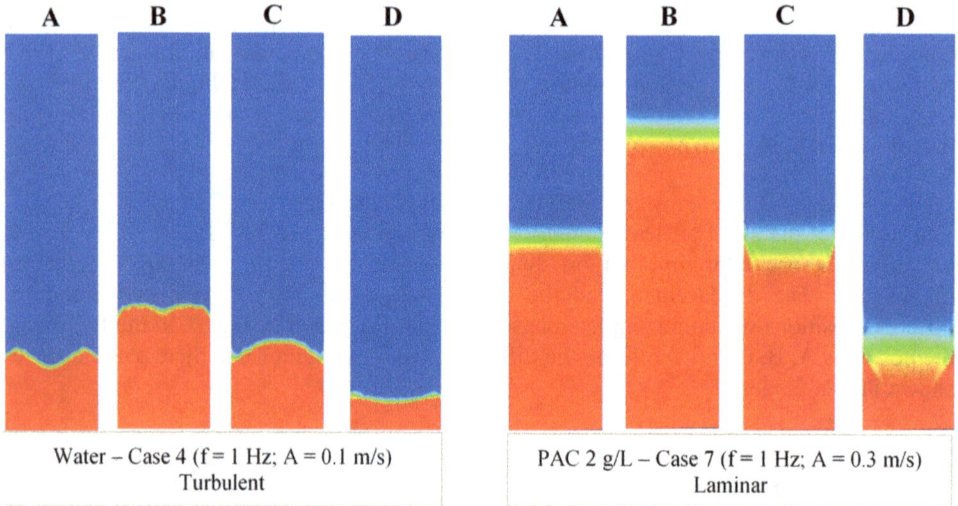

Figure 3: Air–liquid interface patterns two different simulation cases representing both Newtonian and non-Newtonian fluids.

When the interface reached its maximum level, it was strongly impacted by the inertia of the fluid column and the shape of this interface is dependent on the frequency and the velocity amplitude. It could be noted that, the air–liquid interface becomes unstable for the cases with water tending to generate quite a lot of disruptive small peaks like fingers at higher velocity amplitudes. This could be attributed to the surface instabilities generated due to the turbulent flow structure and also due to the absence of surface tension data fed to the simulation. For the laminar cases with non-Newtonian fluids, the air–liquid interface seems to be very calm without any disruptive small peaks. The laminar formulation of the flow structure and the increased viscous resistance could be the reason for such plug-type flow behaviour. However, even the flow is laminar for both non-Newtonian fluids; the interface shows an unstable behaviour at higher velocity amplitudes as shown in Fig. 3. This could be due to the destabilization of the interface by the viscous forces when there is a bulk relative motion between the fluid and the wall. At higher simulated frequencies, the air–liquid interface for both PAC fluids appears to be calmer and tends to show more or less plug type behaviour.

3.2 Variation of velocity magnitude

The variation of velocity field within the bulk liquid medium is a better way to understand the impact of the oscillatory motion. The radial velocity profiles illustrated here were extracted along a horizontal line located 200 mm below the initial air–liquid interface.

3.2.1 Effect of oscillation frequency

Fig. 4 shows the dynamic velocity profiles for water, PAC 2 g/L and PAC 4 g/L fluids at different frequencies and at different positions within a phase cycle for a fixed velocity amplitude (Cases 1–5). Note that the velocity values given by Fluent are absolute values, so that stages C and D are downwards flow (negative in "real" values).

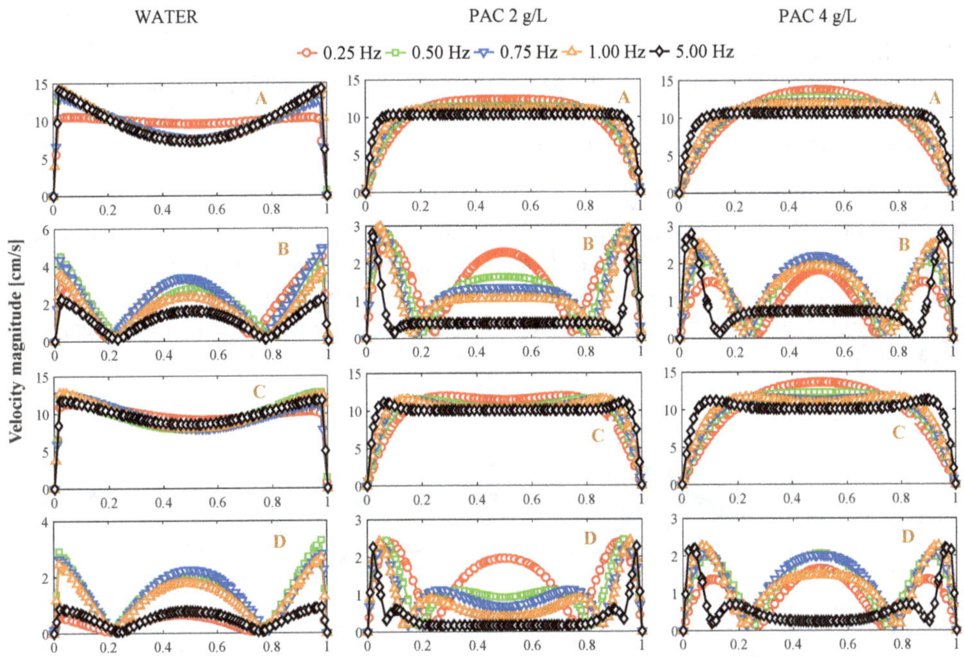

Figure 4: Velocity variation with frequency of oscillation at different phase positions of a complete cycle (velocity amplitude is constant at 0.1 m/s). Non-dimensional horizontal distance (x/D) (-).

The velocity profiles and the corresponding shear rates are calculated in the x–y plane between vertical walls. Thus the symmetry line ("pipe centre" for similar 2D cylindrical simulations, z–r) is at the non-dimensional horizontal distance of 0.5. One should also be aware that since the (x–y) simulation corresponds more to flow between parallel planes, the calculated wall friction in these simulations would be somewhat different from those obtained with a 2-D cylindrical geometry. Furthermore, in real drilling operations, local uneven geometries that might be present within the well can induce flow instabilities. Similarly, different flow conditions may generate different oscillatory motions to induce flow instabilities. The turbulent behaviour of the flow often occurring within a well makes the scenario more complicated from a rheological point of view when the non-Newtonian fluids are circulating. According to Fig. 4, it could be noted that irrespective of the flow condition whether it is laminar or turbulent, the patterns/shapes of the velocity profiles for all the test fluids at different phase positions are slightly qualitatively similar. Specially, at the phase positions A and C for water, the velocity profiles almost coincide each other for all the frequency values. However, there is a considerable variation of the velocity profiles at phase positions B and D, where the piston is in its maximum and minimum levels respectively. However, when the piston is in the middle position of its way (phase positions A and C), the radial velocity profiles for both non-Newtonian fluids show plug-type behaviour with increased frequency, deviating from the usual parabolic shape. The phase positions B and D are the moments where the piston comes to a rest to change its direction and so the bulk liquid medium also lowers its velocity to change the direction. As a result of this, a bell shaped velocity profile could be seen at those phase positions for all the liquids concerned here, where a local velocity maximum (and sometimes a minimum too) can be observed close to the wall region and that propagates towards the centre of the pipe with decreasing amplitude. Furthermore, it is to be noted that, the peak velocity magnitudes achieved by non-Newtonian fluids are lower than that of water, which could be attributed to the higher viscous resistance of the non-Newtonian liquids and also the laminar flow conditions.

3.2.2 Effect of velocity amplitude

Fig. 5 shows the radial velocity profiles for water, PAC 2 g/L and PAC 4 g/L fluids at different velocity amplitudes and at different positions within a phase cycle for a fixed excitation frequency (Cases 4, 6 and 7). Here, again it is to be noted that the velocity values given by Fluent are absolute values, so that stages C and D are downwards flow (negative in "real" values).

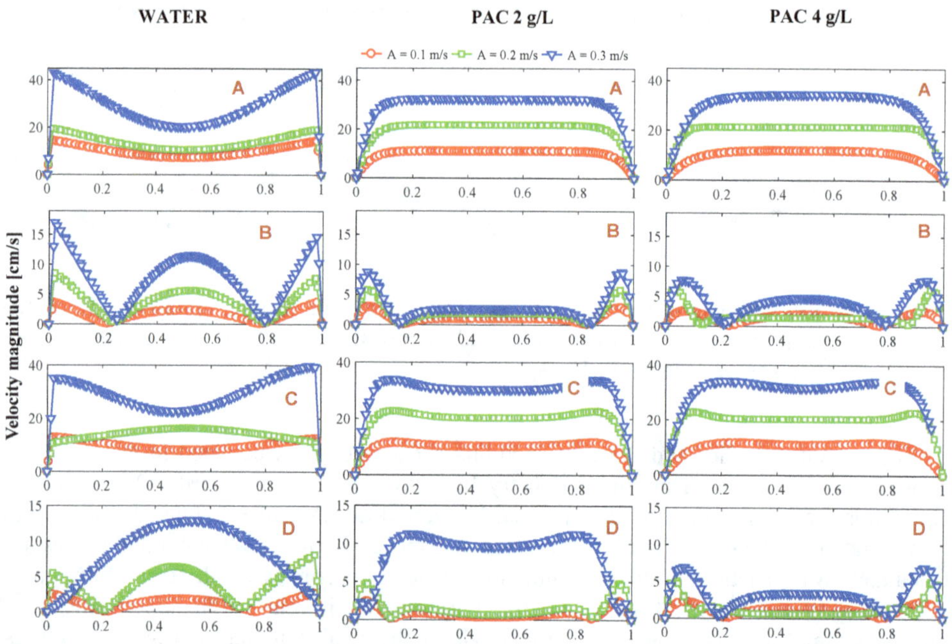

Figure 5: Velocity variation with velocity amplitude at different phase positions of a complete cycle (excitation frequency is constant at 1 Hz). Non-dimensional horizontal distance (x/D) (-).

At the first glance, it can be seen that, when the excitation velocity amplitude is increased, the achieved instantaneous velocity magnitudes are also increased for all the three test fluids. That is just because of the increased kinetic energy provided by the excitation via the piston. For almost all the cases, the velocity gradient close to the wall is greater with compared to that within the middle area of the pipe. In the turbulent case for water, the velocity profiles show significant variation in qualitatively and quantitatively with respect to the excitation velocity amplitude. However, for the laminar cases in non-Newtonian fluids, the radial velocity profiles just 200 mm below the air–liquid interface act more likely to be plug-type specially in the phase positions of A and C. Similar to what we observed in Fig. 4, the velocity gradient increases and suddenly changes its sign near the wall region. After the local velocity reaches its maximum, it tends to travel with a very low velocity gradient close to zero for both the PAC fluids. However, it is to be noted that the velocity gradient for water along the horizontal distance is significant and that could be attributed to turbulent effects within the dynamic flow situation. Therefore, it can be depicted that, the effect of driving velocity amplitude on fluids oscillated vertically lead to significant variations in the radial velocity profiles revealing the presence of an alternating upward/downward motion. This complexity is significant in non-Newtonian fluids since these varying velocity magnitudes and hence the velocity gradients change the shear field of the non-Newtonian fluids.

3.3 Variation of shear rate with frequency of oscillation and velocity amplitude

Fig. 6 shows the variation of shear rate for some selected simulation cases of PAC 2 g/L and PAC 4 g/L along the same horizontal line located 200 mm below the initial air–liquid interface in which the velocity profiles are also drawn. The four cases shown in Fig. 6 have been selected randomly for brevity, just to represent the variation of shear rate. The instantaneous shear rate values along this horizontal position have been calculated using the velocity gradient dy/dx as illustrated by Amaratunga et al. [8]. In that sense, water also provides the values for this velocity gradient, but the importance of discussing the variation in shear rate of the non-Newtonian fluids is because, the shear viscosity of such fluids change with the rate of shear.

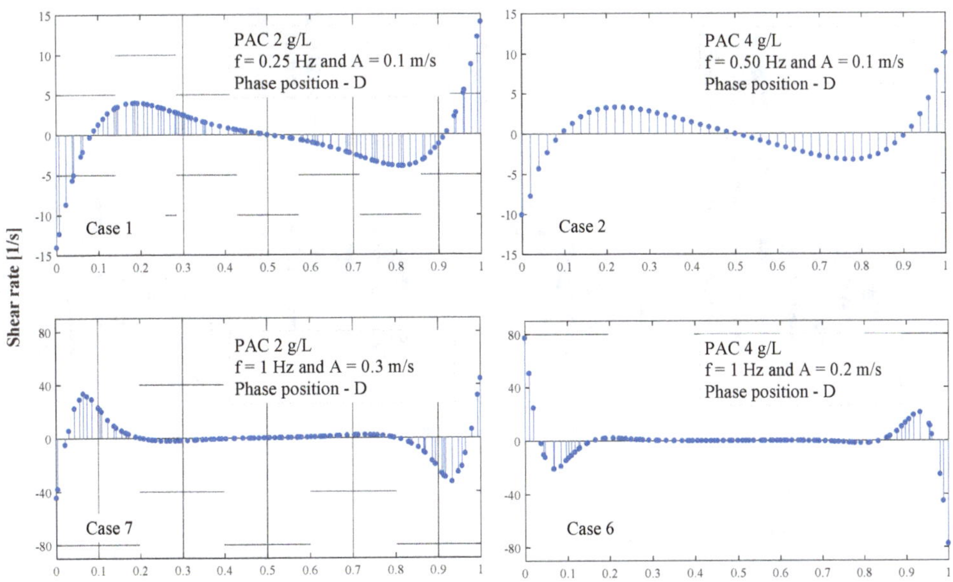

Figure 6: Variation of shear rate along a horizontal line for some selected simulation cases of PAC 2 g/L and PAC 4 g/L. Non-dimensional horizontal distance (x/D) (-).

The transient results are consistent with the interpretation of the phenomenon, in that the shear rate is maximum when the velocity gradient is maximum and vice versa. Shear rate is normally highest close to the pipe wall in both steady state and transient flows with boundary layer build-up. This fact is greatly proven by the profiles of shear rate shown in Fig. 5 for the phase position D of selected cases. However, it could be noted that, the shear rate variation is somewhat significant even within the centre area of the pipe at low frequencies (see Case 1 and 2 in Fig. 6). But, when the frequency is increased, the shear rate variation becomes more flat and align close to zero within the centre area of the pipe since the flow tends to be more plug-type. The cases illustrated in Fig. 6 depict the wall-induced shear variation as a result of the oscillatory motion experienced by the bulk liquid medium within the pipe. This additional shear introduced by the oscillations causes a variation in shear viscosity due to the shear-thinning property of the fluids. Thus, the superimposition of a sinusoidal oscillatory motion on a bulk liquid medium contained in a pipe leads to a change in apparent local shear viscosity of the shear-thinning non-Newtonian fluids. Consequently, the solids carrying capacities and solids separation capacities of such fluid are altered considerably.

It is worth to check how significant the frequency and velocity amplitudes are in terms of shear rate variation in a vertically oscillating liquid medium. Fig. 7 summarizes the maximum amount of shear rate achieved by the two non-Newtonian fluids for all the simulation cases.

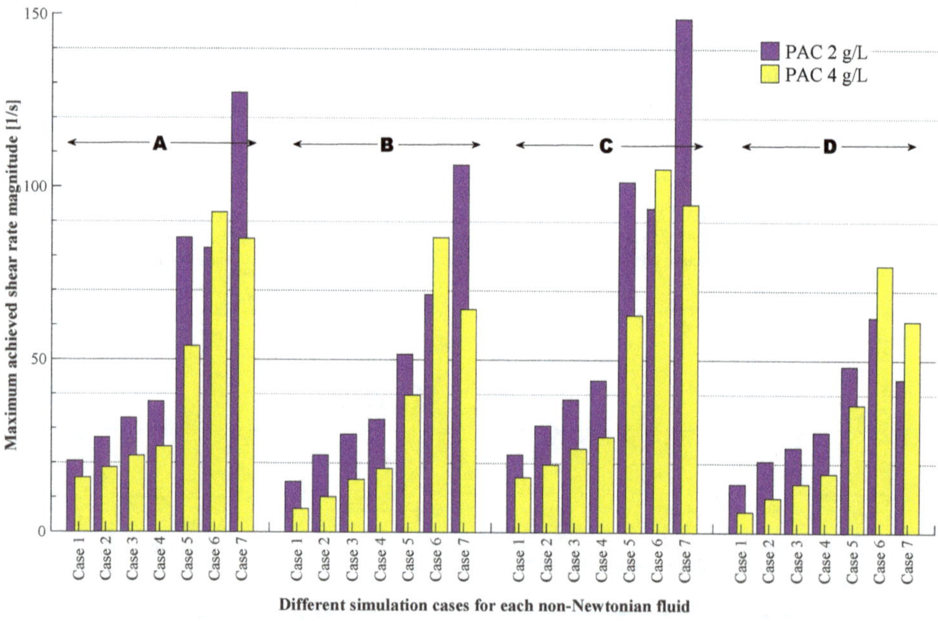

Figure 7: Magnitude of shear rate achieved by the two non-Newtonian fluids for all the simulation cases.

It can be clearly seen from Fig. 7, that for almost all the cases except at higher velocity amplitudes, PAC 2 g/L achieves higher shear variation compared to that of PAC 4 g/L. That could be attributed to the higher viscous resistance of PAC 4 g/L. Due to the higher viscous resistance, PAC 4 g/L is reluctant to be dynamic as PAC 2 g/L within the pipe even the frequency is increased. Therefore, the achieved velocity magnitudes and also the velocity gradients are lower compared to that of PAC 2 g/L. However, for the cases with fixed oscillation frequency and higher velocity amplitudes, the increased kinetic energy of the system is more significant than the viscous resistance.

Out of the four phase positions, A and C are the levels with highest kinetic energy within the system and that is confirmed by the higher shear magnitudes depicts by the figure. It is obvious that, when the frequency is increased the additional shear induced by the walls are also increased. This indicates that, the tendency of shear field variation is significant at higher oscillation frequencies and higher velocity amplitudes. According to the values shown in Fig. 7, out of all the simulation cases and positions of the phase cycle, let us consider the maximum change in shear rate for PAC 2 g/L is approximately 150 1/s and it is in the range of 110 1/s for PAC 4 g/L for the concerned geometry. Based on the K and n values mentioned in the Table 1 for the power-law modelling of the two liquids, the maximum percentage decrement of the shear viscosity could be calculated to be approximately 14% for PAC 2 g/L and 18% for PAC 4 g/L respectively. It is obvious that, a reduction in shear viscosity for a shear-thinning non-Newtonian fluid means an increment of solids separation rate.

Therefore, it should be paid considerable attention when employing shear-thinning non-Newtonian fluids in applications where they undergo high frequency and high amplitude oscillations because they could namely and greatly affect the non-Newtonian rheology, which eventually alters the solids separation and carrying capacities.

4 CONCLUSION

The influence of oscillation frequency and velocity amplitude on dynamic velocity profiles and resulting shear field has been numerically investigated in this study. Seven different cases with varying frequency and velocity amplitudes were simulated using ANSYS Fluent 18.0 version. The air–liquid interface becomes unstable for the turbulent Newtonian cases at higher velocity amplitudes due to the surface instabilities generated by the turbulent flow structure and the absence of surface tension data. For the laminar cases with non-Newtonian fluids, the air–liquid interface seems to be very calm due to the laminar flow behaviour and the increased viscous resistance. When the piston is in the middle position of its way, the radial velocity profiles for both non-Newtonian fluids show plug-type behaviour with increased frequency, deviating from the usual parabolic shape. The peak velocity magnitudes achieved by non-Newtonian fluids are lower than that of water, which could be attributed to the higher viscous resistance of the non-Newtonian liquids and also the laminar flow conditions. When the excitation velocity amplitude is increased, the achieved instantaneous velocity magnitudes are also increased for all the three test fluids. For almost all the cases, the velocity gradient close to the wall is greater with compared to that within the middle area of the pipe. Shear rate is normally highest close to the pipe wall in both steady state and transient flows with boundary layer build-up. However, it could be noted that, the shear rate variation is somewhat significant even within the centre area of the pipe at low frequencies and when the frequency is increased, the shear rate variation becomes more flat and align close to zero within the centre area of the pipe since the flow tends to be more plug-type. According to the calculations based on the results obtained by the simulations, the maximum percentage decrement of the shear viscosity could be approximated as 14% for PAC 2 g/L and 18% for PAC 4 g/L respectively. Therefore, the additional shear introduced by the oscillations causes a variation in shear viscosity due to the shear-thinning property of the fluids, which eventually leads to a change in local shear viscosity of the shear-thinning non-Newtonian fluids. For hydraulic conveying purposes, it can be concluded that the for sinusoidal oscillatory motion of shear thinning liquid medium contained in a pipe leads to an alteration of fluid's solids carrying capacity and solids separation capacity.

ACKNOWLEDGEMENTS

Authors gratefully acknowledge the Norwegian Research Council for funding this study under the project "NFR Improved Model Support in Drilling Automation". We are also grateful to Stein-Tore Johansen at SINTEF-NTNU for valuable discussions on CFD simulations.

REFERENCES

[1] Time, R.W. & Rabenjafimanantsoa, A.H., On the relevance of laboratory scale rheometric measurements for calculation of complex large scale flows in well drilling and pipe flows. *Annual Transaction of the Nordic Rheology Society*, **22**, pp. 69–79, 2013.

[2] Barnes, H.A., Townsend, P. & Walters, K., On pulsatile flow of non-Newtonian liquids. *Rheologica Acta,* **10**(4), pp. 517–527, 1971.

[3] Manero, O., Mena, B. & Valenzuela, R., Further developments on non-Newtonian flow in oscillating pipes. *Rheologica Acta,* **17**(6), pp. 693–697, 1978.

[4] Mena, B., Manero, O. & Binding, D.M., Complex flow of viscoelastic fluids through oscillating pipes: Interesting Effects and Applications. *Journal of non-Newtonian Fluid Mechanics*, **5**, pp. 427–448, 1979.

[5] Chan, W., Lee, S. & Liu, C., Effects of frequency and amplitude of oscillation on low Reynolds number pulsating flow in a circular pipe. *Engineering Computations*, **19**(1), pp. 119–132, 2002.

[6] Eesa, M. & Barigou, M., CFD analysis of viscous non-Newtonian flow under the influence of a superimposed rotational vibration. *Computers & Fluids*, **37**(1), pp. 24–34, 2008.

[7] Torralba, M., Castrejón-Pita, J., Castrejón-Pita, A., Huelsz, G., Del Río, J. & Ortin, J., Measurements of the bulk and interfacial velocity profiles in oscillating Newtonian and Maxwellian fluids. *Physical Review E,* **72**(1), p. 016308, 2005.

[8] Amaratunga, M., Nybø, R. & Time, R.W., PIV analysis of dynamic velocity profiles in non-Newtonian drilling fluids exposed to oscillatory motion. Presented at *OMAE2018*, Madrid, Spain, 17th–22nd June, 2018.

[9] Bird, R.B., Armstrong, R.C. & Hassager, O., *Dynamics of Polymeric Liquids. Vol. 1: Fluid Mechanics*, Wiley, 1987.

[10] Ito, T., Tsuji, Y. & Kukita, Y., Interface waves excited by vertical vibration of stratified fluids in a circular cylinder. *Journal of Nuclear Science and Technology*, **36**(6), pp. 508–521, 1999.

MOLTEN STATE OF THERMOPLASTIC RESIN IN AN INJECTION MOULDING BARREL

SO SHIMOKUSUZONO[1], TATSUYA TANAKA[2], MASAHIRO SASADA[2], YUKA HIRAO[2], KANTA HANAFUSA[2] & AKIRA INOUE[1]
[1]Development Department, TOYO Machinery & Metal Co. Ltd. of Hyogo, Japan
[2]Applied Materials Engineering Laboratory, Doshisha University of Kyoto, Japan

ABSTRACT

In recent years, polymer products have required superior mechanical properties, a particular strength and dimensional accuracy. One of the most popular moulding methods for polymer materials is injection moulding. Compared to other moulding methods, injection moulding is better at making complicated shapes quicker; however, injection moulding suffers from many kinds of moulding faults. Various solutions to reduce the number of faults on a production line have been developed as a result of many years of studies. However, to further decrease the number of faults, researchers need to focus on the basic state of the plasticized resin. In injection moulding, it is important to determine how uniformly a resin is melted because the uniformity of temperature is related to the characteristics of a product. Therefore, melting completion does not mean that resin is uniform. The uniformity of melting must be accurately determined to achieve high quality. First, this study uncovers some effects of screw behaviours on plasticization. The injection moulding machine performs the metering and injection processes with the screw in the barrel. In the metering process, the screw retreats to the metering stroke while maintaining back pressure. In the injection process, the screw advances to the tip of the nozzle. While the screw repeats these movements, the pellets are provided from the hopper to the moving screw. This method of adding the pellets creates differences in the distance that the resin must travel within the screw channel. The longest distance that the resin must travel during an injection cycle is equal to the metering stroke. The viscosity of the resin depends on the barrel temperature, although the viscosity is also influenced by other moulding conditions. An experiment is performed by changing the metering stroke and the barrel temperature. To understand the condition of the resin just before the injection, an original nozzle with a pressure sensor is produced. Thus, the injection pressure waveform can be recorded during each cycle. As changes in the waveform are reflected in the behaviour of the resin, a non-uniform temperature can be visually confirmed by a disturbance in the waveform.

Keywords: injection moulding, molten state, flow analysis.

1 INTRODUCTION

Injection moulding is one of the most popular methods for moulding thermoplastic resins. Compared to other moulding methods, injection moulding is better at making a complicated shape in a short takt time; however, injection moulding has many kinds of moulding faults. These faults are caused by the injection moulding machine system. The plasticizing process is not visually captured due to the heating barrel, and the resin, which has viscoelastic properties, is kneaded under high-temperature and high-pressure conditions. Therefore, it has been difficult to accurately obtain the molten state of resin during injection moulding. Reducing the number of faults while considering the inability to view the plasticizing process is very important to improve productivity in injection moulding.

The study and advancement of moulding processing began in the 1500s, and the plasticization phenomenon of resin for extrusion moulding was studied earlier than that for injection moulding. Plasticization phenomena in black-boxed barrels were proposed by Maddock [1] and Street [2]. In this method, the barrel is rapidly cooled during plasticization to solidify the resin, and then the screw is pulled out to analyse the solidified resin in the screw groove. Additionally, based on this experimental result, a plasticization model was

WIT Transactions on Engineering Sciences, Vol 125, © 2019 WIT Press
www.witpress.com, ISSN 1743-3533 (on-line)
doi:10.2495/CMEM190051

formulated by Tadmor [3], Tadmor et al. [4] and Tadmor and Klein [5]. In injection moulding, Yokoi et al. [6], [7] carried out a method of attaching a glass window to a barrel or mould to visually determine the behaviour of the resin. This method proved to be a melting model similar to an extruder. Furthermore, the pressure and temperature in both the barrel and the mould were measured by sensors, and a numerical analysis using these experimental data was performed by Guerrier et al. [8] and Atinkaynak et al. [9], [10] to predict resin flow. Many studies have confirmed whether the resin is melted in a barrel and determined the flow behaviour of resin in a metal mould. As a result of these studies, some techniques for reducing the number of moulding faults have been established. However, no complete solution has yet been reached. Few specific research studies have evaluated the state of the molten resin in injection moulding. This evaluation is necessary to produce high-quality mouldings because the molten state at the completion of plasticization is greatly related to the production quality. Researchers should consider whether the molten state is uniform as well as whether the resin is molten before it is moulded. Chen et al. [11] reported how the viscosity of the molten resin in injection moulding affects the moulded article. Therefore, this study establishes a method to evaluate whether the parameters of temperature, density and viscosity of a resin are uniform immediately before being injected into a metal mould. This method is then compared to the energy obtained by a flow analysis.

The injection moulding machine performs the metering and injection processes by using the screw in the barrel. During the metering process, the screw retreats to the metering stroke while maintaining back pressure. During the injection process, the screw advances to the tip of the nozzle. While the screw repeats these movements, the pellets are provided from the hopper to the moving screw. This method of adding the pellets creates differences in the distance that the resin must travel within the screw channel. The longest distance that the resin must travel during an injection cycle is equal to the metering stroke. This study was conducted to verify the influence that the difference in screw length, which is caused by the retraction behaviour of the screw, has on the resin.

2 EXPERIMENT

2.1 Flow simulation [12]

It is thought that the amount of energy that the resin receives from shear stress during a plasticization process influences the molten state of the resin in injection moulding. In this study, to investigate the molten state of plastic resin, a flow simulation was performed using POLYFLOW (ANSYS, Inc.).

2.1.1 Simulation conditions

The shear stress and flow path of the molten resin can be calculated by making the following assumptions.

1. The molten resin is an incompressible, non-isothermal, and non-Newtonian fluid.
2. Gravitational forces are ignored during the simulation because the molten resin is a highly viscous fluid.
3. The fluid completely melts, and the flow channel is completely filled.
4. There is no slip on the surface of the solid.

The viscosity of the polymer around the screws was modelled by the Bird-Carreau law (eqn (1) below), and the properties of polypropylene at 180°C, 200°C and 220°C were applied

$$F(\dot{\gamma}) = \eta_{\infty} + (\eta_0 - \eta_{\infty})(1 + \lambda^2\dot{\gamma}^2)^{\frac{n-1}{2}}, \tag{1}$$

where η_0 is the zero-shear-rate viscosity, λ is the time constant, $\dot{\gamma}$ is the shear rate and n is the power law index. The shear viscosity and the shear velocity were measured using a capillary rheometer.

The volumetric flow rate Q was calculated using the following expression:

$$Q = \frac{\pi r^2 h}{t}, \tag{2}$$

where t is the charge time, r is the screw radius and h is the injection stroke. Table 1 shows the volumetric flow rate Q in the flow channel for each temperature condition.

Table 1: Volumetric flow rate in the flow channel of the storage temperature condition.

Temperature (°C)	Volumetric flow rate (mm³/sec)
180	2083
200	2121
220	2106

The plasticization process in each screw design was evaluated by a particle tracking method. The residence time or plasticization time could be used for this purpose. The particle tracking was conducted using a Lagrangian method with a second-order accuracy in a space consisting of two steps

$$\text{1st step:} \quad x_i^{1/2} = x_i^0 + \frac{1}{2}u_i^0\Delta t, \tag{3}$$

$$\text{2nd step:} \quad x_i^1 = x_i^{1/2} + \Delta t\left(u_i^{1/2} - \frac{1}{2}u_i^0\right), \tag{4}$$

where x_i is the position of the particle, u_i is the velocity vector and Δt is the time increment.

To determine the residence time distribution of the particles, 1000 particles were charged via the inflow port of the hopper for 100 s. The trajectory of each particle throughout the screw was calculated, and the histories of the individual particle coordinates, the stress and the strain rates were recorded by particle tracking. The amount of energy that the resin receives from shear stress during plasticization was calculated using eqn (5), and this value was used for one index to evaluate the uniformity of the molten resin

$$U = \int_0^{t_p} \tau \cdot \dot{\gamma}\mathrm{dt}, \tag{5}$$

where τ is the shear stress and t_p is the particle residence time.

The initial temperature of the fluid was assumed to be 180°C, 200°C or 220°C. The screw rotation speed was 100 rpm. The screw during a plasticization process retreats smoothly. However, in this study, the analysed retreat distances of the screw were 0 mm, 24 mm, 48 mm, and 72 mm, and the state was changed by the length of the screw; the results were obtained as a fragmentary state of 1 cycle.

2.2 Evaluation of the molten resin state

2.2.1 Materials and equipment

For this investigation, polypropylene (grade J108M) was selected. The employed injection moulding machine was a 40-ton clamping force model (TOYO PLASTAR ET-40V) with an all-electric drive and a screw diameter of 24 mm. A spiral-shaped metal mould was used to avoid needing to be changed for some metering strokes. The moulding conditions used for the fabrication of the test pieces are shown in Table 2. These conditions are standard for the applied mould. Ten shots were used for the experimental evaluation. This experiment was performed by changing the metering stroke and the barrel temperature.

Table 2: Moulding condition.

Screw rotation speed (rpm)	100		
Injection speed (mm/s)	10		
Back pressure (MPa)	5		
Metering stroke (mm)	24	48	72
Barrel temperature (°C)	180	200	220
Tool temperature (°C)	40		
Cooling time (sec)	25		

2.2.2 Evaluation method

To judge the molten state of the resin, the maximum injection pressure was recorded. This pressure is the value of the resistance that the screw receives at the moment of injection. Additionally, the flowing distance of the resin was determined by putting the spiral product on a protractor and adding up the number of turns and the remaining angle.

3 RESULTS AND DISCUSSION

3.1 Flow simulation

3.1.1 Simulation of the full-length screw

In the case of 200°C, the energy that each particle received between the inflow and outflow was 6.05 MJ/m^3. Fig. 1 shows the energy that the particles received in 3 different zones: the feed zone, compression zone and metering zone. This figure shows that the energy that the particles receive from the screw is highest in the feed zone. If the shear rate becomes large, the energy that the particles receive also becomes large, as shown in eqn (5). The shear rate is a geometric parameter that is influenced by the screw shape and screw rotation speed. The shear rate increases when the ditch of the screw is shallow. Therefore, the shear rate usually becomes highest in the metering zone. However, the energy in the feed zone is the highest. The feed zone accounts for approximately half of the whole screw, and the residence time in the feed zone is longer than that in the two other zones. For the analysis, the resin is supposed to completely melt at the inflow. In actuality, after a solid resin pellet is placed in the hopper, the pellet is transported forward by the frictional force of the screw and barrel while receiving heat from the hot barrel. After fusion begins, the energy the resin receives changes to viscous shear stress. The temperature of the resin is low near the hopper, and the resin viscosity is in a high state. The shear stress that is added to the materials becomes large when the resin viscosity is high. As a result, the energy that the resin receives substantially increases. From this phenomenon, the ratio of mechanical energy that the resin receives in the feed zone is very large. In other words, it is thought that the decrease in the length of the screw in the feed zone from the retreating behaviour has a large influence on the plasticization and melting of the resin.

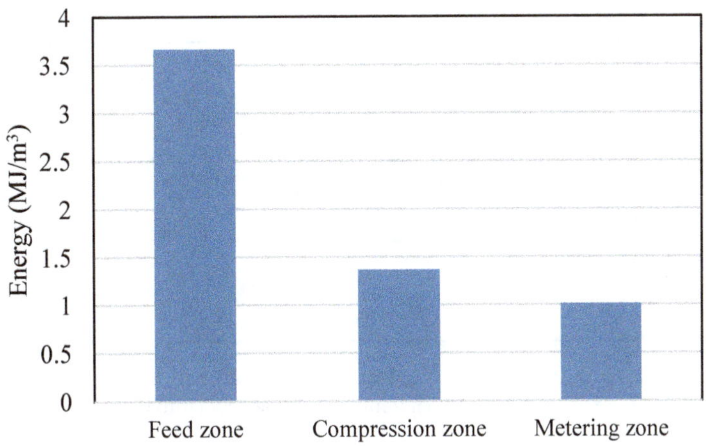

Figure 1: Increase in energy within each zone.

3.1.2 Simulation of the screw in each retreat distance condition

An analysis was performed that varied the length of the screw according to the distance that the screw retreated. A result provided by particle tracking is shown in Fig. 2. The horizontal axis is the value of mechanical energy that each particle received, and the vertical axis is the number of particles in the energy zone. The results confirmed that the particles received less energy when the retreat distance of the screw was large. Therefore, the distribution of the energy was shown to become wider when the retreat distance of the screw was larger. From the actual moulding, it was expected that an irregularity occurred in the uniformity of the molten resin. The mean value of the energy for each screw length is shown in Fig. 3. From this figure, it was confirmed again that the energy decreased when the retreat distance of the screw increased.

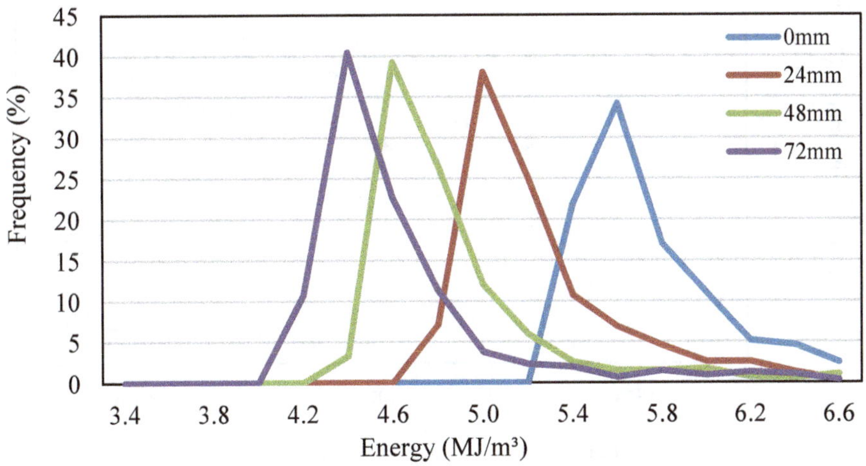

Figure 2: Histogram of each retreat distance condition.

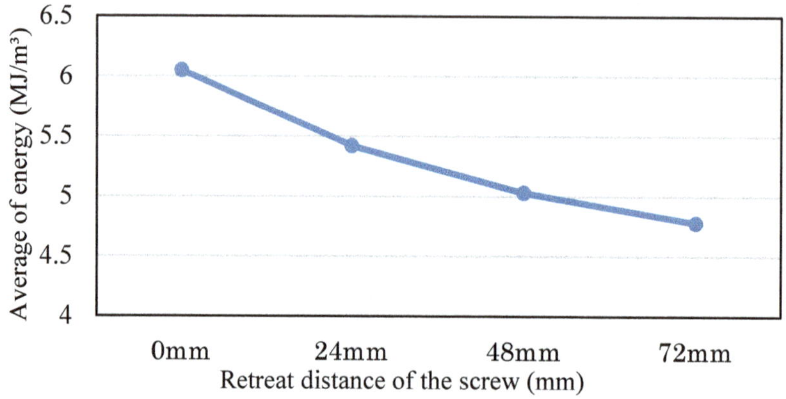

Figure 3: Average energy value at each retreat distance condition.

3.1.3 Simulation of each temperature condition

The difference in the amount of energy when the temperature was changed is shown in Fig. 4. The retreat distance of the screw is shown on the horizontal axis, and the decrease in the amount of energy is shown on the vertical axis. The results confirmed that the difference in the energy increased when the temperature was lower, and it was confirmed that the difference in the energy increased when the retreat distance was larger. This phenomenon occurs because the mechanical energy that the resin receives increases as a result of the shear stress, and the resin with lower temperature has a larger effect from the shear stress. Additionally, as the retreat distance of the screw increased, there was a larger change in the energy.

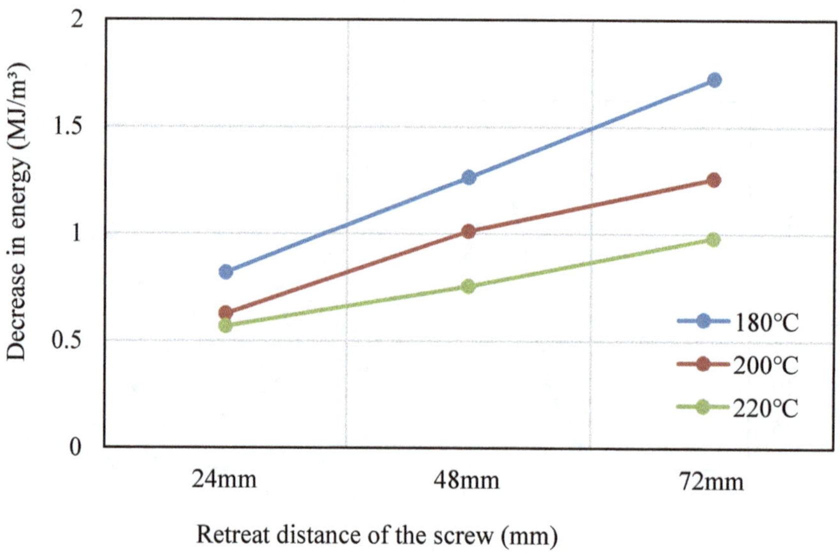

Figure 4: Energy difference from the full-length screw at each retreat distance.

3.2 Evaluation of the molten resin state

The average maximum injection pressure is shown in Fig. 5. The relationship between the barrel temperature, metering stroke and maximum injection pressure was evaluated. According to the results, the maximum injection pressure decreased as the barrel temperature increased, and the maximum injection pressure increased as the metering stroke increased. The cause of the former is that the viscosity decreases as the barrel temperature increases. For the cause of the latter, the influence of the mould on the flow resistance increases due to the increase in the metering stroke. Therefore, the standard deviation is calculated as the degree of variation per shot.

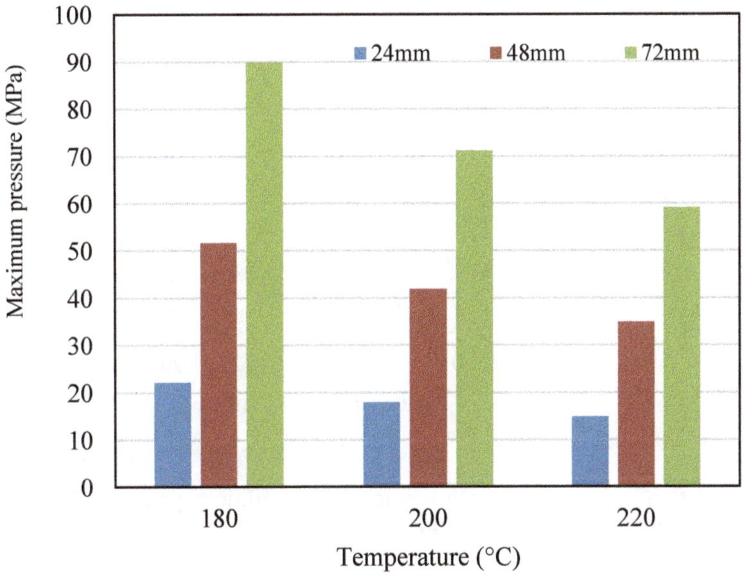

Figure 5: Average maximum injection pressure with respect to the barrel temperature and metering stroke.

The standard deviation of the maximum injection pressure for each barrel temperature is shown in Fig. 6(a), (b) and (c). At the barrel temperature of 180°C, the standard deviation of the maximum injection pressure increases as the metering stroke increases. However, at 220°C, the state of the resin is considered stable because the standard deviations were approximately the same for all metering stroke values. This tendency was similarly seen in the standard deviation of the flowing distance of the moulded particle. In other words, the molten state of the resin is affected by the metering stroke as the barrel temperature decreases. Before we conclude that the standard deviation becomes larger when the metering stroke increases, it is necessary to consider the residence time. This consideration of residence time is important because when the metering stroke is increased, the amount of plasticization increases, and the time required for the resin to be ejected from the hopper and injected is shortened. Therefore, we need to find a way to analyse the standard deviation that is not influenced by the residence time.

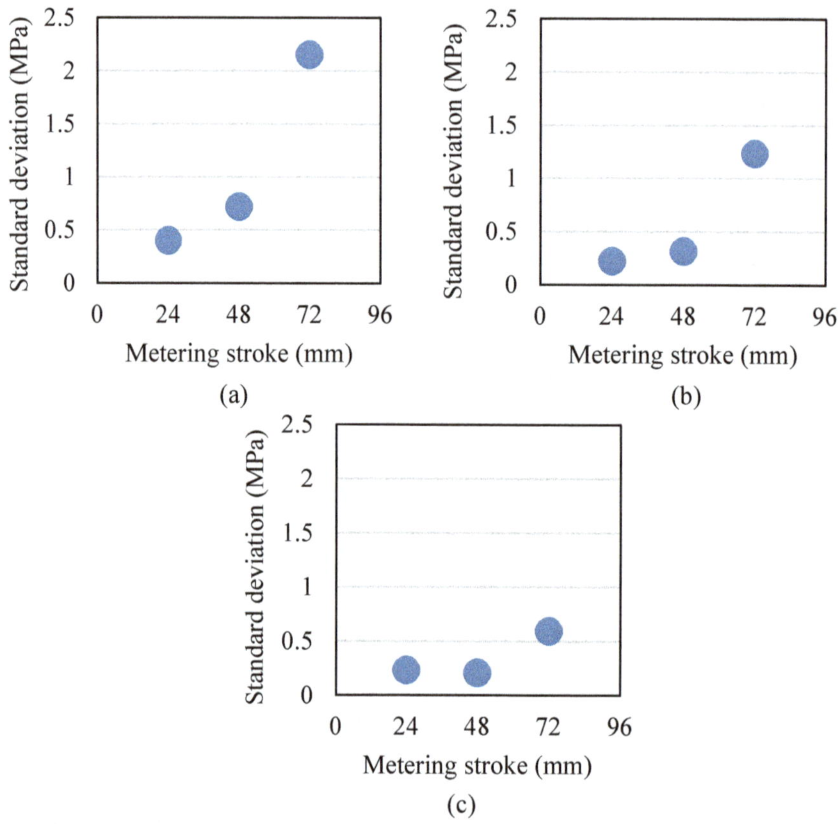

Figure 6: Deviation of the maximum injection moulding pressure for each barrel temperature. (a) 180°C; (b) 200°C; and (c) 220°C.

It may be difficult to compare the molten state of the resin because the injection pressure is influenced by the shape of the metal mould. Therefore, we tried an experiment wherein the resin was injected into the air, similar to a purge, instead of injecting into a mould. A purge is an operation for replacing the resin in the barrel when changing materials or removing the remaining resin in the barrel. However, the back pressure is very important because it largely relates to the molten state of the resin. In this experiment, the measuring process was performed by touching the nozzle and applying back pressure before injecting. Additionally, a pressure sensor (NP465XL-1/2-20MPa of DYNISCO JAPAN) was installed at the tip of the nozzle to determine the state of the resin from a closer position. The pressure was measured every 100 ms. We obtained a waveform by connecting the pressure points. The pressure waveform obtained in one cycle is shown in Fig. 7. There are two places where the pressure increases. The first pressure increase occurs at the time of measuring, which exhibits the same pressure as the back pressure. The second pressure increase occurs at the time of injection. This pressure waveform was extracted for each condition and evaluated. The pressure waveforms at the injection section are shown in Fig. 8(a), (b) and (c). The resin flow immediately after the start of injection was the same regardless of the metering stroke; the pressure waveforms at 24 mm and 72 mm were similarly shaped, but the pressure waveform at 24 mm was half as long as the waveform at 72 mm. Therefore, the waveform just before the end of injection is the focus of evaluating the difference in the molten state of the resin by the metering stroke. However, in this experiment, the nonuniformity of the resin did not appear in the disturbance of the waveform; as the nozzle in this study is generally used in injection moulding, the pressure may not be sufficient to indicate resin nonuniformity.

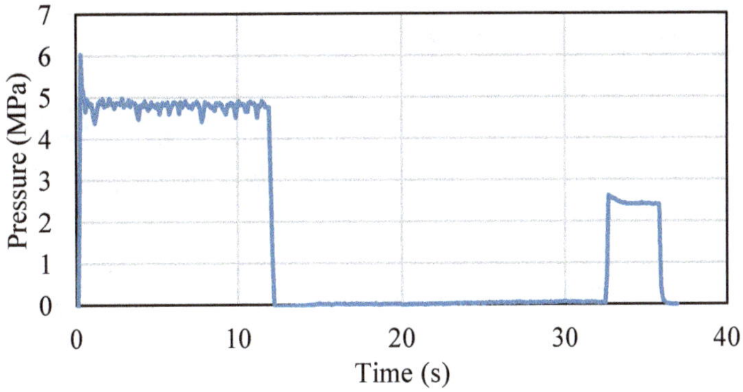

Figure 7: Sample of the pressure waveform in a cycle.

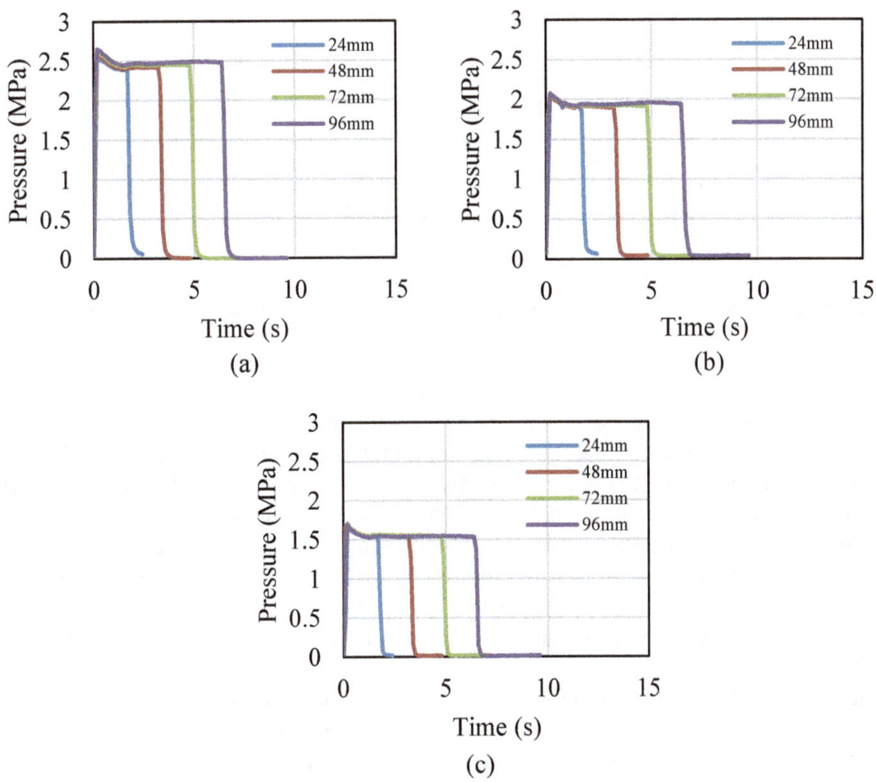

Figure 8: Pressure waveform in the injection section. (a) 180°C; (b) 200°C; and (c) 220°C.

Therefore, we investigated the unevenness of the data under each condition. Fig. 9 shows a result provided under conditions of 72 mm at each temperature. As a result, it was confirmed that the unevenness of the pressure decreased because the temperature of the barrel rose. This observation was well consistent with a result provided by analysis.

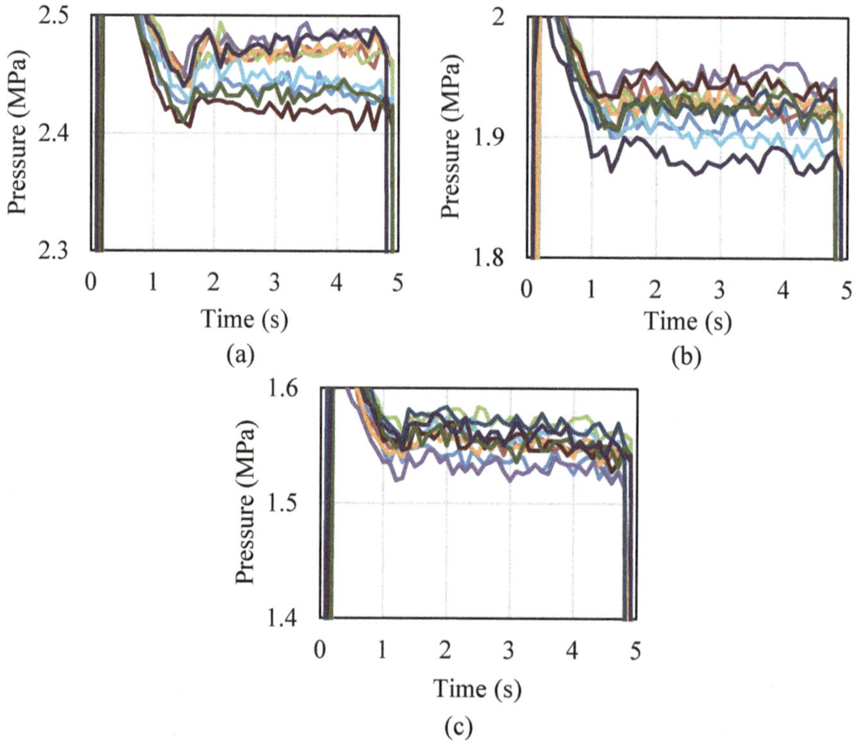

Figure 9: Results provided under conditions of 72 mm. (a) 180°C; (b) 200°C; (c) 220°C.

4 CONCLUSIONS

In this paper, we conducted experiments to verify how the difference in screw length (i.e., where the resin passes by the retracting behaviour of the screw) affects the molten state. This experimentation aims to establish a method that can evaluate whether the molten state of the resin is uniform just before being injected into the metal mould.

A flow analysis was performed, and the results showed that the difference in the energy that the resin received grew large when the retreat distance of the screw was large. It was thought that this difference in energy had an influence on the uniformity of the resin. Furthermore, it was confirmed that the influence was more substantial when the temperature was lower.

Since the data trend of the flowing distance of the moulded product coincides with that obtained from injection moulding, it can be said that it is effective to evaluate the pressure at the time of injection to stabilize the molten state of the resin. Although there was a phenomenon in which the standard deviation of the maximum injection pressure became larger as the metering stroke increased, at the present stage, it cannot be concluded that the increased standard deviation was caused by the variations in the metering stroke. It was found that the residence time must be considered in future experiments. We tried an experiment to inject resin into the air, similar to a purge, instead of injecting the resin into a mould. As a result, the state of the resin was not confirmed even if we compared the experimental means. However, we were able to grasp the unevenness of the resin in a molten state when we investigated the state of the pressure under every condition in detail. This finding was consistent with an analytical result.

REFERENCES

[1] Maddock, B.H., A Visual Analysis of Flow and Mixing in Extruder Screws. *SPE Journal*, **15**, 1959.
[2] Street, L.F., Plasticating extrusion. *International Plastics Engineering*, **1,** 1961.
[3] Tadmor, Z., Fundamentals of plasticating extrusion. I. A theoretical model for melting. *Polymer Engineering and Science,* **6**(3), pp. 185–190, 1966.
[4] Tadmor, Z., Duvdevani, H.J. & Klein, I., Melting in plasticating extruders theory and experiments. *Polymer Engineering and Science*, **7**(3), pp. 198–217, 1967.
[5] Tadmor, Z. & Klein, I., *Engineering Principles of Plasticating Extrusion*, Van Nostrand Reinhold Co., 1970.
[6] Yokoi, H., Takatsugi, S. & Shiraishi, W., Development of a small-sized glass-inserted heating cylinder for visualization and temperature distribution measurement in injection molding. *Institute of Industrial Science,* **52**(9), pp. 410–413, 2000.
[7] Yokoi, H., Hayasaki, S., Kuroda, A. & Tatsuno, M., Visualization of screw-plasticating process of plastics pellets by glass-inserted heating cylinder. *Journal of Flow Visualization and Image Processing*, **7**, pp. 231–241, 2000.
[8] Guerrier, P., Tisello, G. & Hattel, J.H., Flow visualization and simulation of the filling process during injection molding. *CIRP Journal of Manufacturing Science and Technology*, **16**, pp. 12–20, 2017.
[9] Altinkaynak, A., Gupta, M., Spalding, M.A. & Crabtree, S.L., Melting in a single screw extruder: experiments and 3d finite element simulations. *International Polymer Processing*, **26**(2), pp. 182–196, 2011.
[10] Altinkaynak, A., Gupta, M., Spalding M.A. & Crabtree, S.L., An investigation of melting and metering sections in a single-screw extruder. *SPE ANTEC The. Papers*, **60,** pp. 1135–1141, 2014.
[11] Chen, J.Y., Yang, K.J. & Huang, M.S., Online quality monitoring of molten resin in injection molding. *International Journal of Heat and Mass Transfer*, **122,** pp. 681–693, 2018.
[12] Inoue, A. Morita, K., Tanaka, T., Arao Y. & Sawada, Y., Effect of screw design on fiber breakage and dispersion in injection-molded long-glass-fiber-reinforced polypropylene. *Journal of Composite Materials*, **49**(1), pp. 75–84, 2015.

DEVICES FOR ENVIRONMENTAL OBSERVATIONS

GIUSEPPE ZAPPALÀ[1,2], GABRIELLA CARUSO[1], VIVIANA PIERMATTEI[2], ALICE MADONIA[2], SIMONE BONAMANO[2], ALESSANDRA PETRI[2] & MARCO MARCELLI[2]
[1]Consiglio Nazionale delle Ricerche (CNR), IRBIM – MESSINA, Italy
[2]Laboratory of Experimental Oceanology and Marine Ecology (DEB, Civitavecchia), Tuscia University, Italy

ABSTRACT
This paper describes some systems, designed and built by CNR in Messina and by Viterbo Tuscia University in its Civitavecchia laboratory, that fill the gap between traditional offshore oceanography and modern operational oceanography. The evolution of the main parts of such devices is analysed together with the pros and cons of design choices, not excluding the economic point of view that, together with instrument precision and reliability, is one of the main design constraints. Technological and scientific advancements were obtained in the state of the art, offering scientists new low cost instruments to perform research both in coastal and in harsh environment areas, exploiting the progress both in elaboration and in transmission electronic devices. The systems were designed to operate from fixed and mobile platforms, like Ships of Opportunity (SOOP) or towed vehicles. To this aim, CNR Messina set up a device able to automatically release up to eight expendable probes. Ruggedized versions of systems were built to be installed on small towed vehicles in harsh environments (Arctica). Sites at the sea-ice interface where blocks fall from glaciers due to ice melting, make it too difficult and dangerous to operate by traditional ships. During the June 2017 Svalbard campaign, a new generation water sampler, able to fill eight 500 ml bottles, was tested on field and showed a good operational performance. Various measuring instruments can be connected to the data acquisition and transmission system according to the research needs. They include multiparametric probes, expendable probes launched from SOOPs, water samplers for subsequent laboratory analysis, and any kind of device having a standard interface or providing a voltage-current signal. The land-based data assimilation can integrate the acquired in-situ data with satellite observation and the output of mathematical models.
Keywords: environment monitoring, data acquisition and transmissions systems, marine measurements.

1 INTRODUCTION

Modern operational oceanography needs a big quantity of environmental data to perform quality assessment, nowcasting activities, feeding of forecasting models and sea-truth validation of satellite acquired data.

The high cost of traditional oceanographic cruises, together with the low space-time coverage, has stimulated the development of new instruments, systems and methods able to provide scientists with quality measurements.

Among the scientific Institutions who worked to set up new low-cost instruments and devices, the Italian National Research Council (CNR) and the DEB Laboratory of Civitavecchia, Tuscia University of Viterbo must be mentioned. Both these Institutions developed some devices for environmental monitoring and experimented on field their possible use in several marine ecosystems, including temperate (Mediterranean regions such as Sicily, Apulia, Latium) and polar (Arctic Svalbard Islands) areas.

2 SOME HISTORY

The most common industrial devices in use at the beginning of the '1990s were conceived to operate from oceanographic ships and offshore buoys and often constituted a "black box", connectable only with other devices by the same manufacturer; few companies (like Sea Bird

WIT Transactions on Engineering Sciences, Vol 125, © 2019 WIT Press
www.witpress.com, ISSN 1743-3533 (on-line)
doi:10.2495/CMEM190061

Electronics) offered – and still offer – very good performances both in shallow and in deep water, associated to the disadvantage of being quite expensive.

Other companies offered lower precision and cost devices, unsuitable for traditional oceanography and hardly acceptable also for what was going to become a great scientific challenge: Operational Oceanography.

In USA a range of applications for water monitoring were set up by main research institutions (Monterey Bay Aquarium Research Institute – MBARI, Scripps Institution of Oceanography, University of San Diego, California, Woods Hole Oceanographic Institution-WHOI, and others) and governmental agencies (National Oceanic and Atmospheric Administration – NOAA, National Data Buoy Center – NDBC) [1], [2].

In northern Europe single buoys and monitoring networks were set up to control water quality in aquaculture plants against the arrival of pollutants (i.e. oil spills) or to assess pollution in some particular zones like the estuary of the Elbe river, where a nuclear power plant was operating [3], while in Mediterranean (northern Tyrrhenian Sea) the French–Italian ARCOBLEU network was funded by MAST (Marine Science and Technology) Program as a pilot project [4].

In Italy, the most important installations were the platform “Acqua Alta” in Venice Lagoon and the ODAS 1 buoy, offshore the Tuscany-Ligurian coast; at that time, both the monitoring stations were more devoted to the study of meteo-marine phenomena than to water composition and quality [5].

Basically, these installations measured the main parameters characterizing sea water such as temperature, conductivity, depth, dissolved oxygen, pH, turbidity, often using a ruggedized version of laboratory sensors.

Other measurements (like nutrients, bacteriological parameters, more complex chemical analyses) were performed in laboratory on water samples often collected during oceanographic cruises or in coincidence with equipment maintenance.

3 AND IN MESSINA?

Founded in 1916, the Istituto Sperimentale Talassografico CNR of Messina (Sicily, Italy) was one of the oldest marine research institutions, mainly devoted, at that time, to biological oceanography.

At the end of 1980s, the National Research Council launched the Strategical Program for Coastal Pollution Monitoring, composed of 10 operating units co-ordinated by the Istituto, and a new research line focusing on marine monitoring started, opening new perspectives in this field where further advancements are still in progress [6].

After the first experiences made using commercial devices, the basic idea was to set up new low-cost instruments, easy to deploy and maintain, offering good reliability and quality measurements; therefore, exploiting the technological progress in the field of computers and radio transmission, various systems were designed, built and used in national and European research programs [7].

The starting point was to analyze the problems encountered with some commercial devices. In fact, these equipments, basically, consisted in a group of sensors connected to an Analog to Digital converter; the base station started the measurement sending a radio signal to activate the “squelch” line on the remote radio, so powering up the subsystem that performed the measurements, sent the results using a simple analogic modem on the radio link and went back to a "sleep" state with only the radio powered.

In these systems, the measurement resolution was not very high (12 to 16 bit) and the exact regulation of the squelch level was vital for the good system operation; the used RTX equipments were commercial radios, working in VHF or in UHF.

Among the facilities, some systems were able to reply remote interrogation for a pre-defined number of times, until the reception of the required measurements or the expiration of the time-out limit.

The absence of recharging systems (e.g. solar panels) made it necessary to change the batteries of the remote station every two weeks, with the risk of stopping the measurements if, for some reason (e.g. bad meteo-marine conditions) the operation could not be performed.

Moreover, the availability of a set of full-charged batteries always ready was also necessary, so doubling the costs for these items.

To overcome the encountered shortcomings, a new system architecture was designed, complying to the following constraints:

- Availability of battery recharging facilities on the buoy
- Low-power consumption of the equipments
- Good modularity and flexibility
- Easy and reliable radio-link management

The final architecture was defined (Fig. 1), comprising:
Land station:

- A computer, usually a PC-like one, able to interact with the remote station and to store in a local or remote data base the acquired data

Sea station:

- Power supply system with batteries and recharging devices (solar panels and/or wind generator)
- Measuring system (one or more sensors or measuring devices with analogue or digital (usually RS232 or similar) interfaces
- Data acquisition system (a computer performing timed pre-defined operations), interacting with the measuring system
- Data transmission system (a radio device with modem)
- Position control system (a GPS monitoring the position of the station)

3.1 The power supply system

Depending on the connected devices, the final version consists of up to four 12 V sealed lead batteries, each with its own solar panel and charge regulator; a wind generator with separated charge regulator contributes to the energetic balance of the system; the charge regulators and

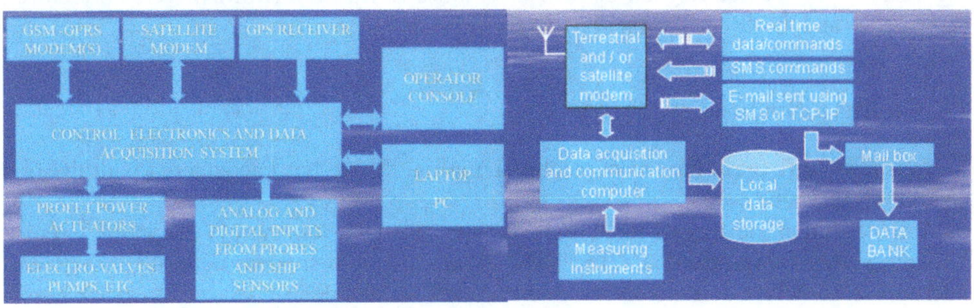

Figure 1: The general system architecture (left) and the data flow (right).

the parallel diodes are low voltage drop. The resulting output goes into one or more wide input voltage DC/DC converters, that supply the appropriate voltages and currents to the other systems. The combined use of solar panels and wind generator enables to recharge batteries also during the night and in bad weather conditions.

3.2 The measuring system

Both marine and meteorological sensors were used, all connected using serial interfaces.

The marine sensors (stand alone or assembled in a multiparametric probe) included water temperature, conductivity, dissolved oxygen, fluorescence, turbidity, pH, while the meteo sensors were connected to an intermediate conditioning and controlling system, in turn communicating with the main data acquisition system (Fig. 1).

3.3 The data acquisition system

Almost all of the developed data acquisition systems were based on a PC-like architecture, that evolved from the first version based on a 286 CPU mounted on a double eurocard board, up to a recent one, based on Pentium CPU mounted on PC-104 (IEEE 696 standard) boards, and to the most advanced version based on Arduino boards. Additional boards that were mounted as needed included analogue interface boards (A/D converters), multi-serial RS232 boards, solid state disks, digital I/O boards. The system was controlled using a specially created set of macro-commands, that will be presented in a subsequent chapter (Fig. 2).

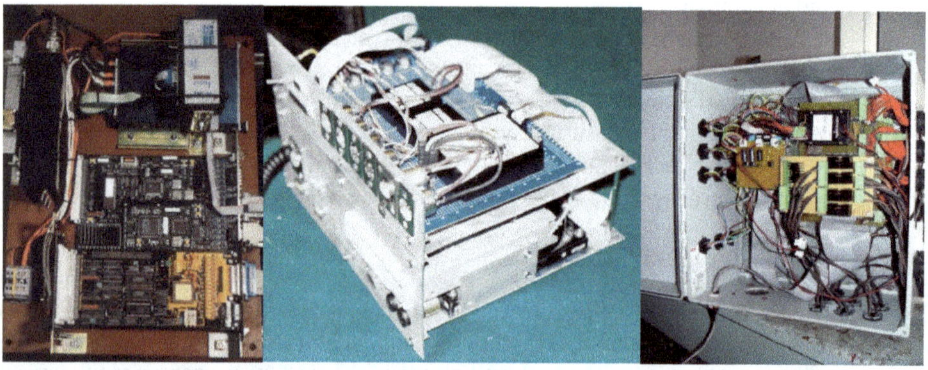

Figure 2: The Venice Lagoon System electronics (left), the PRISMA2 version (centre), and the “SAM” version (right).

3.4 The data transmission system

Various tests were performed to overcome the initial difficulties encountered with the first commercial systems used.

A first version, developed to be used for the “Sistema Lagunare Veneziano” (Venice Lagoon System), assembled a CTE TNC 24 packet radio modem controlled by a Z80 microprocessor with a quartz frequency controlled 1 W UHF RTX connected to an RF power amplifier and to a directional antenna for the remote station and an omnidirectional one for the base station [8].

Some considerations are associated to the use of this kind of accommodation:

- First of all, the use of radio frequency must be authorized by the competent Ministry.
- It is necessary to have a land base station in a suitable location, possibly high on the sea, and free from obstructions towards the remote station.
- The antennas must see each other.
- The additional attenuation due to the marine water "spray" must be taken into account.
- To minimize the required transmission power it should be necessary to use directional antennas having a good gain, but this is not always possible for two reasons: the size of the buoy and its movement caused by wind and streams.
- The maximum theoretical distance between the antennas is a function of their height over sea, and it is limited by the buoy size and the land base station location, where municipal laws could limit the presence and height of antennas.
- Antennas pointed low on horizon receive a lot of unwanted noise.

The use of "intelligent" packet modem has allowed to overcome some of the above limitations, using intermediate buoy stations as "repeaters".

A significant progress was reached with the introduction in Italy of the GSM mobile telephony. A first application, designed for the "PRISMA2" (Research and Experimentation Program for the Protection of the Adriatic Sea) program, used a Motorola Cellular Phone, driven by a specific PCMCIA data interface card; this equipment was then substituted by Nokia products, offering a better performance and simpler programming [9].

The most recent implementations (such as the Automatic Multiple Launcher for Expendable Probes funded by the European Union within the FP5- MFSTEP program; the buoys designed for the "SAM" project funded by the Italian Ministry for University and Research [10], [11] and for Civitavecchia (Latium) C-CEMS buoy [12]) use integrated modems (external or PC104 boards), able to work also in GPRS, so permitting low cost data transmission, as described in a subsequent paragraph.

3.4.1 The data transmission methods

The first commercial systems devoted all the (very rudimentary) controls to the same data acquisition software; the introduction of packet radio modem embedded in the modem itself all the controls, including latency, number of retransmission trials, error management.

The first GSM modems enabled to send SMS between the base and the buoys, so transmitting in a bi-directional way both commands (from a specifically written set of macro-commands) and data (coded to fit in the 160 characters length of SMS standard); a special service offered by the Telecom provider transformed the SMS's into e-mail messages. Of course it was possible also to make point to point connections, for instance for system maintenance, paying the service on a time basis.

The arrival of GPRS and new modems, embedding Internet connection routines, enabled to significantly lower the communication costs, sending directly e-mail messages paying on a data volume basis.

The use of mobile telephony networks allowed to avoid problems arising with private radio links, including the availability of a base station on site, allowing to receive data and to control buoys from anywhere in the world.

3.5 The data acquisition system

3.5.1 The hardware

The Data Acquisition System is based on one or more computer boards, interfacing with measuring devices (sensors or analytical instruments either using serial interfaces, or analogue 0–5 V interfaces), GPS, telecommunication devices (terrestrial or satellite modems...), mass memory devices (usually solid state disks) [13], [14].

I/O interfaces are included to control the connected devices power supply.

The first systems used a double eurocard size 286 CPU board, plugged with serial and analogue interfaces boards, the most recent mount a Pentium family CPU assembled on a PC 104 (IEEE 696) standard board, complemented with all the interfaces and communication devices boards, including GPS. Cheaper Arduino boards are also in use, generally to control single devices or to coordinate data acquisition and transmission activities not requiring high elaboration speeds.

3.5.2 The firmware

The software is based on a set of macro-commands, enabling to fully control the computer and the connected devices using the local keyboard, the base station computer, or simply a mobile phone.

All the macro-commands start with the letter "Z", followed by a character identifying the kind of command and, if necessary, by the command parameters identifying, for instance, the sensor or I/O port involved, the switching on or off of an alimentation line, and so on.

The macro-commands can be combined in "sequences" using a simple text editor and stored in the remote station, where an "event machine" routine runs in an endless loop waiting for the arrival of commands from the local keyboard or from the communication devices, or for the occurrence of an event like a timed sequence or the reach of a geographical point.

Timed events can be programmed on a 10-minute interval basis and can be each different from the others.

Each of the above conditions starts a "parser" module, that reads the sequence file line by line, interpreting and executing the pre-coded commands.

During the time between operations, the system can enter a low power consumption "sleep" state, waiting to be awakened from commands or sequences execution.

The commands include:

- System control and diagnostic commands
- Instrument management commands
- Data acquisition commands
- Telecommunication management commands.

The firmware was written in compiled Microsoft Basic, with some routines in Assembly, running in ROM-DOS (an operating system very similar to MS-DOS) environment; the most recent version uses Arduino's embedded development system language.

The systems continuously monitor the position, launching an alarm message in case of buoy unmooring.

The choice of the above software architecture makes the system really modular, simplifying the integration of new devices and the "mission programming".

3.6 Special devices

An interesting development was obtained in the frame of the FP5 EU-funded MFSTEP (Mediterranean Forecasting System Towards Environmental Prediction) program (2003–2006) [15]. Two devices for the use on Ships of Opportunity (SOOP, also called VOS, Voluntary Observing Ships) were designed and built. The first was an expendable probe measuring temperature and fluorescence, the other a device conceived to manage its release.

3.6.1 The T-FLAP and the ECO BLUBOX

The original idea of expendable probes, measuring water temperature and sending data on board using a couple of very thin wires, was developed in the '60s and was initially used in submarine warfare and later in scientific applications. These devices, still in use, called XBT, contained a temperature sensible device (thermistor), that was excited by a small current generated in a control box on board the ship; to allow the free sinking two coils of wire, one included in the XBT tail and the other in the on board XBT case, de-reeled and the depth was calculated as a function of time [16].

The new expendable probe, T-FLAP (Temperature Fluorescence Launchable Probe) – designed by the Marcelli's research team at the DEB Laboratory of Civitavecchia of the Tuscia University – integrates a temperature sensor with a fluorometer, to measure chlorophyll 'a' for the determination of the phytoplankton biomass. Differently from the XBT that included no active parts, an integrated electronic system manages all the functions of the probe, including the serial data transmission on board the ship [17].

Starting from this device, a compact system for continuous surface water monitoring was obtained (The Eco Blue Box), that comprises, in a waterproof container, the T-FLAP core, a GPS receiver, a water pump to supply the measurement cell and the needed accessories, the data acquisition and storage system, the connection for the on demand communication system (GSM-GPRS or Satellite modem). The small size and low electrical consumption of this arrangement make it suitable also for installation on small sail boats (Fig. 3).

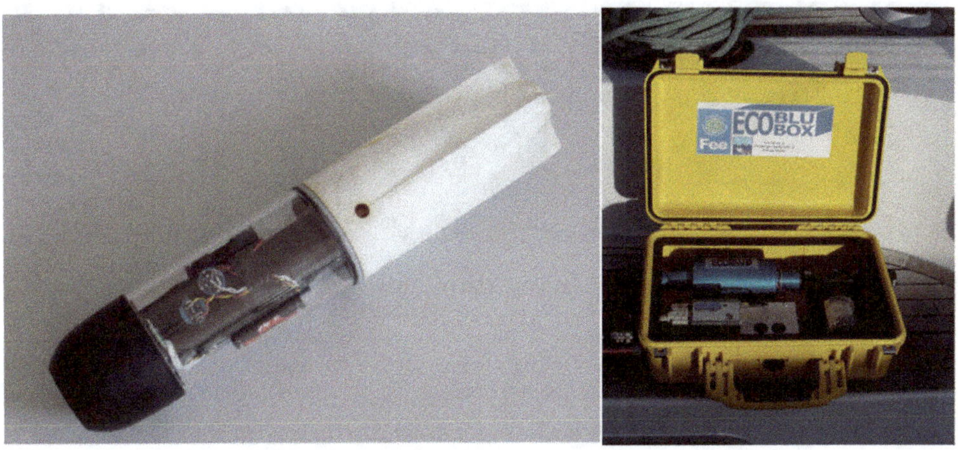

Figure 3: The T-FLAP (left) and the ECO BLUBOX (right).

3.6.2 The expendable probes launcher

This device was designed in the frame of the MFSTEP program to automatically deploy from SOOPs up to eight expendable probes (either commercial standard XBTs or T-FLAPs) [18], [19].

It uses the standard electronics already described, integrated with a special XBT driving board.

The embedded GPS continuously monitors the ship position; when a predefined point (northern of, southern of, eastern of, western of) or time is reached, a deployment routine starts, opening the security lock of the selected launch tube, then the tube door, and starting a data acquisition routine.

The device uses pneumatic actuators controlled by electro-pneumatic valves to lock and move the launch tube doors (Fig. 4).

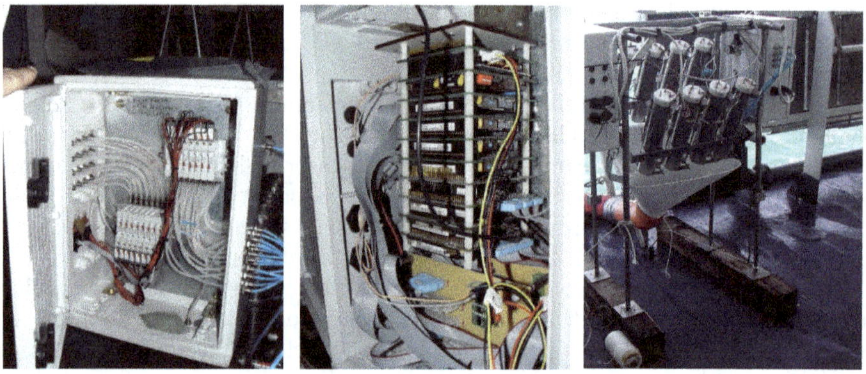

Figure 4: The pneumatic valves (left), the control electronics (centre), and the whole multiple launcher for expendable probes (right).

Figure 5: The “SAM” water sampler (left), the Arctic version (centre), and its electronics (right).

3.6.3 Water samplers

To perform water quality laboratory analysis, for instance to detect the presence of biological contaminants (such as *Escherichia coli* or bacterial pathogens), it is necessary to have devices able to collect and preserve water samplers for further examination in laboratory (Fig. 5).

The first device of this kind was built at CNR in the ‘90s of the past century. It consisted in a frame, hosting six 250 ml bottles, filled via a rotating distributor moved by a step motor, directing the water flux given by a peristaltic pump either into a bottle or to a circuit washing position. To preserve the samples, the bottles contained 10 ml of formalin.

This device worked on a first version of a buoy moored in Messina (Sicily), under the control of the buoy computer [20].

A more sophisticated version of water sampler was developed at the beginning of the current century, to be used on new buoys. It hosted eight 250 ml bottles, connected to the water circuit by electro-valves opening the water intake and the air outlet of each bottle; three peristaltic pumps provided the sea water, the preservant, the fresh circuit washing water. Also this version was controlled by the buoy computer [21].

Another version, derived from the previous, but customized to work in Arctic harsh environment and able to fill eight 500 ml bottles, was developed in 2015 and used during oceanographic cruises at the Svalbard Islands, hosted on a small catamaran towed by an unmanned surface vehicle [22], [23]. The use of a towed vehicle enabled to study the ice-water interface going very near to glaciers, where falling ice blocks would be dangerous to manned boats [24]. This version of water sampler incorporates a control computer, similar to those above described.

3.6.4 Sliding devices

The need to perform measurements at various (subsuperficial) depths, withdrawing also water samples for laboratory analyses, led to the development of two sliding devices: the T-FISH (Towed FISH), able to be towed from small boat and to pump on board a continuous flow of marine water [8] and the SAVE (Sliding Advanced Vehicle) collecting 50 ml of sea water in small cavities in an appositely designed water sampler [25]. Both the vehicles were able to measure main physico-chemical water parameters (Fig. 6).

Figure 6: The T-FISH (upper) and the SAVE with its water sampler (lower).

4 CONCLUSIONS

During the last two decades, substantial technological and scientific advancements in the state of the art have been reached, offering scientists new low-cost instruments to be operative both in coastal and in harsh environment areas. All the above-reported technologies offer new tools fundamental to improve the reliability and the performance of currently available systems for both environmental monitoring as well as operational and forecasting oceanography.

An important result was obtained by the integration of the above described systems, together with the combined use of satellite observations and mathematical modelling, into the C-CEMS (Civitavecchia Coastal Environmental Monitoring System [26]–[29]).

Studies are still continuing to further enhance their performance and to develop new low cost, high precision, measuring systems.

REFERENCES

[1] Griffiths, G., Davis, R., Eriksen, C., Frye, D., Marchand, P. & Dickey, T., Towards new platform technology for sustained observations. *Proceedings of OceanObs 99*. www.bom.gov.au/OceanObs99/Papers/Griffiths.pdf. Accessed on: 23 Feb. 2007.

[2] Lessing, P., Henderson, D. & Edwards, B., Architecture and development of an environmental acquisition and reporting system. *Proceedings of MTS-IEEE OCEANS '99 Conference*, **II**, pp. 785–788, 1999.

[3] Grisard, K., Eight years experience with the Elbe Estuary environmental survey net. *Proceedings of OES-IEEE OCEANS '94 Conference*, **I**, pp. 38–43, 1994.

[4] Sauzade, D., Hénocque, Y. & Carof, A.H., Arcobleu, an integrated surveillance system for chronic and accidental pollution. *Water Science and Technology*, **32**(9–10), pp. 25–31, 1995.

[5] Canepa, E., Pensieri, S., Bozzano, R., Faimali, M., Traverso, P. & Cavaleri, L., The ODAS Italia 1 buoy: More than forty years of activity in the Ligurian Sea. *Progress in Oceanography*, **135**, pp. 48–63, 2015.

[6] Zappalà, G., Caruso, G. & Crisafi, E., Design and use of advanced technology devices for sea water monitoring. *Elsevier Oceanography Series*, **66**, pp. 273–280, 2002. DOI: 10.1016/S0422-9894(02)80032-0

[7] Zappalà, G., Caruso, G., Azzaro, F. & Crisafi, E., Marine environment monitoring in coastal Sicilian waters. *WIT Transactions on Ecology and the Environment*, vol. 95, WIT Press: Southampton and Boston, pp. 337–346, 2006.

[8] Crisafi, E., Azzaro, F., Zappalà, G. & Magazzù, G., Integrated automatic systems for oceanographic research: Some applications. *Proceedings of OES-IEEE OCEANS '94 Conference*, **I**, pp. 455–460, 1994.

[9] Zappalà, G., Alberotanza, L. & Crisafi, E., Assessment of environmental conditions using automatic monitoring systems. *Proceedings of MTS-IEEE OCEANS '99 Conference*, **II**, pp. 796–800, 1999.

[10] Zappalà, G., Caruso, G. & Crisafi, E., The "SAM" integrated system for coastal monitoring. *Proceedings of the Fourth International Conference On Environmental Problems in Coastal Regions, Coastal Environment IV*, ed. C.A. Brebbia, WIT Press: Southampton and Boston, pp. 341–350, 2002.

[11] Zappalà, G. & Azzaro, F., A new generation of coastal monitoring platforms. *Chemistry and Ecology*, **20**(5), pp. 387–398, 2004.

[12] Bonamano, S. et al., The Civitavecchia Coastal Environment Monitoring System (C-CEMS): A new tool to analyze the conflicts between coastal pressures and sensitivity areas. *Ocean Science*, **12**(1), pp. 87–100, 2016.

[13] Zappalà, G., A software set for environment monitoring networks. *Proceedings of the International Conference on Development and Application of Computer Techniques to Environmental Studies X, Envirosoft 2004*, eds G. Latini, G. Passerini & C.A. Brebbia, WIT Press: Southampton and Boston, pp. 3–12, 2004.

[14] Zappalà, G., A versatile software-hardare system for environmental data acquisition and transmission. *WIT Transactions on Modelling and Simulation*, vol. 48, WIT Press: Southampton and Boston, pp. 283–294, 2009.

[15] Pinardi, N. et al., The Mediterranean Ocean Forecasting System: First phase of implementation (1998–2001). *Annales Geophysicae*, **21**(1), pp. 3–20, 2003.
[16] Marcelli, M., Piermattei, V. & Zappalà, G., Advances in low cost marine technologies. *WIT Transactions on Modelling and Simulation,* vol. 51, WIT Press: Southampton and Boston, pp. 497–507, 2011.
[17] Marcelli, M., Di Maio, A., Donis, D., Mainardi, U. & Manzella, G.M.R., Development of a new expendable probe for the study of pelagic ecosystems from voluntary observing ships. *Ocean Science*, **3**(2), pp. 311–320, 2007.
[18] Zappalà, G., Reseghetti, F. & Manzella, G., Development of an automatic multiple launcher for expendable probes. *Ocean Science*, **3**, pp. 173–178, 2007.
[19] Zappalà, G., Development of advanced instrumentation for operational oceanography. *WIT Transactions on Modelling and Simulation*, vol. 46, WIT Press: Southampton and Boston, pp. 841–850, 2007.
[20] Zappalà, G., Crisafi, E., Caruso, G., Azzaro, F. & Magazzù, G., Coastal monitoring by an advanced technology platform. *Proceedings of Oceanology International 98, The Global Ocean*, **1**, pp. 69–84, 1998.
[21] Zappalà, G., Caruso, G. & Crisafi, E., Coastal pollution monitoring by an automatic multisampler coupled with a fluorescent antibody assay. *Environmental Studies*, **10**, pp. 125–133, 2004.
[22] Zappalà, G., Bruzzone, G., Caruso, G. & Azzaro, M., Development of an automatic sampler for extreme polar environments: first in situ application in Svalbard Islands. *Rendiconti Lincei*, **27**(1), pp. 251–259, 2016.
[23] Zappalà, G., Bruzzone, G., Azzaro, M. & Caruso, G., New advanced technology devices for operational oceanography in extreme conditions. *International Journal of Sustainable Development and Planning*, **12**(1), pp. 61–70, 2017.
[24] Piermattei, V. et al., Cost-effective technologies to study the arctic ocean environment. *Sensors*, **18**(7), p. 2257, 2018.
[25] Zappalà, G., Marcelli, M. & Piermattei, V., Development of a sliding device for extended measurements in coastal waters. *WIT Transactions on Ecology and the Environment*, vol. 111, WIT Press: Southampton and Boston, pp. 187–195, 2008.
[26] Zappalà, G., Bonamano, S., Madonia, A., Caruso, G. & Marcelli, M., Microbiological risk assessment in a coastal marine environment through the use of mathematical models. *WIT Transactions on Ecology and the Environment,* vol. 164, WIT Press: Southampton and Boston, pp. 3–14, 2012.
[27] Zappalà, G. et al., Integrated marine measurements in Civitavecchia, near Rome. *WIT Transactions on Modelling and Simulation*, vol. 55, WIT Press: Southampton and Boston, pp. 221–235, 2013.
[28] Zappalà, G. et al., Assessment of environmental conditions in Civitavecchia (Rome, Italy) harbour. *WIT Transactions on Ecology and the Environment,* vol. 182, WIT Press: Southampton and Boston, pp. 271–282, 2014.
[29] Zappalà, G. et al., A multi-platform approach to marine environment assessment in the Civitavecchia (Rome) area. *Journal of Operational Oceanography*, **9**, pp. s131–s143, 2016.

UNCERTAINTY OF MEASUREMENT IN HPM PULSE TESTING

RAFAŁ PRZESMYCKI, MAREK BUGAJ & MARIAN WNUK
Military University of Technology, Poland

ABSTRACT
Immunity measurement of informatic (IT) equipment and specialized military equipment on high power microwave (HPM) pulses are taking on more and more importance. In quite a short time, these tests can become a standard procedure for military and IT systems. The article presents a laboratory stand for the measurement of the immunity of military equipment to HPM pulses along with a description of the measurement procedure. In the research laboratory, one should strive to accredit each of the measurement procedures being carried out. At the Electromagnetic Compatibility Laboratory in the Military University of Technology, the measurement procedure discussed above was prepared for accreditation. One of the basic aspects of accreditation is the determination and assurance of the lowest measurement of uncertainty. The main goal of this article is to present how to determine and calculate the uncertainty budget for the measurement of the immunity of IT equipment and specialized military equipment to HPM pulses.
Keywords: electromagnetic pulse, immunity, uncertainty, electromagnetic field, EMC, HPEM, HPM.

1 INTRODUCTION

In a high-energy electromagnetic environment of high power, microwave sources are a relatively new kind of threat systems for both military and civilians. Operation using this type of source is usually referred to as intentional electromagnetic disturbance. The High Power Electromagnetic Environments (HPEM), which is inherently harmful, can also occur at frequencies up to several dozen MHz.

Technological development in the field of high-power microwaves has made it possible to produce more efficient microwave sources to irradiate large battle installations. At the same time, it is possible to generate relative high fields strength with the use of compact systems that can fit, for example in a suitcase. That is why the potential threat from High Power Microwave (HPM) has increased. In addition, when the HPM system is applied, especially in combat conditions, by a member of the North Atlantic Treaty Organization (NATO), there may be a need to stage joint defence systems against generated HPM fields. However, in the event of any scenario occurring, the real threat may occur in both friendly and hostile sources and has not yet been fully defined by any NATO member state. For this reason, the information provided below is general and public. National authorities should consult to determine if the HPM environment should be defined for a given system or device [1].

HPM type disruptions can be emitted in the air or be conducted through wires/cables until they reach their intended destination. There are two main possible ways of transmitting electromagnetic energy radiated to electronic systems. This can be done through the "front door" (antenna) or "back door" (cables and openings). Induced currents and voltages can definitively cause damage to electronic systems/circuits to which they reach [1].

The possibility of damaging unprotected electronic systems with disastrous consequences is considered as a realistic scenario for the development of the operational situation, including the critical elements of civilian infrastructure, where the possibility of sabotage and terrorism is a serious problem. A potential attack and additional action using currently and generally available technologies can be taken into account in any situation where a failure of electronic

WIT Transactions on Engineering Sciences, Vol 125, © 2019 WIT Press
www.witpress.com, ISSN 1743-3533 (on-line)
doi:10.2495/CMEM190071

systems can have fatal consequences for these systems. The examples concern such issues as: computer security, maintenance-free electronically opened/closed valves, damage to computer data, loss of connectivity and power loss. The result of the disturbance can be even permanent damage to the device. On the other hand, HPM sources belong to the general class of non-lethal weapons that can be used to eliminate hostile capabilities while maintaining a minimum of losses and own damages [1].

2 HPEM ENVIRONMENT

Although HPEM effects such as electromagnetic pulse due to lightning (LEMP) and electromagnetic pulse due to the discharge of a nuclear explosion (NEMP) have been studied in the past and the effects of these tests are relatively easy to access, HPM is a relatively new area of research.

Characteristics of hostile sources are unknown and difficult to predict, and possible “friendly” sources, due to the fact that they have not yet been sufficiently defined, should be the subject of security issues. All this makes the determination of the HPM environment very difficult.

Potential attackers can be: criminals, terrorists, dissatisfied employees/clients, soldiers of hostile armed forces/special forces and competitors. Potential scenarios of events using HPM are similar. A dissatisfied employee can bring a small, improvised device to the workplace, and by starting this device may cause numerous problems related to the functioning of electrical, electronic and other infrastructure. To create a threat, a terrorist may use the HPM source installed on the vehicle. In the case of a military approach, it is also possible to use airborne sources.

The characteristics of the threat environment are limited by the attacker's technical skills, the level of technology and access to technological means. The radiated power of HPM sources is from kilowatts to gigawatts (in impulse).

The frequency range is from tens of MHz to several GHz. Fig. 1 shows the frequency ranges of sources: LEMP, NEMP and HPM sources taken from IEC 61000-2-13 [2].

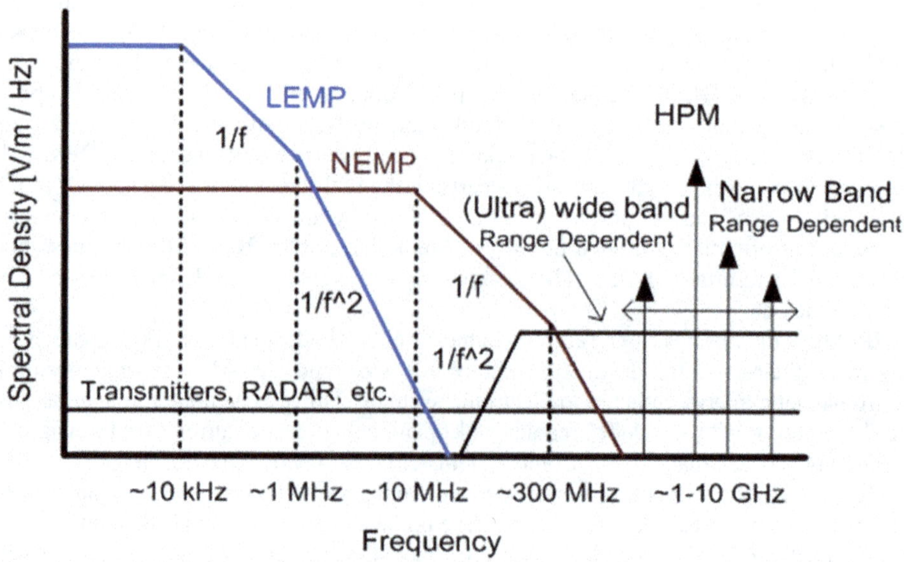

Figure 1: Typical high power electromagnetic environments (HPEM) [1], [2].

Depending on the type of objective and the knowledge of the target, to achieve the desired effect and cause the maximum effects, the attacker will choose the types of HPM sources with the most effective course. The possible types of HPM sources and waveforms are described in the next item of the paper. The level of electromagnetic energy at the place of application of the HPM source depends on many parameters [2].

3 HIGH POWER MICROWAVE PULSE

HPM sources can be divided into four categories. These are, among others:

a) Mobile HPM sources installed on the platforms – typically HPM sources radiate fields towards the equipment of the facility's infrastructure from areas that are not controlled by devices of the technical protection system, such as areas outside the fence. In this situation, it is more likely to use truck-mounted sources than man-borne sources, because the power required to generate effective levels of electromagnetic fields to radiate the intended target from a considerable distance requires the use of sources characterized by large geometric dimensions and high mass. It is also possible to use sources carried by air.
b) A portable HPM source – a portable HPM source can be brought inside the target – to the site of the object being attacked, hidden under clothing or inside a briefcase, purse or even in a can. The proximity of the electronics subject to attack may cause the "small" source to become even more dangerous than the "larger" source used at a greater distance, because the effect of damping through the screens and distance protections are eliminated.
c) An HPM source that generates conductive disturbances – this type of HPM source can cause energy to be injected directly through the power cords into the facility's infrastructure, such as power lines and communication lines. This type of threat can be caused both from the outside and from the inside of the object.
d) HPM source based on a projectile – an HPM source of this type radiates the entire facility with a generated electromagnetic pulse. This impulse generates a large current impact in the external cables of the facility's infrastructure, such as power lines or communication lines, or in water pipes, etc. [3].

In general, the number of types of HPM waveforms is limited to four. Waveform parameters such as frequency, power, etc. depend on the type of source.

a) The continuous wave (CW) – a HPM system, which generates a CW waveform, generally contains a magnetron, microwave resonance lamp. The CW waveform signal is most effective when its frequency is tuned to the most susceptible frequency from the point of view of the target. A continuous wave can be characterized by frequency and power [3].
b) Narrowband pulsed CW – narrowband pulsed continuous waveform CW is a typical HPM waveform. In the Fourier spectrum of this type of wave there is a narrow envelope around the "center frequency". This type of waveform can be characterized by the carrier frequency, peak power, pulse width and pulse repetition frequency and derivative parameters such as duty cycle, average power, etc. The advantage of the shape NB is the fact that a small amount of medium power is used to generate an effective high-energy pulse, which results in the use of this type of high-energy impulses to disturb the work or damage to the electronic equipment. It is assumed that narrow-band sources are characterized by the highest impulse power at the peak. The generated signal is most effective when the target is susceptible to the peak signal at its

most sensitive frequency. This signal can be designed to engage antennas or antenna structures (coupling through the "front door") to cause damage to the target. Narrowband sources can also be designed to engage with the target through unintentional tracks ("rear door" coupling) causing damage to electronics [3].

c) Dumped sinewave (DS) – DS waveform is characterized by: carrier frequency, peak power, decrement attenuation and energy value in the pulse. The DS waveform is characterized by a "center frequency" and some frequency bands that can increase the likelihood of a "rear door" coupling. The advantage of the DS waveform is a wider band than the CW or NB waveforms; usually one octave [3].

d) Ultra-wide band (UWB) – UWB waveform is characterized by: peak power, rise time, decay time, bandwidth and repetition frequency. UWB pulses are characterized by the ability to excite many resonant modes in a facility or target. Due to the fact that the power that occurs in a small part of the frequency range is relatively small, and also because the total energy is contained in one pulse is small, it is difficult to generate a large field strength in resonance to destruction or damage to the target. In order for the UWB pulses to be effective, more power is required than for the narrowband frequency with the appropriate NB frequency [3].

4 THE LABORATORY STAND FOR GENERATION AND MEASUREMENT OF HPM PULSE

The main element responsible for the generation of a HPM pulse is a generator Diehl DS110, which is shown in Fig. 2. It is compact enclosed equipment, looks like a suitcase. Components of the generator are high-voltage power supply, 300 kV Marx generator and a resonant antenna. High voltage power supply system supplies the Marx generator by 50 kV voltage (symmetrical voltage ± 25 kV), he is charged until the threshold set by the switch that sets pressure gas N_2 in the generator. When changing the position of the switch, antenna is fed a 300 kV voltage, the sparker contained in the air is isolated by the pressure gas N_2 [2],[4]–[6]. The basic specification of the HPM DS110F generator is shown in Table 1.

Figure 2: View of the generator Diehl DS 110.

Table 1: DS110F specifications.

Parameter	Value
Size	500 x 410 x 200 mm
Weight	24 kg
Peak radiated power	160 MW
Radiation pattern (without reflector)	Dipole
Pulse duration	4 ns
Repetition rate	> 5 Hz (10 Hz)
Centre frequency	350 MHz
3 dB bandwidth	100 MHz
Operating time (without reflector)	> 1 h

The Diehl Measurement System incorporates a fast digitizing oscilloscope running from an uninterruptible power supply (UPS). The UPS is used in a different manner than originally intended. The UPS operates the oscilloscope for up to 3 h without mains connection before recharging the batteries of the UPS is required.

Oscilloscope and UPS are installed in a shock damped and electromagnetically shielded housing. This allows not only outdoor measurements without mains available, but also shielded measurements without any connection to outer environment except the field measuring probe.

The Diehl Measurement System combines a state-of-the-art Tektronix Digital Phosphor Oscilloscope with a computer UPS and an additional battery pack within a shielded and shock damped housing. Due to this combination the Diehl Measurement System can be operated e.g. in free field without mains, and it can be operated with closed shielding lids when in close distance to radiating HPM systems. The Diehl Measurement System includes a so-called D-dot probe and a Balun which are designed to measure pulsed electromagnetic fields. Within the shielded housing a laptop computer is stored usable to calculate measured electromagnetic fields and to further process measured data [2], [4]–[6].

Fig. 3 shows an appearance of HPM pulse in the time domain, while Fig. 4 shows the frequency spectrum of the pulse. The red color presented the pulse generated by generator with reflector, the blue color presented the pulse generated by generator without reflector.

4.1 D-dot field probe

For measuring fast HPM pulses with the Diehl Measurement System a combination of D-dot probe Prodyn AD-70 and a Balun Prodyn BIB-170F was chosen as the optimum free field sensor. The probe is mounted on a plastic stand to reduce measurement influences to a minimum. Free field sensors as the Prodyn AD-70 are dual sensors. They have two antenna elements that are sensitive to the same field but measure in opposite directions. They have their own ground reference between the two elements. Electric (D-dot) sensors have capacitive sensing elements in opposite to magnetic (B-dot) sensors having inductive sensing elements or loops. The D-dot sensors measure time derivative of electric displacement. The

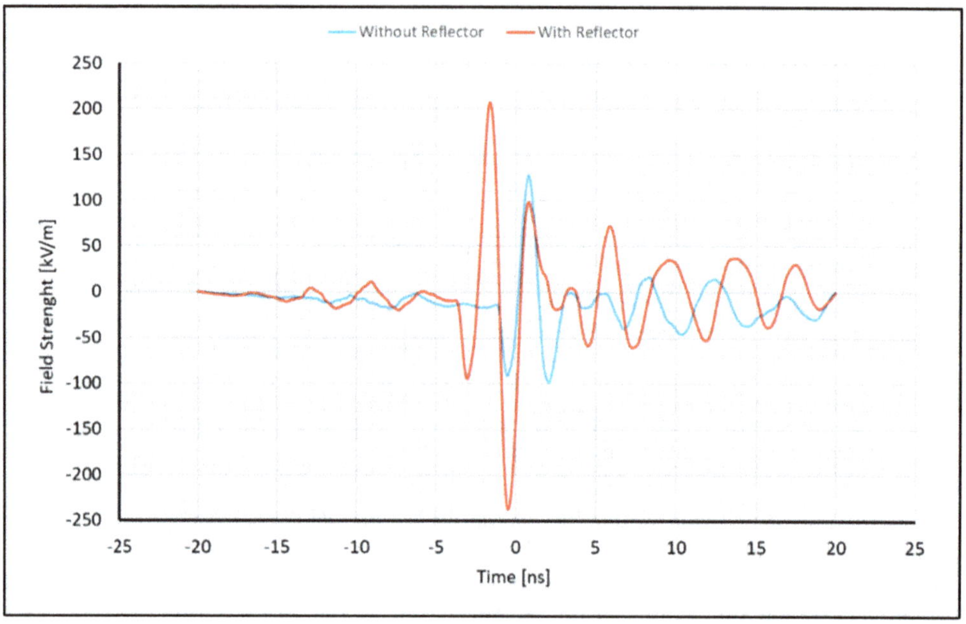

Figure 3: An example of the time domain for a sample HPM pulse.

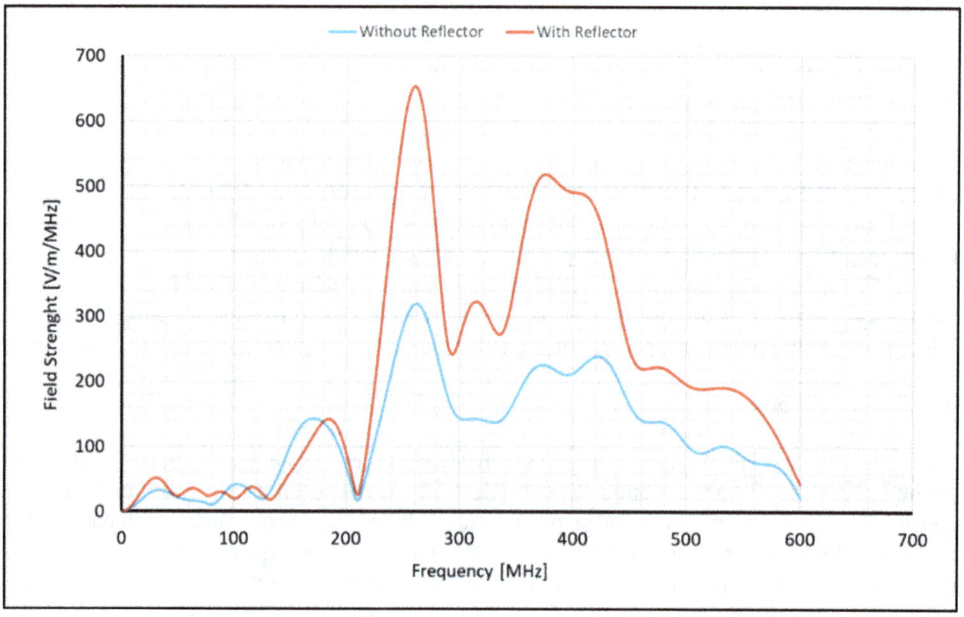

Figure 4: An example of the frequency spectrum for a sample HPM pulse.

sensor is passive devices and does not need external power supply. Since the sensor is symmetric a Balun is required. Balun is the abbreviation for BALanced and UNbalanced and describes a unit which converts the symmetric probe signals to a common, non-symmetric signal as usual in coaxial cables. The D-dot effective antenna size of 0.001 m^2 as well as the Balun insertion loss of 8 dB are included as standard parameters within the Excel macros. These values can be changed in Microsoft Excel worksheet when using probes other than the original ones delivered with the Diehl Measurement System [6].

4.2 Oscilloscope

The main component of Diehl Measurement System is a state-of-the-art Tektronix Digital Phosphor Oscilloscope DPO70404. It contains four analogue channels each having 4 GHz bandwidth and 25 GS/s real time sampling rate. For pulsed field measurement of Diehl Measurement System an interpolated higher sampling rate is not usable, the real time sampling method is always to be used. The Tektronix DPO70404 incorporates an own computer board for operating the equipment. It would be possible to install Microsoft Excel with processing macros but Diehl experience is that using a separate computer makes the oscilloscope more stable and data analysis more independent. The oscilloscope allows to record so-called FastFrame formats. FastFrame is able to capture many trigger events respectively many HPM pulses as sequential records in a larger single record. It is possible to view and measure each record individually. Time stamps display the absolute trigger time for a specific frame and the relative time between triggers of two specified frames. This means that several HPM pulses can be stored within one file and can be processed in the computer in one step. Data measured with the oscilloscope can be stored either on internal hard disk, on USB sticks connected at back or front USB ports or directly on network drives when connected with other computers. Diehl recommends storing measured data first on internal hard disk. To copy them onto the laptop computer use either USB memory sticks or network connection. This ensures that only a duplicate copy is onto the laptop computer when processing measured data [6].

The output of the D-dot sensor is proportional to the first time derivative of the electric displacement D(t), that is:

$$\text{D-dot} = dD/dt, \tag{1}$$

and, rewriting equation:

$$Vo = R\ Aeq\ dD/dt, \text{ or } dD/dt = Vo/R\ Aeq, \tag{2}$$

then, recalling that $D = \varepsilon_o E$:

$$dE/dt = Vo/R\ Aeq\ \varepsilon_o. \tag{3}$$

The output of the D-dot sensor is measured in volts that are proportional to units the first time derivative of electric displacement dD/dt, coulombs per sq m. The constant of proportionality between the sensor output and the first time derivative of the electric field dE/dt (= dD/dt ε_o) is the permittivity constant ε_o. The constant of proportionality between the sensor output Vo and the first time derivative of the electric field dE/dt is the product of the output impedance R and the equivalent area Aeq of the sensor, and the permittivity constant ε_o. This is the calibration factor, but it is applied to the first time derivative of the electric field, dE/dt, not to the electric field itself [7].

The sensor output resulting is a cosine function with amplitude V_{op} and frequency f, where:

$$Ep = 1/2\ f\,(dE/dt)p = V_{op}/2\ f\,R\ Aeq\ \varepsilon_o. \tag{4}$$

4.3 Complete measurement chain

Since the output signal of the D-dot sensor is balanced, but the input of an oscilloscope is singled-ended, a balanced-to-unbalanced transformer (Balun) is directly connected to the sensor. This device adds a specific insertion loss to the sensor signal and performs an impedance transformation from the balanced sensor output (typically 100 Ω = 2·50 Ω to a coaxial output (typically 50 Ω) [8].

Before the output signal from the D-dot probe is registered by the oscilloscope, it must be sent. There are two types of data transfer from the D-dot probe to the oscilloscope. The first of them is based on high-frequency tracks (concentric lines) that directly connect the D-dot probe, Balun and oscilloscope. Due to high amplitudes of HPM pulses, shielding of RF lines however, it can be a big problem. It is possible to use a second type of data transfer using a short coaxial connection that is used from the Balun to the optical transmitter. The optical transmitter converts this signal into an optical signal. The optical signal is sent via optical fiber to the receiver, which converts back to an electrical signal. In practice, the D-dot probe is directly connected to the Balun. In developed laboratory stand there are only RF paths. This is a first type of data transfer from the D-dot probe to the oscilloscope.

The Fig. 5 has been presented a block diagram of the laboratory stand to generation and measurement of HPM pulse. The main elements of this laboratory stand are the Diehl generator and field probe, which they are placed in an anechoic chamber so that the probability of damage other electronic devices located near the generator is very small. In addition, they have been placed on the metal floor with ferrite absorbers thus ensuring that a common reference mass. Devices in the measurement room allow their safe operation by authorized persons. Fig. 6 shows a schematic with only every component used for the field measurement.

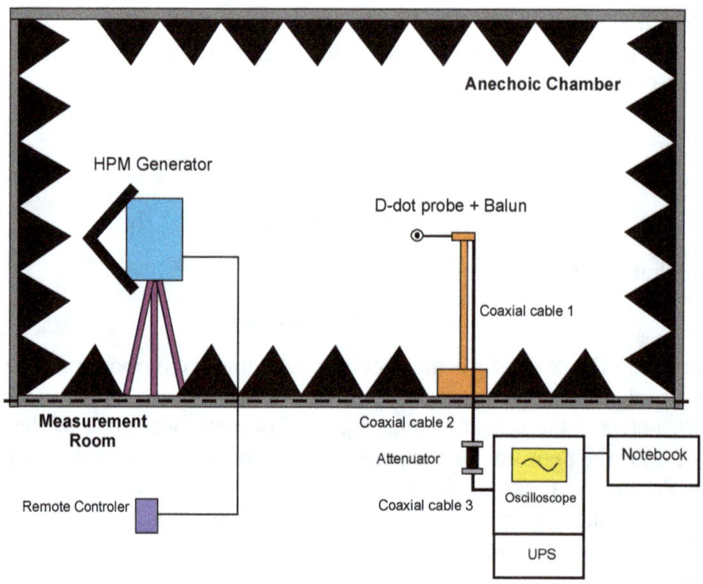

Figure 5: The block diagram of the laboratory stand for the generating and measurement of a HPM pulse.

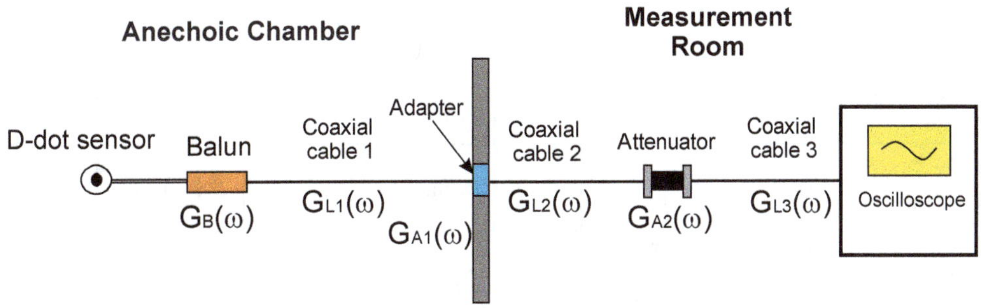

Figure 6: Schematic of the measurement chain for the field measurement in the laboratory stand for the measurement of a HPM pulse.

The complete signal path of course must be considered non-ideal, because each component modifies the original signal from the output of the D-dot sensor. The influence of each component on the value of the received signal can be described by means of complex transfer functions in which G_B (ω) is for Balun, G_{L1} (ω), G_{L2} (ω), G_{L3} (ω) for coaxial cables, G_{A1} (ω), G_{A2} (ω) for the adapter in the chamber wall and silencer. These complex transfer functions describe the amplitude response and linear phase difference of each element depending on the angular frequency ω as a linear system. The data for each element of the measurement chain can easily be obtained by measuring S_{21} using a vector analyzer VNA [8].

5 UNCERTAINTY OF MEASUREMENT FOR LABORATORY STAND TO GENERATION AND MEASUREMENT OF HPM PULSE

The traceability of measurement quantities is very important for an accredited test procedure. Accuracy and reliability of measurement results are based on the uncertainty budget. Starting from the model equation, you can calculate the uncertainty budget for the presented measurement station. In this type of measurements very often logarithmic values are used for particular quantities, which simplifies the final formulas for uncertainty budget calculations. The equation for the measurement chain described and shown in Fig. 6 is as follows:

$$E(t) = E_{calc}(t) + dV_{scope}(t) + dV_{sensor}(t) + P_S + \sum dER_x, \tag{5}$$

where Table 2 describes the components included in the uncertainty budget. Due to the complete consideration of the complex measuring chain, the transmission functions of each element depending on the angular frequency reduce their contribution to the uncertainty budget only to components derived from S_{21} measurements for each of them [8], [9]:

$$\sum dER_x = dER_B + dER_{L1} + dER_{L2} + dER_{L3} + dER_{A1} + dER_{A2}. \tag{6}$$

Table 3 shows the full calculation of the expanded measurement uncertainty for the measuring stand discussed, including source uncertainty and assumed distributions. The calculations were based on the literature [10], [11].

For the presented test method for measuring the amplitude of HPM pulses, the expanded measurement uncertainty can be determined to be ± 3.08 dB. The expansion factor k = 2.0 is dictated by additional degrees of freedom due to type A errors during the analysis of all components derived from measurements S_{21}.

Table 2: Uncertainty budget.

Symbol	Source of uncertainty	Value	Probability distribution	Divisior	c_i	$u_i(y)$	$(u_i(y))^2$
$E_{calc}(t)$ [dB]	Estimation	0.50	Rectangular	1.732	1	0.289	0.083
$dV_{scope}(t)$ [dB]	Calibration certificate or documentation and the conformity certificate of the oscilloscope	0.65	Rectangular	1.732	1	0.375	0.141
$dV_{sensor}(t)$ [dB]	Estimation from reference measurement within an Anechoic chamber	1.9	Normal, k = 2	2.000	1	0.950	0.903
P_S [dB]	Verification measurement before each test	2.0	Normal, k = 2	2.000	1	1.000	1.000
dER_B [dB]	Calibration certificate or documentation and the conformity certificate of the vector network analyzer	0.35	Rectangular	1.732	1	0.202	0.041
dER_{L1} [dB]	Calibration certificate or documentation and the conformity certificate of the vector network analyzer	0.35	Rectangular	1.732	1	0.202	0.041
dER_{L2} [dB]	Calibration certificate or documentation and the conformity certificate of the vector network analyzer	0.35	Rectangular	1.732	1	0.202	0.041
dER_{L3} [dB]	Calibration certificate or documentation and the conformity certificate of the vector network analyzer	0.35	Rectangular	1.732	1	0.202	0.041
dER_{A1} [dB]	Calibration certificate or documentation and the conformity certificate of the vector network analyzer	0.35	Rectangular	1.732	1	0.202	0.041
dER_{A2} [dB]	Calibration certificate or documentation and the conformity certificate of the vector network analyzer	0.35	Rectangular	1.732	1	0.202	0.041
$u_c(E(t))$ [dB]	Combined standard uncertainty		Normal			1.540	2.372
$U(E(t))$ [dB]	Expanded uncertainty		Normal k = 2			3.08	

Table 3: Uncertainty contributions.

Symbol	Quantity	Unit	Description
$E(t)$	Electrical field strength	dBμV/m	This is a result of measurements
$E_{calc}(t)$	Calculated electrical field strength	dBμV/m	This is a value calculated from the sample data in oscilloscope
$dV_{scope}(t)$	Amplitude error of the oscilloscope	dB	Amplitude error of the oscilloscope
$dV_{sensor}(t)$	Output error of D-dot probe	dB	This value summarizes all the imperfections of the field probe
P_S	Measurement repeatability	dB	This value is an estimation for the repeatability of the measurement, e.g. the placement of the sensor
$\sum dER_x$	Sum of errors from the measurement chain	dB	This value is the overall error from the S_{21} measurement of the measurement chain's elements

6 CONCLUSIONS

The study of the resistance of specialist military equipment to the pulsed electromagnetic field is getting more and more important. In quite a short time, these tests can become a standard procedure for military and IT systems. The testing procedures in standardization documents defining risk levels and testing procedures for HPM exposure have existed for several years. This article presents the measurement procedure and the laboratory stand for immunity test of military equipment to pulsed electromagnetic fields.

In the research laboratory one should strive to accredit each of the performed measurement procedures. At the Electromagnetic Compatibility Laboratory in Military University of Technology, the measurement procedure discussed above was prepared for accreditation. One of the basic aspects of accreditation is the determination and assurance of the least possible measurement uncertainty.

The main purpose of this article, however, was to present the method of determining and calculating the budget of uncertainty for immunity test of specialist military equipment to pulsed electromagnetic fields. Due to mathematical relations, which take into account all the information of the measuring chain from the point of view of electrical parameters, the expanded uncertainty of measurement is 3.08 dB for the discussed measuring stand. Using this test method, you can offer reliable and repeatable qualification tests for HPM compromising military equipment, which rely on the accuracy and reliability of such tests.

REFERENCES

[1] AECTP 250 Ed.2, 2011: Electrical and electromagnetic environmental conditions. NATO International Staff – Defence Investment Division, 2011.

[2] Przesmycki, R. & Wnuk, M., Susceptibility of IT devices to HPM pulse. *International Journal of Safety and Security Engineering*, **8**(2), pp. 223–233, 2018.

[3] NO-06-A218-4, 2014 – Technika wojskowa, Warunki Środowiskowe Część 4: Elektryczne i elektromagnetyczne oddziaływania środowiskowe, 2014.

[4] Przesmycki, R., Wnuk, M. & Bugaj, M., HPEM susceptibility assessment of data storage devices. *21st International Conference on Microwave, Radar and Wireless Communications (MIKON)*, Kraków, Poland, 2016. DOI: 10.1109/MIKON.2016.7492113.

[5] Przesmycki, R. & Wnuk, M., HPEM susceptibility assessment of informatic devices. *Progress in Electromagnetic Research Symposium (PIERS)*, Shanghai, China, 2016. DOI: 10.1109/PIERS.2016.7734568.

[6] DIEHL BGT Defence, *Operating and Maintenance Manual for DIEHL DS110.KS2.F.MP1.B (HPM Case System DS110)*, document No BDB8118952.

[7] Reed Edgel, W., *The Next Best Thing to Calibration, Functional Testing,* Prodyn Aplication note.

[8] Kreitlow, M., Schmidt, G. & Sabath, F., Accreditation of a NEMP test procedure: Approach, measurement technique, uncertainty. *2016 IEEE International Symposium on Electromagnetic Compatibility (EMC)*, Ottawa, ON, pp. 841–846, 2016.

[9] Kreitlow, M., Schmidt, G. & Sabath, F., *The Uncertainty of Measurement in NEMP Testing,* July 2016.

[10] BSI, *General Requirements for the Competence of Testing and Calibration Laboratories,* International Organization for Standardization (ISO) Std. ISO 17 025:2005, 2005.

[11] BIPM, Evaluation of measurement data – Guide to the expression of uncertainty in measurement (GUM), Joint Committee for Guides in Metrology (JCGM/WG 1), 2008. www.bipm.org/utils/common/documents/jcgm/JCGM 100 2008 E.pdf.

ROBUST DOUBLE ACTIVE CONTROL SYSTEM DESIGN FOR DISTURBANCE REJECTION

SOUMAYYA CHAKIR, VAN TRONG NGUYEN, TAE-WAN KIM,
DONG-HUN LEE, JIN-HO SUH & YOUNG-BOK KIM
Department of Mechanical System Engineering, Pukyong National University, South Korea

ABSTRACT

This paper introduces a new disturbance decoupling and rejection method based on robust control framework for a robot system with multiple arms. While in action, the movement of one individual arm of the robot system affects the motion of the other joint, which brings the control performance down. In order to compensate and keep the desirable control performance, the authors designed a double active control system which can effectively reject the direct mutual disturbances. The designed control system consists of two controllers, the first one is an inner loop controller and works as a disturbance observer that attenuates the disturbances. The second one, on the other hand, is an outer loop controller, which was designed based on H_∞ control theory to maintain the system stability and a robust control performance under the uncertainty. The combined control system is applied, simulation and experimental results show that the proposed control system effectively suppress the mutual disturbances, and an enhanced tracking performance is obtained.

Keywords: robust H_∞ control, feedback control, PID control, 2DOF robotic arm.

1 INTRODUCTION

In the midst of the transition to the fourth industrial revolution, the use of automated machines and robots in the industry is of a great importance. Robotic arms, in particular, imitating the human arm movement, they can execute simple, repetitive or difficult tasks on several degrees of freedom (DOF). Thanks to their efficiency and accuracy, they are widely applied in many aspects of the industry, and the marine industry is no exception.

Therefore, in order to control industrial robots to generate accurate motion, multiple control methods were designed. A robust control approach [1], sliding mode control [2], and adaptive control [3] approaches have been developed. However, the mutual disturbances of the joints were not studied thoroughly and had not been directly controlled. Thereupon, an approach to suppress these types of perturbations was developed based on the generation of feedforward control signal computed in advance [4]. This method has two major drawbacks, first, the perturbation suppression is done off-line. Second, there was no room for tackling the unpredicted disturbances that may occur in real time action. Likewise, another approach can be listed, a perturbation observer that has been designed to control the disturbances [5], the observer system, however, is based on estimations and assumed disturbances, which does not often give the desired response.

In the light of all the above, and for the purpose of this paper, the disturbance rejection of the robot arms on each other is achieved by two types of feedback compensators that have been designed, notably a double active control system. First, a real time inner loop controller, designed based on the system configuration, and its objective is to cancel mutual disturbances of the robot arms. However, the input of this controller can be considered as disturbance for the plant which requires the second control system to be designed. Therefore, the outer loop controller was designed based on the robust control theory for better control performances.

In order to verify the efficacy of the proposed control system, a robot system of two-arms, illustrated in Fig. 1, is being considered, and a series of numerical simulations with commercial software MATLAB and Simulink have been investigated. Furthermore, using

WIT Transactions on Engineering Sciences, Vol 125, © 2019 WIT Press
www.witpress.com, ISSN 1743-3533 (on-line)
doi:10.2495/CMEM190081

the graphical programming software LabVIEW, experimental studies have been carried out. Comparative experiments using a PID controller and the proposed control system were performed and analysed, and the results show that the robust double active controller was effective in suppressing disturbances, as well as it gave enhanced and better tracking performances.

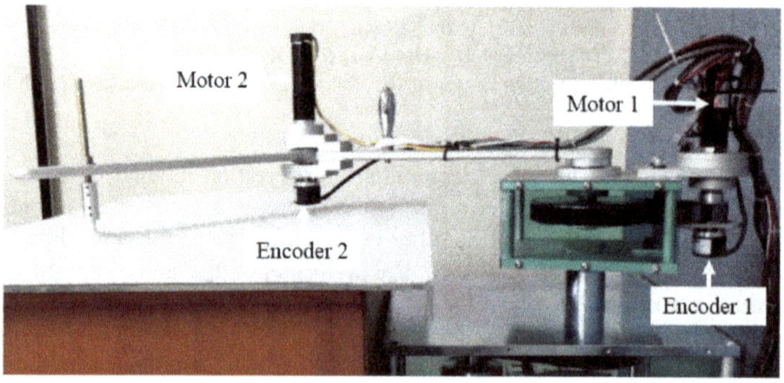

Figure 1: A two-arm robot system.

2 SYSTEM MODELLING AND CONTROLLER DESIGN

2.1 System identification process

In order to design the control system, having a mathematical model that describes the dynamics of the system is compulsory. The Lagrange equations and the identification process using a software are the main methods to obtain a mathematical model of a system. However, the presence of different types of frictions, and difficulties to clearly calculate the different parameters, for instance the inertia moment, determining the system dynamics will be difficult to obtain using Lagrange equations. Therefore, the method adopted provides an individual identification of the system dynamics; each arm of the robot system has been active, one at a time. While the first joint is moving, the transfer function of the active joint is obtained based on the angle of rotation. The inactive joint thus moves and based on its angle of rotation, we obtain the mathematical representation of the active joint disturbance on the inactive one, thus noted $G_d(s)$, and vice versa.

2.2 Disturbance rejection system design

Every real system is subjected to different types of disturbances, whether they occur within the components of the system, or they are some extern disruptions. In fact, let us consider the system with disturbance input depicted in Fig. 2.

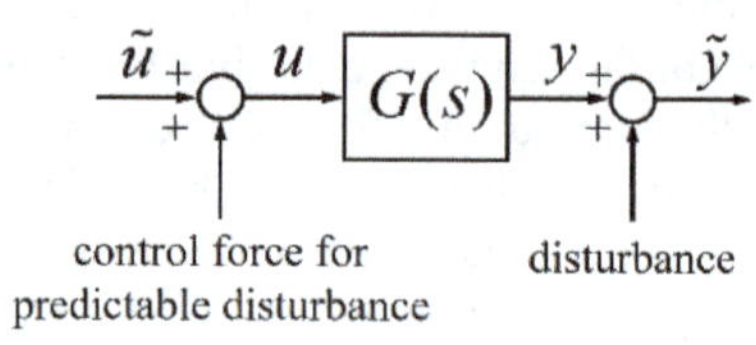

Figure 2: Bloc diagram of a controlled system with input and output signals.

The overall output denoted $\tilde{y}$ is the summation of the output signal y of the plant, and y_d the direct disturbance input signal, which in a muti-joint robot system mainly represents the mutual perurbation.

$$\begin{aligned}\tilde{y}(s) &= y(s) + y_d(s) \\ &= G(s)\cdot u(s) + y_d(s),\end{aligned} \tag{1}$$

where $G(s)$ is the mathematical model of the plant. The objective is to reject the direct disturbances input signal $y_d(s)$. Therefore, if the control input $u(s)$ is to be designed as in the eqn (2), the $y_d(s)$ signal will be eliminated, under the condition that $G(s)$ is proper.

$$u(s) = -G(s)^{-1}\cdot y_d(s). \tag{2}$$

The disturbance input can be expressed as in eqn (3):

$$\begin{aligned}u(s) &= -G(s)^{-1}\cdot G_d(s)\cdot y(s) \\ &= -\bar{G}(s)\cdot y(s),\end{aligned} \tag{3}$$

with, $\bar{G}(s)=G(s)^{-1}\cdot G_d(s)$ as the mathematical representation of the inner loop controller. This input signal may be computed in advance and inserted in the specified time, which will help cancel the mutual disturbances [4]. However, the presence of the inner loop compensator, and the mutual disturbances are drawbacks to the system. Therefore, a robust and optimal H_∞ controller is to be designed to provide better performances. Instead of the classic representation of this controller, is illustrated in Fig. 3(a), where $\Delta(s)$ represents the uncertainties, $G(s)$ is the generalized plant, and $K(s)$ is the controller to be designed.

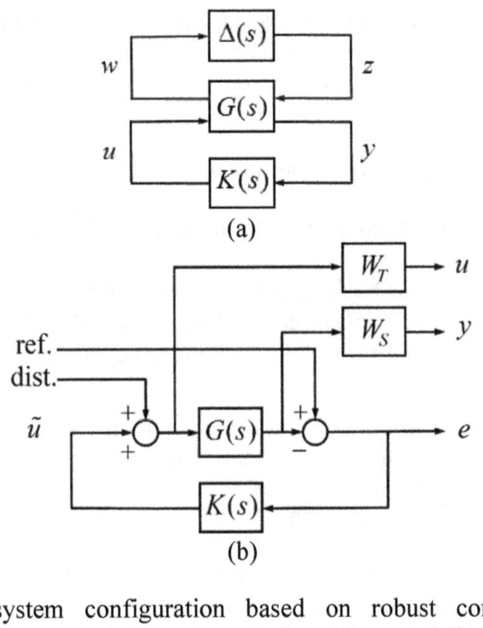

Figure 3: Control system configuration based on robust control framework. (a) Configuration of standard H_∞ control problem; and (b) Mixed sensitivity H_∞ control optimization.

The new configuration of H_∞ control, has a double objective [6]. First, enhancing the tracking performances requires the design of a stabilizing feedback controller *K(s)*, the outer loop controller. Second, the design of an inner loop controller is necessary in order to achieve a robust stability. The weighting functions W_T and W_S are to help minimize the control input u, the plant output y, as well as the error signal e.

For instance, let us consider the two arms robot system illustrated in Fig. 4 with the control system. Simulations and real experiments were conducted based on the identified system and the control system designed.

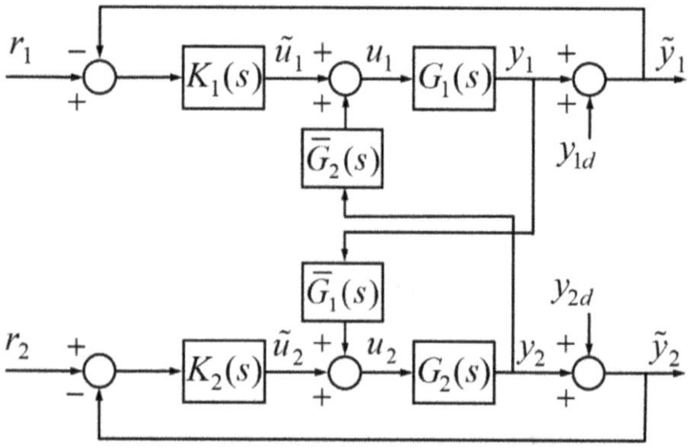

Figure 4: Block diagram of two-arm robot control system.

3 SIMULATION RESULTS

In this section, a comparative study was conducted in order to evaluate the robustness and the control performances of the system. The simulation results were obtained using the designed control system on one hand, and on the other hand experiments using the PID controller were executed.

3.1 Simulation results while the second arm is active

In this experiment, the control input signal to the plant, as well as the overall output of the system are being analysed.

In order to maintain a better tracking performance while the inner loop controller is not active, the control input u_2 in Fig. 5(a) has to be very big to obtain better results. On the other hand, the same control u_1 input is needed, but better tracking performances of the inactive arm, when $\bar{G}_i$ is active.

Using the H_∞ controller and activating the inner loop controller, Fig. 5(b), brings the control input signal to half the amount while it was not activated.

3.2 Simulation results while the first arm is active

Similar to the previous case, while the inner loop controller is not active even greater control input signal u_1 is required when the first arm is active, Fig. 6(a). However, the tracking performance improved once the inner loop controller is active, even though the same amount of energy is required for the control input.

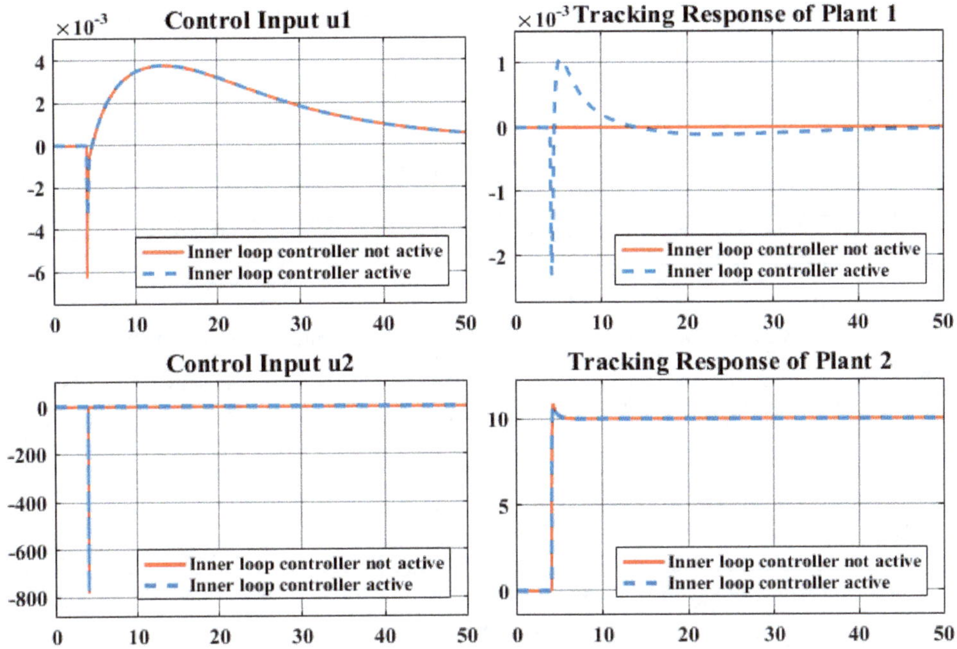

(a)

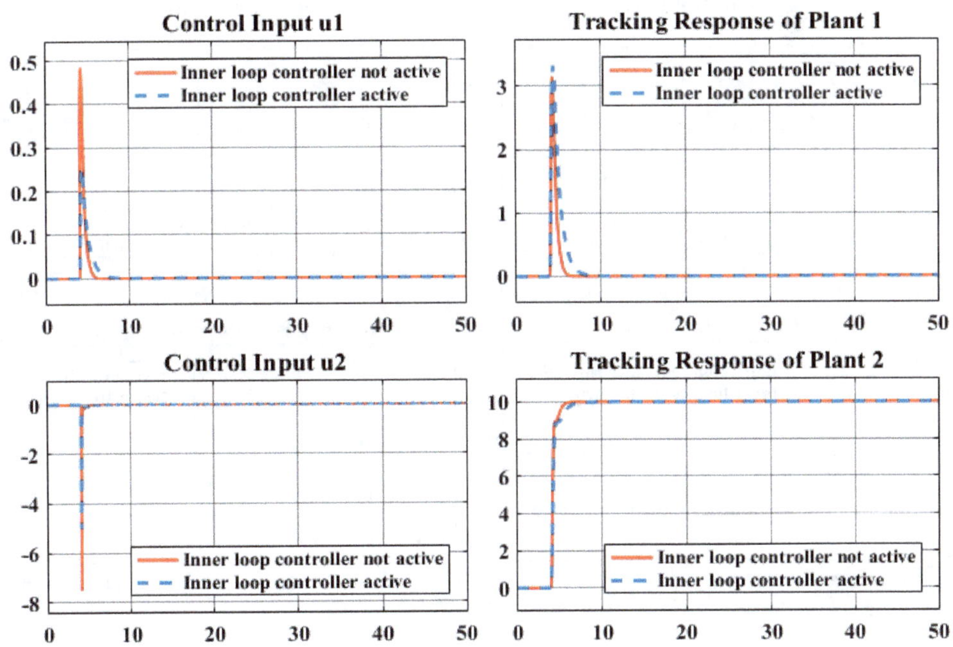

(b)

Figure 5: Step responses while the second arm is active (the first arm is inactive). (a) Response of the PID control system; and (b) Response of the H_∞ control system.

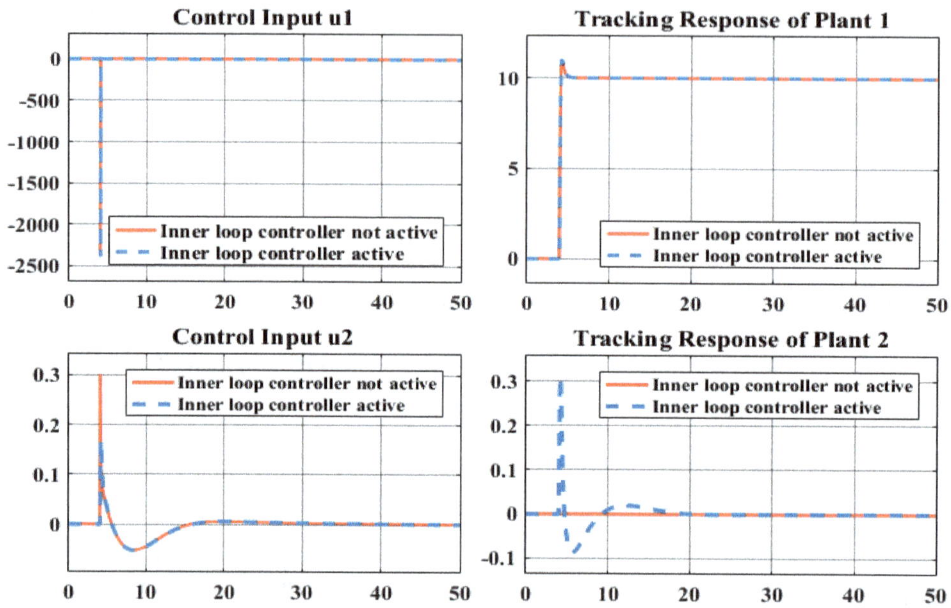

(a)

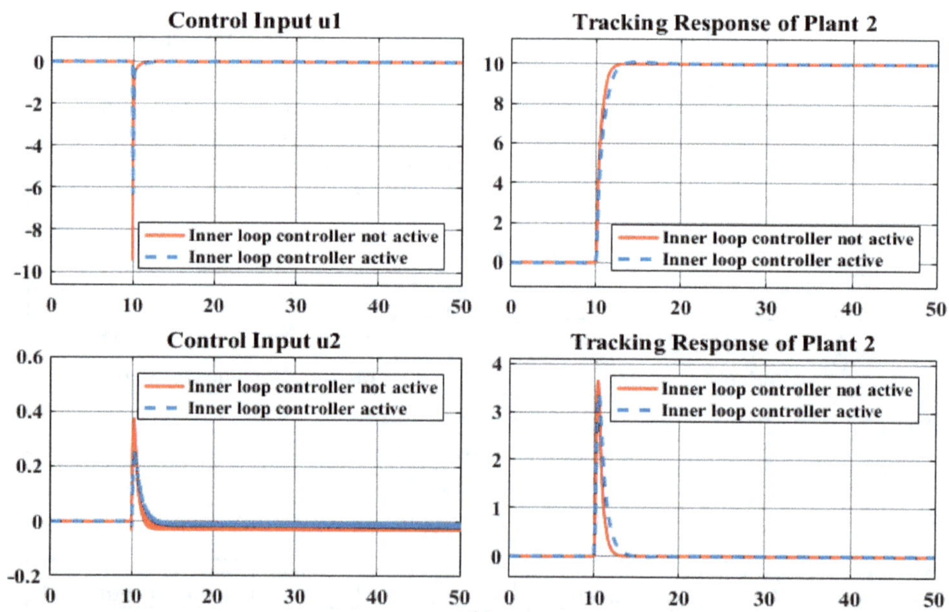

(b)

Figure 6: Step responses while the first arm is active (the second arm is inactive). (a) Response of PID control system; and (b) Response of H_∞ control system.

Likewise, using the designed controller and activating the inner loop controller Fig. 6(b) brings the control input signal to half the amount while it was deactivated. Moreover, better tracking performances are provided with much less control input signals.

3.3 Simulation results using the pulse signal

The experiment executed using the PID controller shows good tracking performances, however, a high control input signal is required, which does not change despite the inner loop controller being active, Fig. 7.

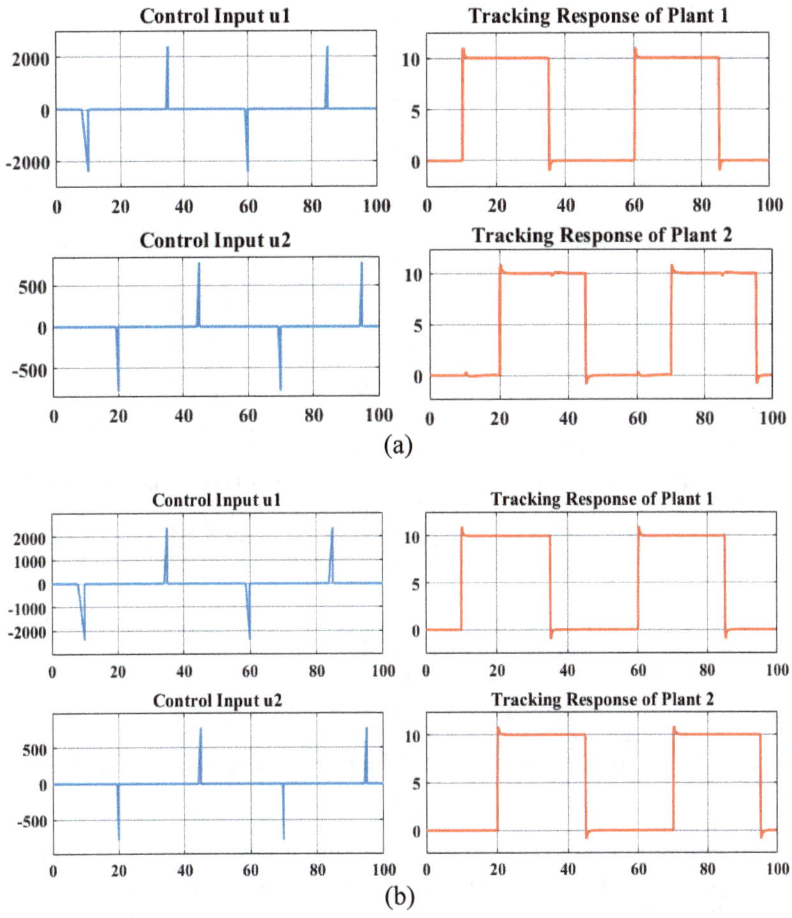

Figure 7: Pulse type system responses with PID controller. (a) $\overline{G}_i = 0$; and (b) $\overline{G}_i \neq 0$.

The pulse response using the designed control system shows better results than the PID controller, especially, regarding the control input signal, which is the main objective of the H_∞ based control system, Fig. 8(a). Moreover, when the inner loop controller is active, not only the overall output of the plant is better, but also less control input signal is required to maintain the desired response to the refrence, Fig. 8(b).

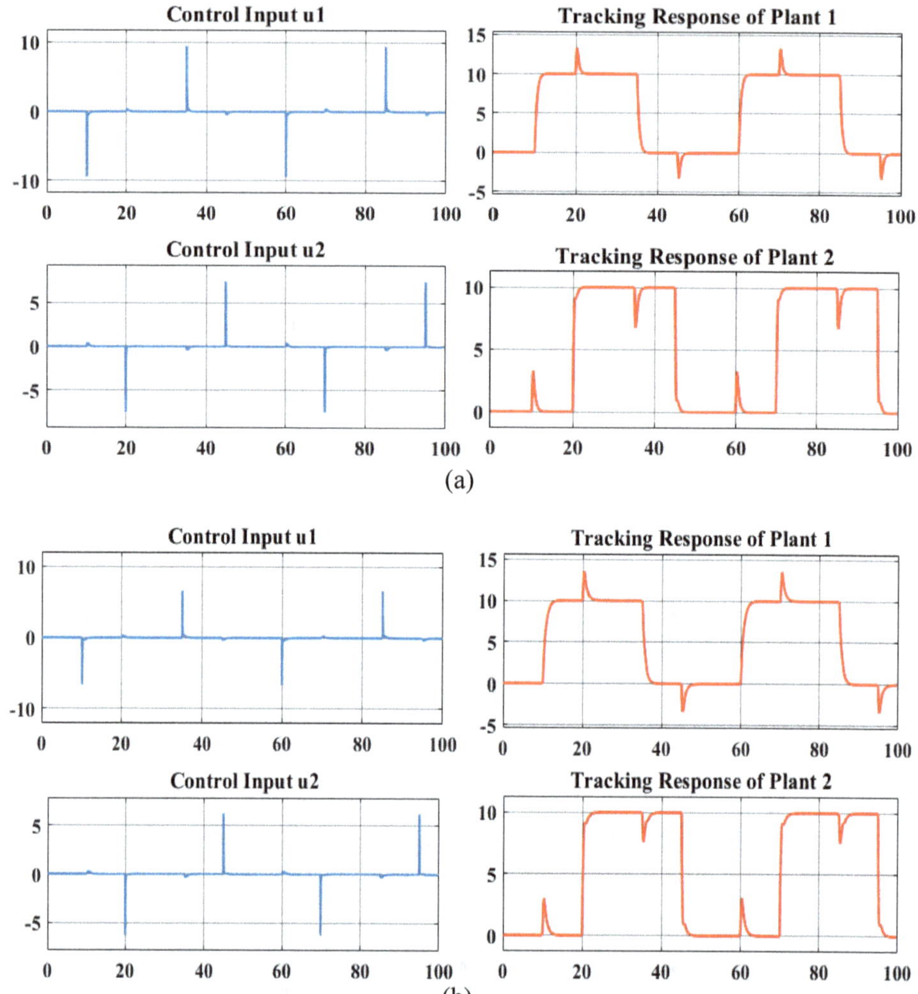

Figure 8: Pulse type system responses with H_∞ controller. (a) $\overline{G}_i = 0$; (b) $\overline{G}_i \neq 0$.

Finally, from the previous simulation results, the designed control system appears to provide better performances, an internal stability, as well as an improved system robustness. Therefore, in the next section, this designed control system will be put to the test with the real system illustrated in Fig. 1.

4 EXPERIMENT RESULTS

In a similar fashion to the simulation tests, a comparison study of the PID and the designed control system responses will be presented. The two arms robot system has been subjected to a pulse reference, and the rotation angles were measured by two incremental encoders.

The experiment results support perfectly the simulation results, and sustain the fact that the proposed control system performs well. The following figures illustrate the results obtained using the PID controller, Fig. 9(a), and using the designed control system, Fig. 9(b).

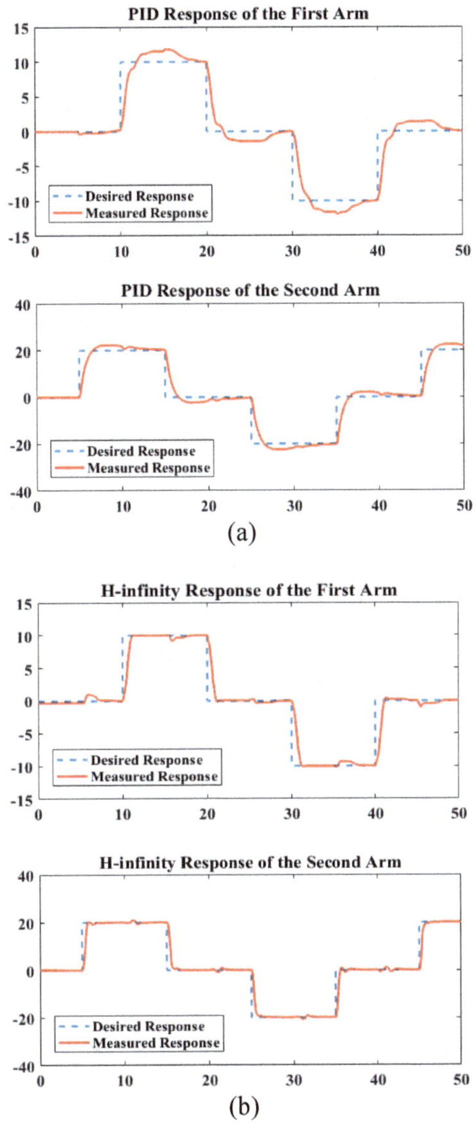

Figure 9: Experimental responses. (a) Responses of PID control system; and (b) Responses of H_∞ control system.

5 CONCLUSION

Throughout this study, the authors proposed a new control method to assure better performances, mutual disturbances rejection, and a robust system stability. The designed control system constitutes of two controllers, an inner loop controller, which improves the internal stability, and an outer loop control system that was designed based on the H_∞ control theory, in order to maintain a better tracking response to the reference signal. The simulation and the experimental results prove the efficacy of the proposed controller, and provides, to a higher standard, a better control for the studied system.

ACKNOWLEDGEMENTS

This research was a part of the project titled "Development of technology to commercialize flatfish automatic grader", funded by the Ministry of Oceans and Fisheries, Korea. This research was also part of the project titled "Development of the ICT convergence smart sliding docking system", funded by the Ministry of SMEs and Start-Ups, Korea.

REFERENCES

[1] Dang, D.C. & Kim, Y.B., Accurate motion regeneration technique with robust control approach. *Science and Technology Development*, **18**(K6), pp. 183–191, 2015.

[2] Mohammed, A.A. & Eltayeb, A., Dynamics and trajectory tracking control of a two-link robot manipulator using PID and sliding mode control. *Proceedings of the 2018 International Conference on Computer, Control, Electrical, and Electronics Engineering (ICCCEEE)*, Sudan, 2018.

[3] Tomei, P., Adaptive PD controller for robot manipulators. *IEEE Transactions on Robotics and Automation*, **7**, pp. 565–570, 1991.

[4] Yasuda, M., Osaka, T. & Ikeda, M., Feedforward control of a vibration isolation system for disturbance rejection. *Proceedings of the 35th Conference on Decision and Control*, Kobe, Japan, pp. 1229–1233, 1996.

[5] Song, Y.E., Kim, C.Y. & Lee, M.C., Sliding-mode control with sliding perturbation observer for surgical robots. *IEEE International Symposium on Industrial Electronics*, Seoul, South Korea, 2009.

[6] Gu, D.W., Petkov, P.H. & Konstantinov, M.M., *H-infinity Design: Robust Control Design with MATLAB®*, 2nd ed., Springer, pp. 31–48, 2013.

ANALYSIS OF JAPANESE FOLDING FAN'S CHARACTERISTIC STRUCTURE AND TRADITIONAL TECHNICAL SKILL

FUJIKO ABE[1], YOSHIFUMI OHBUCHI[1], TSUTAO KATAYAMA[2] & HIDETOSHI SAKAMOTO[2]
[1]Kumamoto University, Japan
[2]Doshisha University, Japan

ABSTRACT

Japanese folding fans were born during the Heian Period (794–1185) and have been used in every situation. Many Japanese fans are used widely at present as a portable fan, a tool of Japanese dance and tea ceremony. The making of folding fans in Japan has a history of 1,000 years or more and the production technology of the Japanese fan has been followed up until now. This research has aimed to clarify the characteristics of the traditional Japanese fan through the material of fan components, the fan-making process and the traditional production skill of a Japanese fan.

Keywords: Japanese folding fan, natural material, traditional skill, reproduction, concave–convex effect.

1 INTRODUCTION

The making of folding fans in Japan has a history of 1,000 years or more and the production technology of the Japanese fan has been followed up to now [1]. The function of a folded and portable fan that was an invention of Japan, has spread all over the world, and its materials were changing in various countries. The drawing and production technique of a fan was polished up by the painters who represent Japan in the Edo era (1603–1868) [2], [3]. Moreover, the predecessors' wisdom about the production technique and a lot of the processing technologies are condensed by fan-making craftsmen [4]. In previous papers [5]–[7], the quantitative evaluation of the shaped form, the historical transition and the work–study program for understanding the Japanese folding fan were shown.

The material and structural characteristics of a Japanese fan are examined in detail in this research. Moreover, the function and the artistry arising from those characteristics are considered.

2 MATERIALS AND MANUFACTURING PROCESS

2.1 Materials of Japanese traditional fan

A traditional Japanese folding fan consist of (a) thin bamboo bones and (b) fan-shaped Japanese handmade paper "Washi" which was processed for the fan. "Washi" is a cellulose fiber reinforced paper and the fiber orientation is mainly at right angles to the folding direction as shown in Fig. 1(b). Additionally, "Washi" is a layer structure shown in Fig. 1(c). Fig. 2 shows the assembled bamboo bones and the part's name. The thickness of "Washi" for a folding fan is about 0.13–0.16 mm.

2.2 Manufacturing process

The manufacturing process is shown in Fig. 3.

In this manufacturing process, it is difficult to insert the middle bone "Nakabone" into Washi (Japanese paper) in D-stage and E-stage. Here, the sebum has an important role. The

WIT Transactions on Engineering Sciences, Vol 125, © 2019 WIT Press
www.witpress.com, ISSN 1743-3533 (on-line)
doi:10.2495/CMEM190091

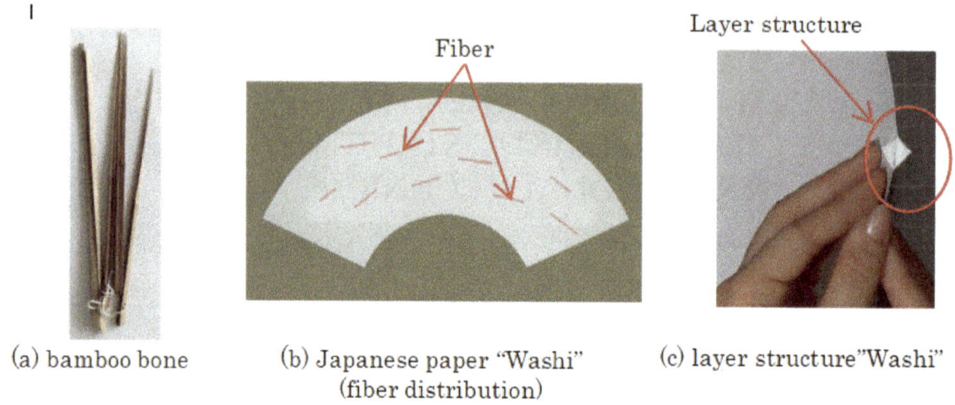

Figure 1: Materials of traditional Japanese folding fan.

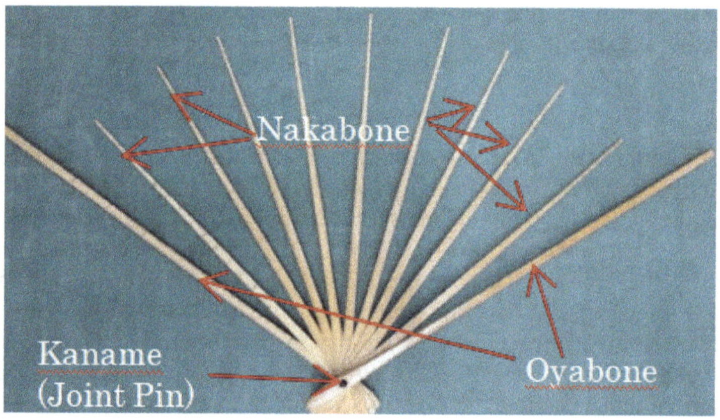

Figure 2: Assembled bamboo bones.

craftsman puts up to the bamboo bone of the fan by touching his/her scalp as shown in Fig. 4 in order to reduce the insert friction of paper and prevent the tearing of paper. After this operation, a small route in Japanese paper for insert has to be made with the bamboo bone in advance before inserting "Nakabone" into paper ("Nakazashi" process) and the inserting process is then carried out. Bamboo bones "Nakabone" are inserted into the direction of thickness of the Washi because of the layer structure shown in Fig. 1(c). The "Nakabone" insertion structure of Japanese folding fan is greatly different from that of fans of other countries.

3 STRUCTURAL AND ARTISTIC CHARACTERISTICS OF JAPANESE FOLDING FANS

In this section, Japanese folding fan's structural and artistic characteristics are shown in comparison with those of other countries' fans [8]. Fig. 5 shows some typical examples of European style folding fans [8]. The themes from literature and daily life are painted on the fan face; in Fig. 5(a), 5(b) and 5(c) is a fan with lace. Fig. 5(d) shows a jour fan which is called "Decoup" and Fig. 5(e) is a fan with slits which is called "Cabriot".

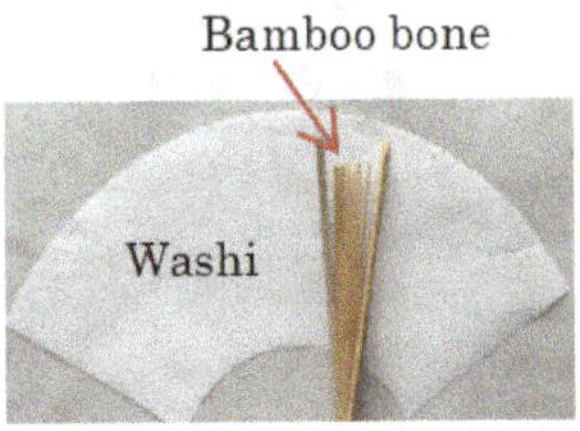

(A) materials

(B) folding process

(C) cut of top and bottom parts

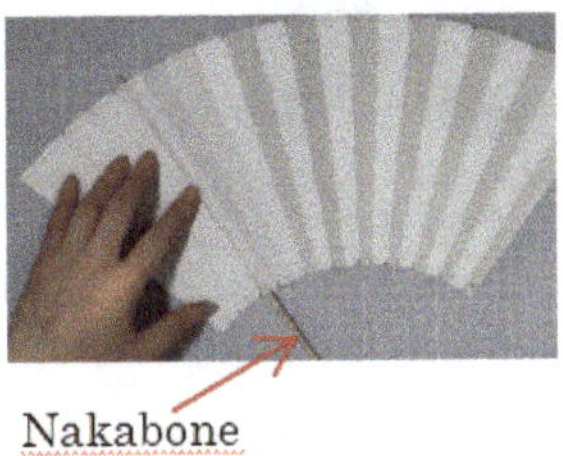

(D) provisional insert process of Nakabone

(E) final insert process of Nakabone

(F) gluing process

(G) finished folding fan

Figure 3: Manufacturing process of folding fan. (A) Arrangement of bamboo bones and the sector of Japanese paper "Washi"; (B) Tracing the folding lines on the fan-shaped paper with a folding guide, which is a pair of accordion folded paper guide; (C) Cutting the top and bottom of the fan-shaped paper; (D) Dispreading the folded fan-shaped paper, and inserting the thinner middle bones (Nakabone) into the middle of the dispread paper ("Nakazashi": provisional insert process); (E) Insertion of the starch-pasted thinner middle bones into the dispread paper; (F) Gluing up the dispread paper and the thinner middle bones; and (G) Adhering two thick bamboo bones (Oyabone) to both ends of the dispread paper.

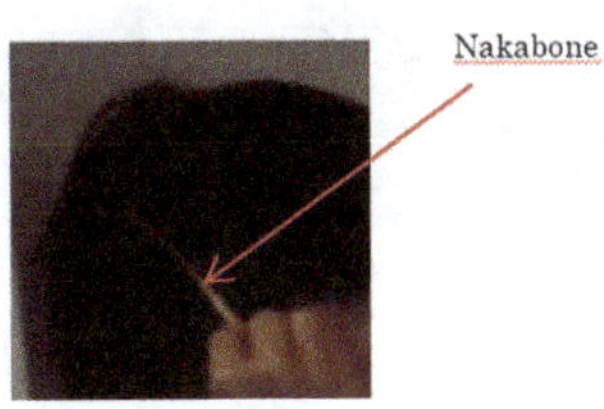

Figure 4: Putting up to the sebum on bamboo bone "Nagabone".

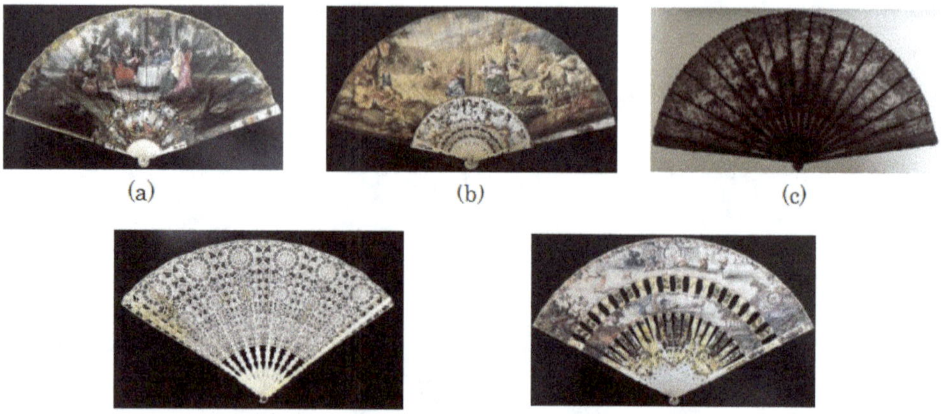

(a) (b) (c)

(d) (e)

Figure 5: Examples of western style folding fan. (a), (b) and (c) Fans with lace; (d) A jour fan which is called "Decoup"; and (e) A fan with slits which is called "Cabriot".

First, the structural characteristics of the fan are considered. In European style fans, the fan consists of the wooden board or the ivory board for the folding fan's bone and the cloth shown in Fig. 5. The fan made only from a wooden board and the board of the ivory is common. In this case, the waving is given to the fan's bone boards. Therefore, these are heavy and bulky in comparison with those of a Japanese fan. As an arbitrary opening angle cannot be adjusted, the fan opens completely and is used.

In contrast to these, a Japanese folding fan has the folding structure of Washi shown in Fig. 6(a). With this structure, it can be adjusted to an arbitrary opening angle. The Japanese folding fan has large strength because the folding structure produces the 3-D structure effect, though thin bamboo bones are used. As a result, the airflow can be effectively created. Fig. 6(b) shows another portable fan "Uchiwa". This has the same function as a folding fan and is made of the same materials. But, this is a more flexible than the folding fan because of its two dimensional structure and it is inconvenient to carry as it cannot fold up.

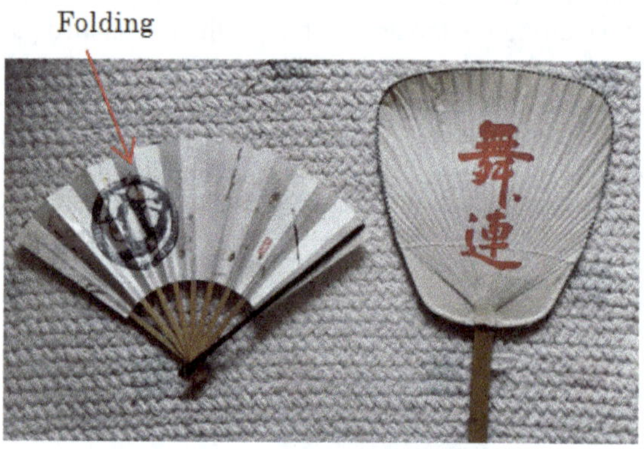

(a) folding fan (b) Uchiwa

Figure 6: Japanese folding fan and 2D-portable fan "Uchiwa".

Next, the artistic characteristics of the fan is considered. The western style folding fan is used while opened completely. Therefore, the pictures drawn on the fan face are in two dimensions, and it is the same as a usual painting. However, a Japanese fan has the folding as shown in Fig. 6(a). Therefore, the picture drawn on the fan face appears different from the usual picture painted in two dimensions. The geometrical shape of a two dimensional picture on the fan face changes by the folding effect shown in Fig. 7. Moreover, the distortion depends on the fan shape. Even if the fan shaped paper sizes are the same, the distance l from joint pin "Kaname" to drawing objects give influences in the distortion of the painted shape as shown in Fig. 8. As just described, the distortion of them is very complicated and depends on the distances [9], [10].

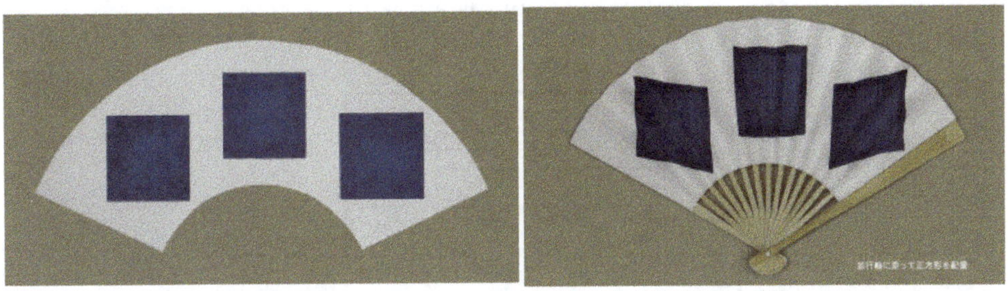

(a) (b)

Figure 7: Geometry shape of picture change. (a) 2-D picture; and (b) Folding fan picture.

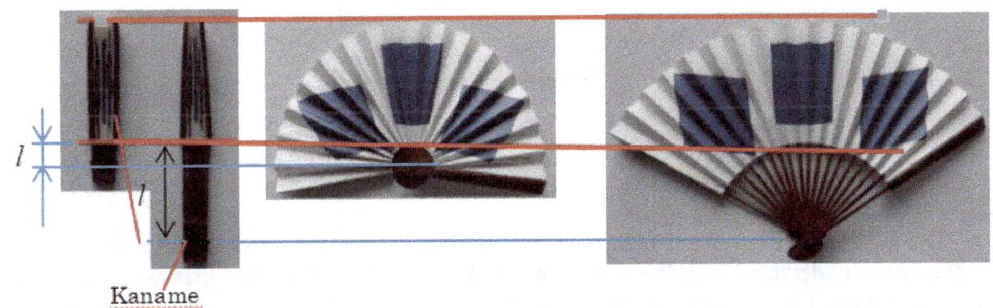

Figure 8: Distance dependency of distortion shape in a folding fan (in the case of a same fan shaped paper).

When the painter draws the folding fan picture, it is important to understand the distortion mechanism. The folding fan picture becomes a picture of a correct composition when the folding fan is opened to final position. An example of the traditional Japanese fan is shown in Fig. 9.

As previously stated, the pictures or the images drawn on the folding fan face are deformed in a complex manner according to the size of the structure. So, an image transformation program was created for mathematically analyzing the deformation. The original image and the experimental and analytical results are shown in Figs 10 and 11, respectively. It is understood that the difference of the distance from "Kaname" greatly influences the distribution of fan face distortion as can be deduced from Fig. 11.

Figure 9: An example of the traditional Japanese fan.

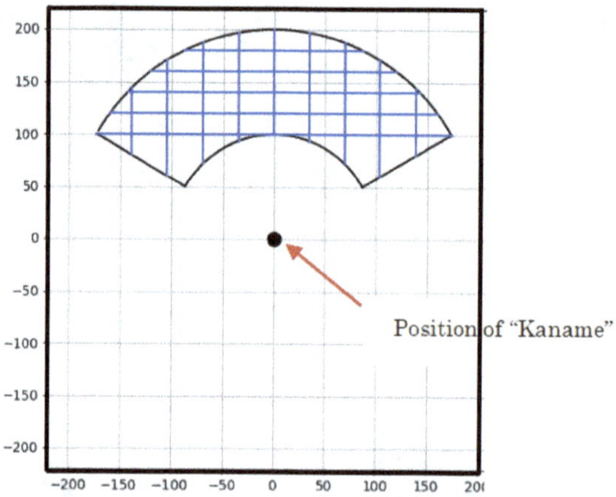

Figure 10: Original image (2-D painting).

The simulation can represent a nearly accurate behavior. This result makes it easier for a successor to understand how the images would be transformed. Hence, they would acquire the technique to find the best way to draw pictures, taking the deformation into consideration. Referring to the result, the successors would be able to understand the traditional Japanese fan deeply.

As another characteristic, there is the concave–convex illusion effect by the folding. The perspective is caused by arranging the drawing object on the mountain segments or valley segments of the folding fan. Furthermore, the Japanese folding fan can be arranged by controlling its opening angle.

4 SUMMARY

Traditional Japanese folding fans consist of thin bamboo bones and "Washi", which is a cellulose fiber reinforced paper, and the fiber orientation being mainly at right angles to folding direction. The 3-dimensional structure of folding fan given by the folding produces the concave and convex illusion effect and the perspective effect. The Japanese fan creates various artistic impressions by changing the opening angle of the folding fan. The complex distortion of the pictures or the images drawn on the folding fan face, was clarified by the proposed image conversion program.

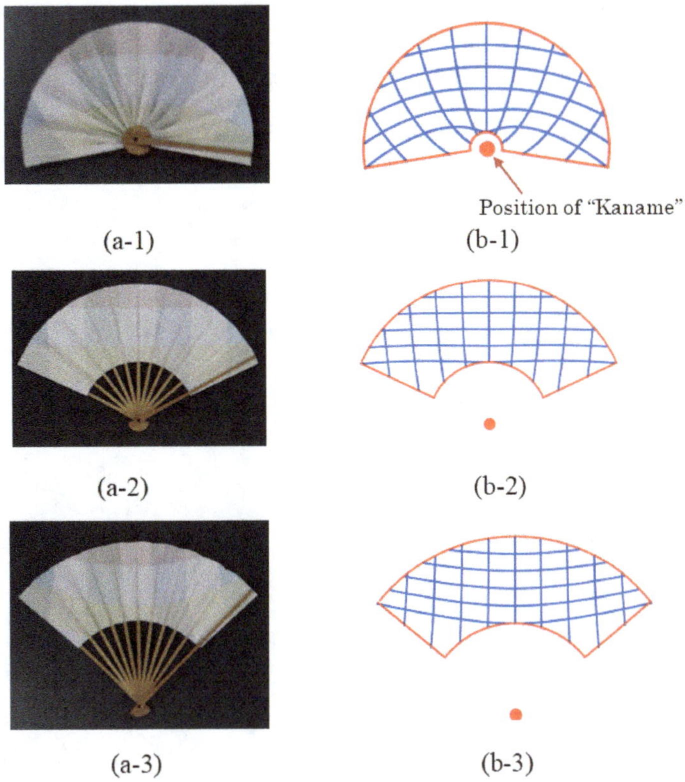

Figure 11: Comparison between experiment and simulation results. (a) Experiment; and (b) Simulation.

REFERENCES

[1] Nakamura, N., *Senmen-ga and Ougi-E*, Kawahara Publishing Co., 1969.

[2] Kobayashi, T., Semmen-ga. *Japanese Art*, **32**, 1993.

[3] Misuo, M., *Summary of Semmen-ga*, Kourin Publishing Co., 1981

[4] Ōta Memorial Museum of Art, www.ukiyoe-ota-muse.jp/exhibition/2011. Accessed on: 20 Dec. 2018.

[5] Abe, F., Ohbuchi, Y. & Sakamoto, H., Quantitative evaluation by measurement for production technology and reproduction of traditional folding fan. *Proceedings of the 13th International Symposium on Advanced Science and Technology in Experimental Mechanics,* A19040, pp. 1–5, 2018.

[6] Abe, F., Ohbuchi, Y. & Sakamoto, H., Transitions of Japanese folding fan for export in Meiji era. *Proceedings of Society for Design and Art Fusing with Science and Technology,* pp. 12–17, 2018.

[7] Abe, F., Ohbuchi, Y. & Sakamoto, H., Learning about Japanese traditional technique by experiencing how to make a "folding fan". *Proceedings of the 7th Asian Conference on Engineering*, pp. 15–16, 2018.

[8] Bennet, G.A., *Unfolding Beauty*: *The Art of the Fan. The Collection of Ester Oldmam and the Museum of Fine Arts*, Thames and Hudson, USA, 1998.

[9] Minde, K., Distortion in pictorial scenes on Japanese fans. *Journal of Graphic Science of Japan,* **43**(2), pp. 3–9, 2007.
[10] Koyama, S., Structure of Semmen-ga. *Journal of Graphic Science of Japan,* **33**, pp. 127–132, 1999.

STUDY ON VESSEL MOTION CONTROL WITH TOWING ROPES AND DAMPERS FOR BERTHING

TAE-WAN KIM[1], DONG-HUN LEE[1], JIN-HO SUH[1], YONG-WOON CHOI[2] & YOUNG-BOK KIM[1]
[1]Department of Mechanical System Engineering, Pukyong National University of Busan, South Korea
[2]School of Engineering, Soka University of Hachioji, Japan

ABSTRACT

The problem of manoeuvring vessels within harbour areas is generating considerable interest in terms of marine cybernetics. The vessel is operated by the pilot and moves at low or zero speeds in shallow water areas. In this case, the vessel is usually assisted by the cooperation of thrusters, main propulsion systems, tugboats and pilots, etc. In this paper we suggest a new vessel berthing technique using dampers and winches as a solution for complex and dangerous berthing work. In the proposed berthing method, we design the controller to counteract the waves and other effects for guaranteeing safety during the berthing process. Experiments are also conducted to verify the proposed vessel berthing method and the effectiveness of the designed control system in which unpredictable external force is applied. Finally, the experiment results show that the vessel could approach the defined position in time with effective control action made from designed controller.

Keywords: vessel, tension, harbour, berthing system, winch, damper.

1 INTRODUCTION

Vessel berthing is widely considered as the most complicated process in the marine control and automation field [1]–[3]. A reason is that the hydrodynamic of the vessel considerably changes, and its controllability is greatly cut down when the vessel moves from deep to shallow water. Furthermore, by only manipulating the main propellers and thrusters, it is very easy for dangerous collisions to appear between the manoeuvred vessel and the moored ones because of unpredictable vessel motion [4].

Various approaches have been tried to make some solutions about this problem. Bui et al. [5], [6] and Bui and Kim [7] conducted automatic ship berthing by using bow and stern thrusters. The authors introduced a steering motion model of a vessel and identified all of the model parameters. An observer based optimal controller was designed to cope with the uncertain ship dynamics to obtain desirable control performance by estimating system states. This approach seems to be a useful method, but the reduced actuator controllability under low-speed conditions has not fully been resolved. In 1994, Hasegawa and Fukutomi [8] suggested an artificial neural network control strategy for automatic vessel berthing.

However, the conventional methods proposed in this time are kinds of automatic manoeuvring technologies or methods which could not give any useful solutions for completing berthing work.

Therefore, the authors propose a new method to complete the final berthing process with preserving working safety. The proposed system configuration is illustrated in Fig. 1. As shown in Fig. 1, the two damper cylinders and two winch systems are prepared for controlling ship motion. Pulling by winch with rope and pushing with damper the vessel, we can move the vessel to the final position more safely. It means that the complicate berthing work could be completed by controlling winches and dampers with designed control system without any supporting system. The effectiveness and usefulness of proposed method are clearly presented by experiment results.

WIT Transactions on Engineering Sciences, Vol 125, © 2019 WIT Press
www.witpress.com, ISSN 1743-3533 (on-line)
doi:10.2495/CMEM190101

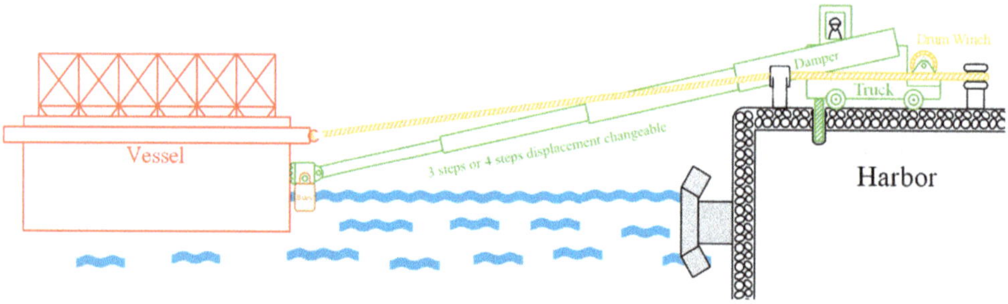

Figure 1: Schematic drawing of the proposed berthing system.

2 SYSTEM DESCRIPTION AND MODELING

The system proposed and intended for experiment is shown in Fig. 2. The authors use a vessel model to execute the berthing experiment, where two pairs of motor and pneumatic cylinder are installed in the harbour side.

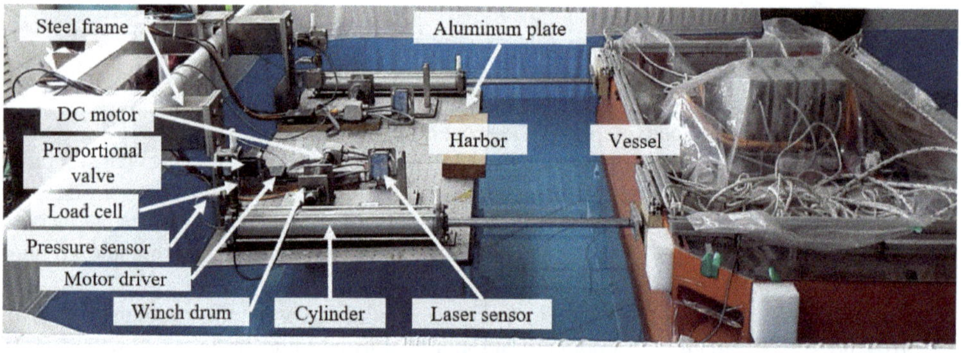

Figure 2: Schematic drawing for the berthing system.

The proposed idea and system configuration are initialized by considering final berthing condition. The most complicate area (or distance) for berthing may be confined in 10 m from ship to quay side. In this area, generally, the tugboats and pilot assistance are needed to occupy safe working process. The object of this study is to show an easier and safer method than conventional technologies. In the proposed method, any additional assistance system such as tugboat or pilot is not necessary except damper and winch systems installed on the land side. The possibility of the proposed method is presented by following experiment.

As previously mentioned, two pairs of winch and damper cylinder are prepared for controlling ship motion. For starting berthing work, at first, the ends of dampers are attached and the winch ropes are connected to the vessel simultaneously. After that, the berthing process is started by controlling pulling and pushing forces made from actuators (see Fig. 5). It means that the desirable control performance and safe berthing work can be achieved by controlling winch and damper systems properly and effectively.

2.1 System modelling

Based on the preparation of experiment apparatus, the authors performed the system identification process to analysis the dynamic characteristics of vessel berthing system.

At first, in order to obtain the transfer function of winch system, 10 V electric power (shown in Fig. 3(a)) was applied to the winch system such that the rope tension force was obtained as shown in Fig. 3(b). In this figure, the dotted line is tension force get from experiment and the solid line is calculated tension force obtained from a transfer function of winch system.

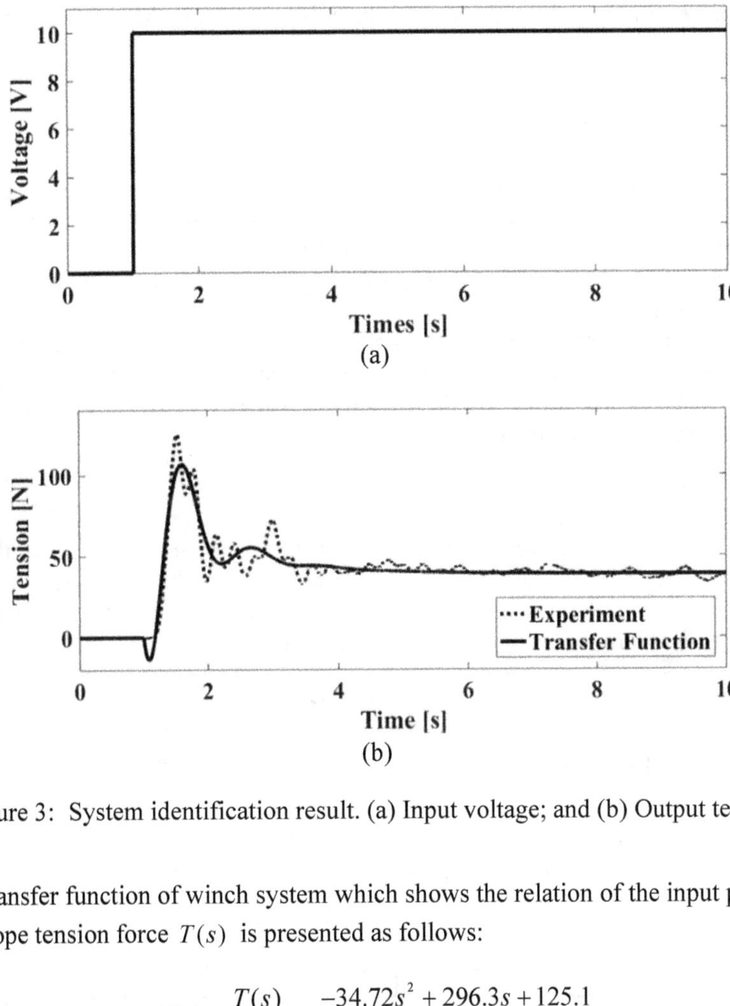

Figure 3: System identification result. (a) Input voltage; and (b) Output tension.

The transfer function of winch system which shows the relation of the input power $V(s)$ and the rope tension force $T(s)$ is presented as follows:

$$G(s) = \frac{T(s)}{V(s)} = \frac{-34.72s^2 + 296.3s + 125.1}{s^3 + 4.75s^2 + 39.77s + 32.4}. \tag{1}$$

We may provide many types of berthing techniques and methods using the proposed system configuration.

In the proposed berthing system, two dampers and two winches are provided and controlled to achieve the given objective.

Here, the authors introduce a considerable simple berthing strategy. It is called a semi-automatic berthing method or technique. In this method, the dampers are manually controlled. On the other hand, the winches are automatically controlled such that the rope tension forces are kept in the defined values.

For example, if the damper control valves are closed, the winches rotate and pull the vessel until the tension forces approach to the target values. It means that the vessel motion or position depends on the expended length of damper cylinder under the appropriately defined rope tension. For accomplishing full-automatic berthing system, two actuator system should be actively controlled. However, the semi-automatic berthing method is most friendly one for real working condition and environment. Based on this idea, the authors tried berthing experiment in several times such that the results will be shown in the next section.

3 EXPERIMENT

3.1 Experimental apparatus

In order to evaluate the usefulness of designed control system, the experimental apparatus is set up and tested in the water basin. The experiment was performed in water basin using a vessel model (weight: 215 kg, length: 2 m, width: 1 m).

The vessel motions are measured and controlled by data acquisition system PCI-6229 (NI) with LabVIEW software.

A pair of winch and damper system is installed on the harbour side. The distance from the vessel to the harbour is measured by the Laser sensor mounted on the plate, the rope tension is measured by Load cell sensor and the cylinder pressure is measured by pressure sensor. All apparatus used for experiment are represented in Fig. 2.

3.2 Controller design

We design a controller based on sliding mode control scheme in order to keep the desired rope tension. Especially, the super-twisting algorithm is introduced for designing sliding mode controller. The configuration of designed feedback control system is illustrated in Fig. 4. Then, the typical form of a sliding surface is given as follows:

$$\sigma = \ddot{e} + G_1\dot{e} + G_2 e, \tag{2}$$

where e is tracking error and $G_1 = 20.67, G_2 = 4.01, \lambda = 0.75, k = 6.2$.

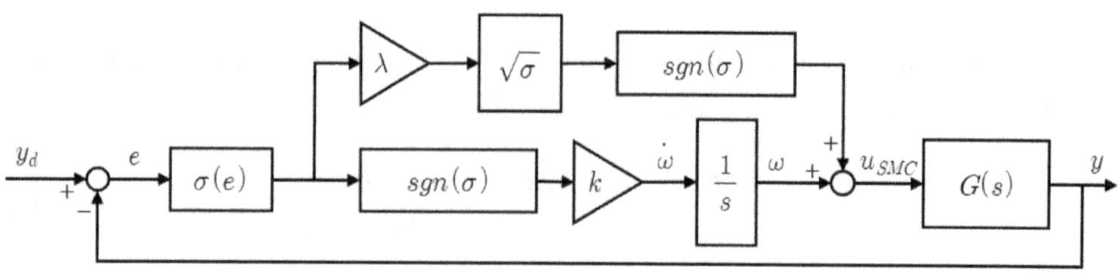

Figure 4: Block diagram for sliding mode control system with super-twisting algorithm.

3.3 Experiment results

The objectives of this study are keeping the target tension of rope and moving the vessel to the desired position simultaneously. Especially, the target rope tension should remain in the defined value throughout entire berthing process.

The authors performed experiment by applying the designed controller. The experiment scenario is illustrated in Fig. 5.

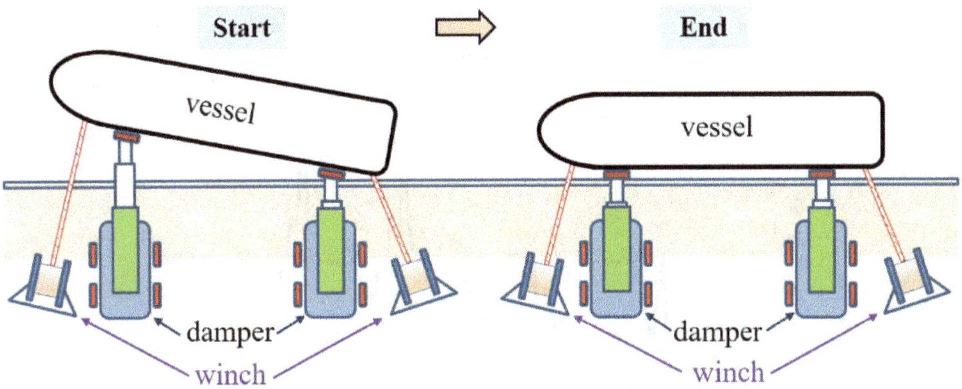

Figure 5: Experiment scenario.

In this study, the target tension was set to 120 N which is shown as the red dotted line in Fig. 6(b).

In Fig. 6(a) shows vessel position during moving to the final position. While the vessel is approaching to the target position (0.57 m), the external disturbance is attacking the vessel in several times. At 3 s, 7 s and 11 s, the vessel was attacked by pushing force toward harbour side. Conversely, at 14 s and 18 s, pulling force attacked the vessel.

With the condition aforementioned, the vessel can approach to the final position in 14 s by controlling two actuators. The controlled vessel motion is illustrated in Fig. 6(a). Also, the rope tension and winch control signal are shown in Fig 6(b) and (c), respectively.

Fig. 6(d) shows the pressure change of damper cylinder during entire berthing process.

As shown in the experiment results, we can find out that the vessel is well controlled and robustly moves to the final position as intended, regardless of attacking of disturbance.

4 CONCLUSION

In the conventional berthing technologies or methods, there exist many limits to the safety of berthing operation where rudders, main propellers, thrusters, tugboats and mooring lines are generally provided. The high technologies have been developed and applied to many industry fields. However, the berthing work strongly still depends on human power and stays on the pre-modern technology level.

To overcome primitive berthing method, the authors proposed a new berthing technique/method by combining dampers and winches. The proposed idea was initialised by considering the real working conditions. As mentioned before, many working apparatuses for berthing should be prepared on land and on the sea side of harbours simultaneously. In this study, the authors proposed a new berthing method only using the apparatuses prepared

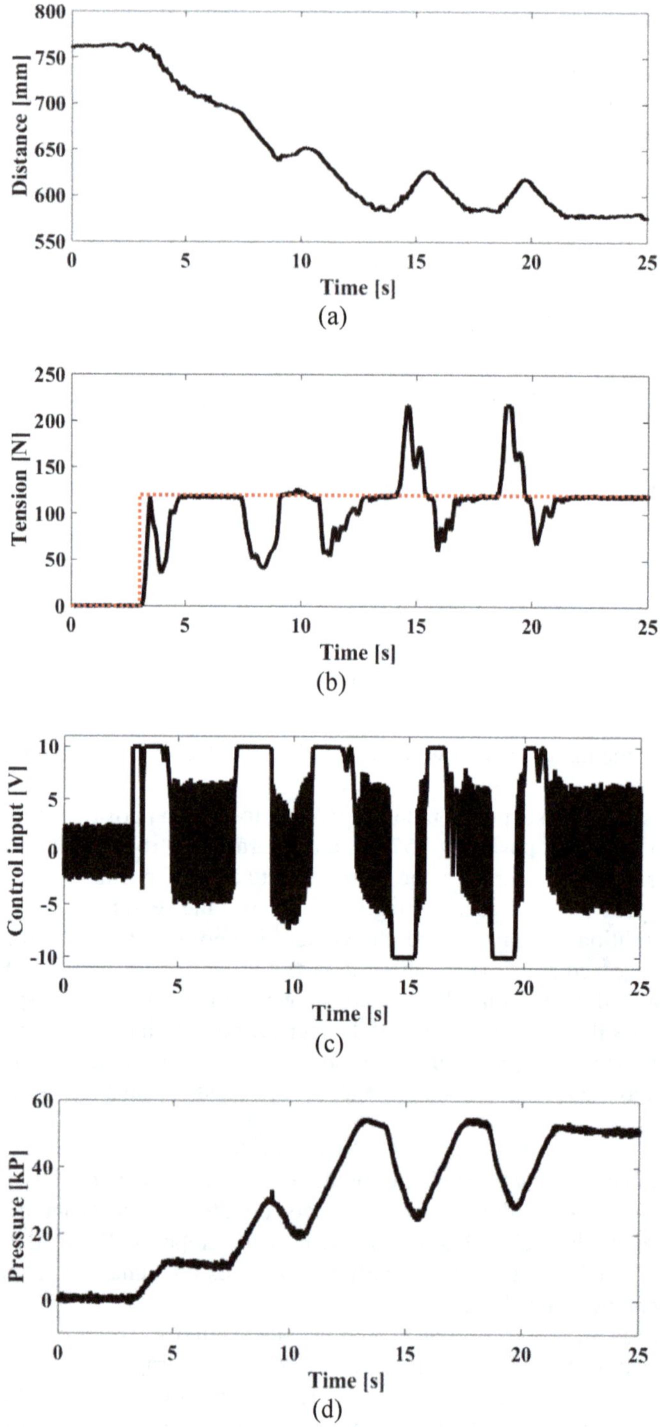

Figure 6: Experiment results. (a) Vessel position; (b) Winch rope tension; (c) Tension control input; and (d) Damper cylinder pressure.

on land side without the sea side apparatuses. It is the novelty of this research such that we could obtain numerous advantages. The representative one is reduction of air pollution get from not using tugboats. Except that, it is expected that we find out uncountable side effects by providing the proposed technique.

ACKNOWLEDGEMENTS

This research was a part of the project titled 'Development of technology to commercialize flatfish automatic grader', funded by the Ministry of Oceans and Fisheries, Korea. This research was also part of the project titled 'Development of the ICT convergence smart sliding docking system', funded by the Ministry of SMEs and Startups, Korea.

REFERENCES

[1] Kim, K.H., Kim, B.G. & Kim, Y.B., A study on the optimal tracking control system design for automatic ship berthing. *Journal of the Korean Society for Power System Engineering,* **22**(4), pp. 72–80, 2018. DOI: 10.9726/kspse.2018.22.4.072.

[2] Kim, Y.B., Choi, Y.W. & Chae, G.H., A study on the development of automatic ship berthing system. *Proceedings of the Korean Society for Power System Engineering Autumn Conference,* pp. 419–423, 2006.

[3] Kim. Y.B., Choi, Y.W., Suh, J.H. & Lee, K.S., A study on the development of the real-time detection technique for automatic ship berthing. *Proceedings of the Korean Society for Power System Engineering Spring Conference,* pp. 306–313, 2006.

[4] Zhang, Y., Hearn, G.E. & Sen, P., A multivariable neural controller for automatic ship berthing. *IEEE Control Systems,* **17**(2), pp. 31–45, 1997.

[5] Bui, V.P., Jeong, J.H., Kim, Y.B. & Kim, D.W., optimal control design for automatic ship berthing by using bow and stern thrusters. *Journal of Ocean Engineering and Technology,* **24**(2), pp. 10–17, 2010.

[6] Bui, V.P., Kawai, H., Kim, Y.B. & Lee, K.S., A ship berthing system design with four tug boats. *Journal of Mechanical Science and Technology,* **25**(5), pp. 1257–1264, 2011. DOI: 10.1007/s12206-011-0215-4.

[7] Bui, V.P. & Kim, Y.B., Development of constrained control allocation for ship berthing by using autonomous tugboats. *International Journal of Control Automation and Systems,* **9**(6), pp. 1203–1208, 2011. DOI: 10.1007/s12555-011-0622-4.

[8] Hasegawa, K. & Fukutomi, T., On harbour maneuvering and neural control system for berthing with tug operation. *Proceedings of the International Conference Maneuvering and Control of Marine Craft,* pp. 197–210, 1994.

DIGITAL EMULATION OF DIELECTRIC RELAXATION FUNCTIONS FOR CAPACITIVE SENSORS OF NON-DESTRUCTIVE DIELECTRIC SPECTROMETRY

ALDIS KALPINSH, VAIRIS SHTRAUSS & ULDIS LOMANOVSKIS
Institute for Material Mechanics, University of Latvia, Latvia

ABSTRACT

This paper addresses a problem of reference means for the adjusting, calibrating and testing of non-destructive dielectric spectrometers employing one-side access capacitive sensors. Due to specific features of these sensors caused by non-homogeneous electric exciting fields, the application of the conventional reference means, particularly the artificial ones, is not always possible. We propose to solve the problem by digital emulation of dielectric relaxation functions for capacitive sensors. A digital emulator is developed for one-side access capacitive sensors with active guarding for the spectrometers, working according to a two-port network measuring model. The test results are provided for digital emulation of the responses of the capacitive sensor system to sinusoidal excitations, corresponding to complex dielectric permittivity described by the Debye relaxation function, with varying of the real part of dielectric permittivity from 1 to 100, over a frequency band from 10 Hz to 0.33 MHz.
Keywords: digital emulator, one-side access, capacitive sensor, dielectric spectrometer, relaxation function.

1 INTRODUCTION

Dielectric measurements [1] have long been of increasing interest for the characterization of non-metallic materials in many areas of science and technology, such as material science, electronics, biology, medicine, agriculture, food processing industry, etc. The potentialities of the dielectric method were significantly increased with developing one-side access (OSA) capacitive sensors [2] allowing non-destructive testing (NDT) and the characterization of materials without preparation of specimens.

Nonetheless, an actual problem for adjusting, calibrating and testing of measuring instruments of dielectric characteristics is regarding materials with the known and stable dielectric characteristics over a wide dynamic and frequency ranges. The problem is extremely urgent for dielectric spectrometers working with capacitive sensors of OSA type for reasons such as: (i) OSA sensors excite test objects by non-homogeneous electric field; (ii) the sensors have low sensing (geometric) capacitances with high stray capacitances; (iii) bulk reference specimens are needed for OSA sensors, whose minimum thickness is limited by the effective depth of penetration of electrical field; and (iv) contrary to parallel-plate capacitive sensors working with homogeneous electric field, there are no simple formulas for calculating sensing capacities of OSA sensors. The mentioned reasons do not allow or noticeably restrict to use the known solutions of artificial reference means. For example, due to small sensing capacitances of the order of hundreds and even tens femto farads, it is impossible to adjust test OSA sensors by connecting the adjustable impedances to the electrodes [3]. The limited possibilities have making artificial reference dielectrics with included electrically conductive layers [4], and producing of synthetic reference dielectrics through mixtures and liquids.

We propose to solve the problem by digital emulation of dielectric relaxation functions. Thus, the goal of this study was the development of a digital emulator for OSA capacitive

WIT Transactions on Engineering Sciences, Vol 125, © 2019 WIT Press
www.witpress.com, ISSN 1743-3533 (on-line)
doi:10.2495/CMEM190111

sensors, which generates the response signals corresponding to the given dielectric relaxation function.

The rest of this work is organized in six sections. In Section 2, NDT dielectric spectrometry is considered for quality testing, in Section 3, operation of OSA capacitive sensors with active guarding is explained, the mathematical model corresponding to an equivalent circuit diagram of CSS is developed in Section 4, the implementation of the digital emulator is described in Section 5, some test results of the digital emulator are demonstrated in Section 6. Section 7 contains conclusions.

2 NDT DIELECTRIC SPECTROMETRY FOR QUALITY TESTING

We have developed several NDT dielectric spectrometers for solving various quality testing problems based on the ideology [2] that a lot of parameters of users' interest correlate with the dielectric characteristics, and, so many quality testing problems can be solved through NDT dielectric spectrometric measurements. Our spectrometers are classical frequency-domain instruments working over relatively low frequency band 10 Hz–1 MHz with exciting test object with electrical field generated by sinusoidal voltage signals at discrete frequencies, usually altered geometrically with the common ratio 2.

Two peculiar features of the spectrometers are as follows: (i) they are equipped with capacitive sensors of OSA type to accomplish non-destructive testing; and (ii) stray-immune capacitance measurements are carried out by using active guarding of the sensing electrode to increase the sensitivity and accuracy of the measurements. The capacitive sensors are protected by patents [5], [6].

A simplified representation of a NDT spectrometer is shown in Fig. 1. The spectrometer consists from three main blocks: (i) multi-frequency excitation generator; (ii) capacitive sensor system (CSS) of OSA type; and (iii) multi-frequency signal processing unit.

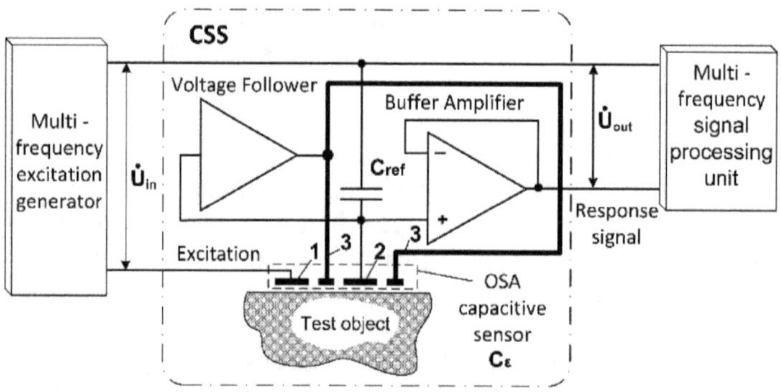

Figure 1: Simplified functional diagram of a NDT spectrometer.

3 OSA CAPACITIVE SENSOR WITH ACTIVE GUARDING

The *capacitive sensor system* (see Fig. 1) works as follows. A *test object* is excited by electrical field through a *driven electrode 1* and a *sensing electrode 2* generated by sinusoidal voltage signals from a *multi-frequency excitation generator*. The output signal of CSS is measured by a *multi-frequency signal processing unit*. A unity gain *buffer amplifier* is used to connect a source with a high impedance to a low-impedance load.

Both the driven electrode 1 and sensing electrode 2 are connected to the multi-frequency excitation generator, though the driven electrode 1 is connected directly, but sensing electrode 2 through the reference capacitor C_{ref}. To carry out stray-immune capacitance measurements, the sensing electrode 2, the reference capacitor C_{ref} and the unity gain buffer amplifier are covered with a screen, which forms a guard electrode 3 on working surface of CSS around sensing electrode. The guard electrode through a *voltage follower* are fed by the same voltage that of the sensing electrode. Thus, the guard electrode suppresses the electric field between the driven electrode and the sensing electrode outside the sensor working surface, so the current in the circuit of sensing electrode is made directly proportional to resultant electric field intensity in the test object. Voltage drop on the reference capacitor C_{ref} created by the current in the circuit of sensing electrode forms a CSS output signal which via the buffer amplifier is passed to the multi-frequency signal processing unit.

By suppressing the electric field between the driven electrode and the sensing electrode outside the sensor working surface, the proposed solution for CSS [5] prevents the influence of stray capacitances on the output signal and ensures that the current through the sensing electrode is proportional to the capacitance between the driven electrode and the sensing electrode. We denote this capacitance as a *sensing capacitance* C_0 when CCS of OSA type is in the air. When the capacitive sensor is put on the test object without a gap, the sensing capacitance C_0 is multiplied by the complex dielectric permittivity of the test object

$$\tilde{\varepsilon}(j\omega) = \varepsilon'(\omega) - j\varepsilon''(\omega)\,, \tag{1}$$

consisting of a real part, $\varepsilon'(\omega)$ and an imaginary part, $\varepsilon''(\omega)$, and produces a *measuring capacitance*

$$\dot{C}_\varepsilon = C_0\tilde{\varepsilon}(j\omega)\,. \tag{2}$$

4 MATHEMATICAL MODEL OF CSS

The measuring principle of NDT dielectric spectrometers are based on the mathematical model of CSS in the form of two-port network, which an equivalent circuit diagram for analysis is demonstrated in Fig. 2.

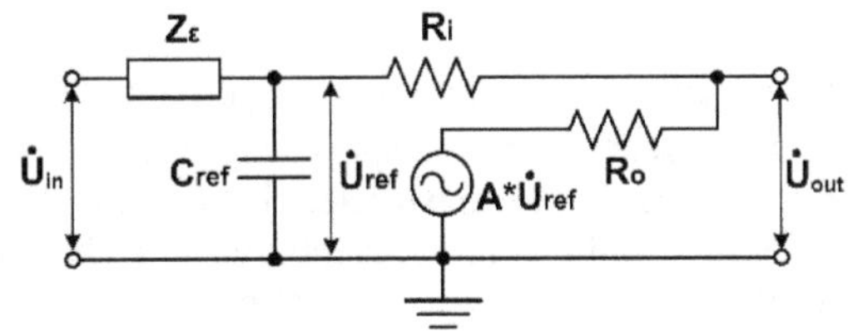

Figure 2: Equivalent circuit diagram of CSS.

The OSA capacitive sensor is represented as impedance Z_ε, while the unity gain buffer amplifier is represented by resistors R_i, Ro and voltage source $A*\dot{U}_{ref}$. Due to deep negative feedback of the buffer amplifier, the input resistance R_i is very large and the output resistance R_o is very small. The voltage gain A of the buffer amplifier is practically equal to 1. Therefore, it can be assumed that the output voltage U_{out} practically repeats the voltage drop on the reference capacitor C_{ref} and, so R_i and R_o can be neglected in the mathematical model of CSS. In Fig. 2, Z_ε is the measuring (sensing) impedance

$$Z_\varepsilon = \frac{1}{j\omega C_0\tilde{\varepsilon}(j\omega)}\,, \tag{3}$$

while C_{ref} establishes a reference reactance

$$X_{ref} = \frac{1}{j\omega C_{ref}}. \tag{4}$$

For CSS in air,

$$Z_{\varepsilon} = Z_0 = \frac{1}{j\omega C_0}. \tag{5}$$

CSS two-port network has the following transfer function

$$K_{\varepsilon}(j\omega) = \frac{\dot{U}_{OUT}}{\dot{U}_{IN}} = \frac{X_{ref}}{X_{rej} + Z_{\varepsilon}}, \tag{6}$$

which results in

$$K_0 = \frac{C_0}{C_0 + C_{ref}} \tag{7}$$

for CSS in air, and in

$$K_{\varepsilon}(j\omega) = 1/\left[1 + \frac{1 - K_0}{K_0 \tilde{\varepsilon}(j\omega)}\right] \tag{8}$$

for CSS put on the test object.

Eqn (7) shows that the transfer function of two-port network for CSS in air does not depend on frequency. To measure the complex dielectric permittivity, two parameters of CSS (sensing capacitance C_0 and reference capacitance C_{ref}) need to be known, and the amplitudes and phases of the sinusoidal input and output signals shall be measured. To emulate the dielectric relaxation functions, only K_0 must be known.

5 IMPLEMENTATION OF DIGITAL EMULATOR

An idea of emulation is to generate the same response signal for the given excitation signal as CSS would generate for the given complex dielectric permittivity of the test object. We are elaborated the emulator as a tool that replaces the sensor in the operations of adjusting, calibrating and testing of the spectrometer. According to eqn (8), only the transfer function for CSS in air (eqn (7)) is necessary to know to emulate the response of CSS for the given $\tilde{\varepsilon}(j\omega)$.

Therefore, for sinusoidal excitation signals, the emulator should execute three operations:

- to fix frequency for each sinusoidal excitation signal,
- to attenuate the excitation signal by a definite amount defined by a value of the complex dielectric permittivity $\tilde{\varepsilon}(j\omega)$ at the fixed frequency in the transfer function (eqn (8)), and
- to delay the phase of the excitation signal by a definite amount corresponding to the same value the complex dielectric permittivity.

In Fig. 3, a functional diagram is shown of a NDT spectrometer in CSS emulation mode, when CCS is replaced by the emulator.

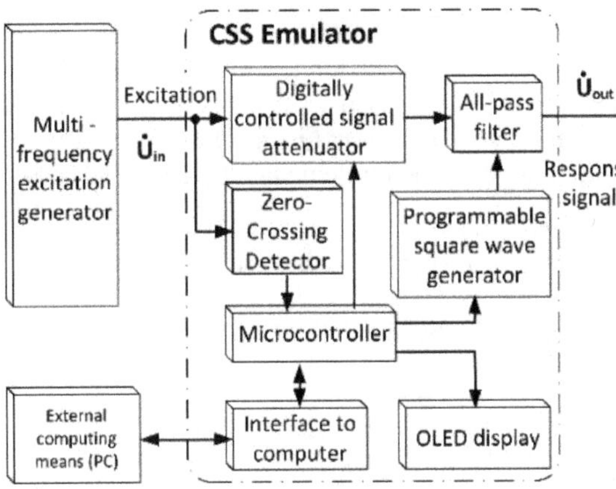

Figure 3: Functional diagram of CSS emulator.

To execute the above-mentioned operations CSS emulator includes:

- units for processing excitation signal – a digitally controlled signal attenuator and an all-pass filter,
- local computing means – a microcontroller and a OLED display;
- auxiliary control units – a programmable square wave generator, a zero-crossing detector and an interface to computer.

Excitation signal processing units generate a response signal from the given excitation signal. Local computing means controls response signal generation and displays emulator status information. The microcontroller provides the control of the operation of the signal processing units by means of auxiliary control units.

The transfer function of CSS according to the chosen dielectric relaxation function is computed by external computing means (PC) and is loaded into the microcontroller memory via the interface to computer.

5.1 Fixing the frequency of excitation signal

To determine the frequency, the harmonic excitation signal is converted into a periodic square waveforms by a zero-crossing detector (see Fig. 3). The duration of the period of square wave corresponds to the period of the harmonic excitation signal. The square waveforms are fed to the peripheral pin of microcontroller. The microcontroller performs measurement of the duration of the square waveform.

5.2 Attenuation of excitation signal

Each excitation signal at the given frequency is attenuated according to an attenuation factor determined by the complex dielectric permittivity in the transfer function (eqn (8))

$$|K_{\varepsilon}(j\omega)| = \frac{|\dot{U}_{OUT}|}{\dot{U}_{IN}}, \tag{9}$$

where $|K_{\varepsilon}(j\omega)|$ indicates a part of the excitation signal to be provided at the output of the emulator.

The proposed processing of analog signals is technically easy to implement by means of a digital-to-analog converter (DAC). In addition, the use of DAC also solves the issue of the digital control. So, a standard DAC [7] is used as a programmable divider. The circuit of the attenuator of the excitation signal is shown in Fig. 4. The digital word D (control code), which determines the signal attenuation to be implemented by the DAC corresponding to the attenuation factor

$$D = Int\{|K_{\varepsilon}(j\omega)| \times 2^{n} + 0.5\}\ , \tag{10}$$

is loaded to the DAC, where n is the number of bits of DAC.

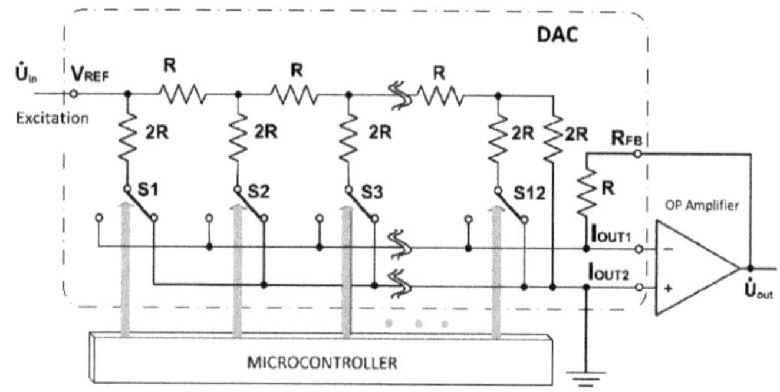

Figure 4: Circuit of digitally controlled signal attenuator.

5.3 Delaying the excitation signal

After attenuation of the excitation signal, phase shift of $Arg\{K_{\varepsilon}(j\omega)\}$ is carried out for the signal, with no effect on the signal amplitude. Such a task is performed through a phase (all-pass) filter. For the programmable control of the procedure, a universal switched capacitor filter [8] is used. Connection diagram of filter is shown in Fig. 5.

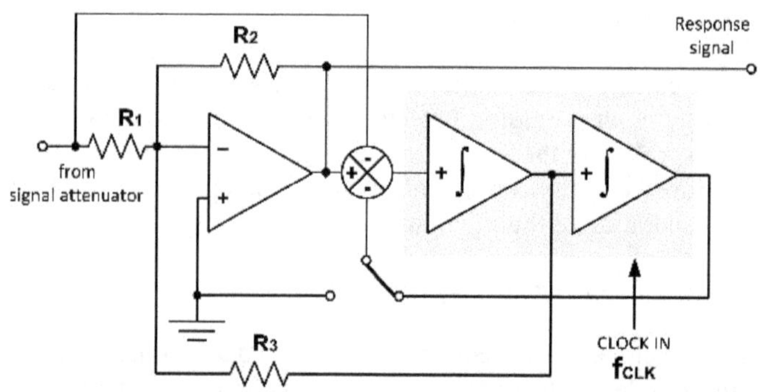

Figure 5: Block diagram of 2nd order universal switched capacitor filter.

The phase responses of all-pass filter corresponding to different center frequencies f_0 are shown in Fig. 6. It can be seen that change of the center frequency f_0 moves the phase response along the frequency axis. This means that it is possible to regulate the phase shift of signal to a specific frequency f_x by adjusting the center frequency.

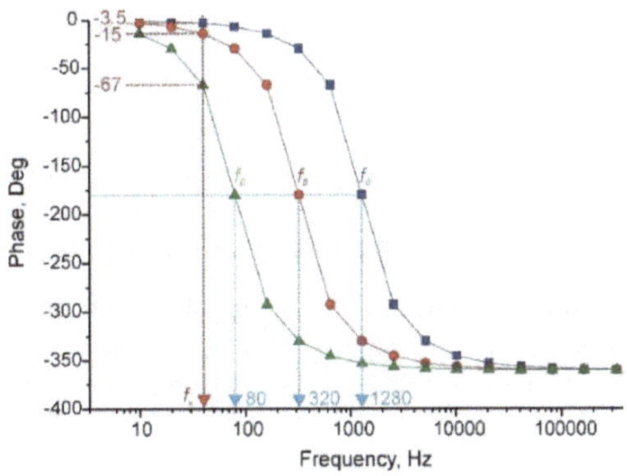

Figure 6: Phase response curves for 2-nd order all-pass filter.

The center frequency of the switched capacitor filter is set by a frequency of an external clock

$$f_{CLK} = 100 f_0 . \tag{11}$$

Thus, a microcontroller (see Fig. 3) can set a calculated phase shift $Arg\{K_\varepsilon(j\omega)\}$ by means of a programmable square wave generator for each excitation signal.

6 TEST RESULTS OF DIGITAL EMULATOR

Below, the test results are provided of the digital emulator to emulate complex dielectric permittivity according to the Debye relaxation function [1]

$$\tilde{\varepsilon}(j\omega) = \varepsilon_\infty + \frac{\Delta\varepsilon}{1 + j\omega\tau_0}, \tag{12}$$

for $\varepsilon_\infty = 2$, $\tau_0 = 10^{-3}$ and three value of $\Delta\varepsilon$ = 1, 30 and 100 over the frequency range 10 Hz to 32 7680 Hz distributed geometrically with common ratio 2. The complex dielectric permittivity (eqn (12)) was emulated for OSA multi-electrode capacitive sensor with active guarding (Fig. 7) with the following technical specification.

- Diameter (mm) 45
- Minimum thickness of test object (mm) 20
- Sensing capacitance (pF) 0.11
- Reference capacitance (pF) 16

Figure 7: Photo of OSA capacitive sensor.

In Fig. 8, relative errors are shown for the attenuation factor

$$\delta = \frac{\left|\hat{K}_{\varepsilon}(j\omega)\right| - \left|K_{\varepsilon}(j\omega)\right|}{\left|K_{\varepsilon}(j\omega)\right|}, \tag{13}$$

where $\left|K_{\varepsilon}(j\omega)\right|$ is the desired attenuation factor according to eqn (9), and $\left|\hat{K}_{\varepsilon}(j\omega)\right|$ is practically measured emulated ratio of the output signal against the excitation signal.

It can be seen a tendency of increasing attenuation error with decrease of the attenuation factor. In the current implementation of the digital emulator, the relative error of signal attenuation did not exceed 1%.

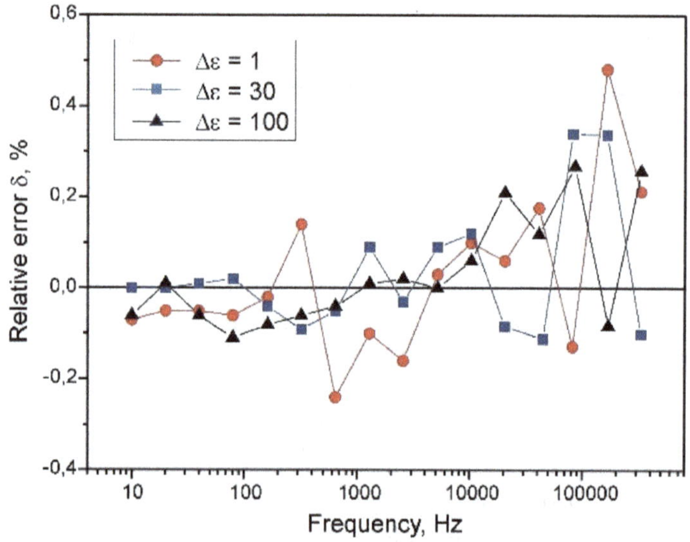

Figure 8: Relative errors of the attenuation factor.

7 CONCLUSIONS

We propose to solve problems of reference means for adjusting, calibrating and testing of non-destructive dielectric spectrometers employing OSA capacitive sensors by digital emulation of dielectric relaxation functions. Based on the developed two-port network measuring model of OSA capacitive sensors with active guarding working, a digital emulator has been elaborated. Tests provided for the digital emulator has witnessed its practical capability to emulate sinusoidal excitation signal attenuation with relative errors that do not exceed 1% over wide frequency band for dielectric constant varying from 1 to 100.

ACKNOWLEDGEMENTS

This work was supported by the European Regional Development Fund under project No. 1.1.1.1/16/A/008 "Development of multi-functional tester for non-destructive quality testing of materials and structures from rigid cellular plastics".

REFERENCES

[1] Kremer, F. & Schonhals, A., (eds) *Broadband Dielectric Spectroscopy*, Springer-Verlag: Berlin Heidelberg, 2003.

[2] Shtrauss, V., Nondestructive relaxation spectrometry of dielectrics. *Mech. Comp. Mat.* **25**(4), pp. 521–534, 1990.
[3] Boschetti, M. & Gottardo, A., Method for calibrating apparatuses for determining the moisture content of products based on capacitive measurements, and device for stimulating the dielectric properties of products, such as wood, for use in this method. Patent document, Patent No.: US 8,410,796 B2, April 2, 2013.
[4] Troxler, R.E., Apparatuses and systems for density gauge calibration and reference emulation. Patent document, Patent No.: US 9,190,183 B2, November 17, 2015.
[5] Matiss, I., Kalpinsh, A., Shtrauss, V., Lomanovskis, U. & Bernavs, I., Capacitance transducer for NDT of dielectric properties of materials. USSR copyright certificate, SU 1638665A1, Bulletin, 12, 1991. (In Russian.)
[6] Kalpinsh A., Shtrauss, V. & Lomanovskis, U., Capacitive probe for non-destructive testing of dielectric materials. Patent Office of Republic of Latvia, 9, p. 1169, 2013.
[7] Analog Devices, 8-/10-/12-bit high bandwidth multiplying dacs with serial interface. Data sheet AD5426/AD/5432/AD5443. https://www.analog.com/media/en/technical-documentation/data-sheets/ad5426_5432_5443.pdf.
[8] Lacanette, K., A basic introduction to filters – active, passive, and switched capacitor. National semiconductor application note 779, April 21, 2010. www.ti.com/lit/an/snoa224a/snoa224a.pdf.

CORRECT SIZING OF REFLECTORS IN ULTRASONIC INSPECTION OF THE FORGING TITANIUM ALLOY

THEODOR TRANCĂ[1] & IULIANA RADU[2]
[1]AROEND, Romania
[2]ZIROM – SA, Romania

ABSTRACT

Commercial Ti-6Al-4V forgings are widely used in the rotating components of aircraft engines. The failure of such parts can be quite catastrophic because of the large amount of kinetic energy. To ensure the safety and longer lifetime of these critical parts working in the hostile environments of high temperature and high stress, the need to detect smaller defects becomes more and more important. Ultrasonic inspection is one of the non-destructive evaluation (NDE) methods widely used by the titanium forging's manufacturers because of its ability to penetrate the interior of a component. Over the last decade, sizing methods were established like DGS (Distance Gain Size) or DAC (Distance Amplitude Correction) for defects smaller than the beam profile. Those methods utilize the echo amplitude and provide results which are proportional to the defect area. In this article, the correct sizing of small defects below one wavelength is investigated. By properly choosing the simulation method, it is ensured that all physical wave modes are included in the simulation and that the discretization error is negligible. A good correspondence between the simulation and classical defect sizing for defects larger than one wavelength is found. In the region between one quarter of a wavelength and one wavelength resonance effects are found, which results in classical defect sizing methods giving conservative results. In the region below one quarter of a wavelength classical DGS and DAC sizing leads to under-sizing.

Keywords: simulation program, grain noise, small defect.

1 INTRODUCTION

With the improvement of material technology and ultrasonic inspection, the necessity to detect and size smaller defects is higher. Therefore, not only for flat bottom holes, but also for the disc shaped reflectors the usability for small irregularities needs to be checked.

Measurement of ultrasonic indication is generally performed by echo dynamic or area amplitude based sizing. For discontinuities larger than the beam spread, echo dynamic sizing (sizing by probe travel, e.g. -6 dB drop method) are practiced and for indications more minor than the beam spread area-amplitude based sizing procedures are applied (like DAC or DGS).

2 CORRECT DIMENSIONING OF THE REFLECTORS SMALLER THAN ONE WAVELENGTH

Area-amplitude based sizing procedures compare the reflection of an indication to the reflection of reflectors with known dimension, e.g. artificial reflectors like flat bottom holes (FBH) or side drilled holes (SDH). The most traditional procedures used from the first days of the ultrasound testing, is the Distance Amplitude Correction (DAC), based on calibration managing multiple flat bottom holes machined into calibration blocks and some extrapolation based on the inverse square law.

2.1 Disagreement with area-amplitude relationship

The area-amplitude relationship is a consequence of using the Kirchhoff (also called physical optics) approximation when solving the equations governing ultrasound scattering by an FBH. This approximation assumes that the motion of the FBH surface when reflecting

WIT Transactions on Engineering Sciences, Vol 125, © 2019 WIT Press
www.witpress.com, ISSN 1743-3533 (on-line)
doi:10.2495/CMEM190121

an ultrasound pulse is identical to the motion that would occur if the pulse were reflecting from an infinite planar surface. A further assumption in obtaining the area-amplitude relationship is that the finite width ultrasound beam can be approximated as an infinite plane wave. Under these assumptions, it is seen that the FBH surface motion will be independent of the size of the FBH. Auld's reciprocity formula [1] states that the voltage received from a void defect is the integral over the defect surface of the product of the traction generated by the incident pulse in the absence of the defect, multiplying the total surface motion of the defect in response to the incident pulse. In the case of the FBH using the Kirchhoff approximation with an incident plane wave, it is readily seen that Auld's formula predicts an output voltage in direct proportion to the area of the FBH, i.e., the area amplitude relationship.

The relationship between echo response and FBHs has been addressed since some of the earliest developments in ultrasonic testing. Krautkrämer [2] referred to these as a disk-shaped reflector (DSR) and developed the famous AVG (English DGS) method of relating amplitude responses from FBHs to curves made for each style of probe. The relationship between echo amplitude and probe and FBH size can be summarised in the form of an equation:

$$\frac{V_f}{V_0} = \frac{SA}{\lambda^2}\frac{1}{T^2}e^{-2T\delta}, \tag{1}$$

where:
V_f = the maximum amplitude of the echo from the target
V_0 = the maximum possible signal amplitude if all energy is returned to the receiver
T = the distance along the beam axis to the target
A = the area of the defect
S = the area of the probe
λ = the wavelength of ultrasound (nominal)
δ = the attenuation coefficient.

From the findings of Krautkrämer [2], it was possible to note that the amplitude change for a FBH was directly proportional to its area. Therefore, having set a response on the Cathodic Ray Tube (CRT) to a specific amplitude (within the linear region of the instrument display) the response from area would produce a signal with half the amplitude and the response from an FBH double the area would produce a signal with double the amplitude.

As part of producing setup standards, a study was performed to determine if smaller FBH could be used by applying the theoretical gain difference in dB based on the area amplitude relationship. There are advantages to using biger FBHs since they can be drilled deeper and provide separation between the back-wall signal and the signal from the bottom of the hole.

The results of the Ultrasonic Testing measurements, however, indicated that a smaller difference was consistently observed rather than the theoretical dB difference (see Fig. 1).

The difference between the signals of Φ 2.0 mm FBH and Φ 1.2 mm FBH is measured at 7.5 dB instead of 8.87 dB, the theoretical value (measured on the Φ 265.0 reference block).

The cause of the discrepancy was determined by modeling the interaction of the beam produced by the specific transducers used and the FBHs.

It was shown that interaction between the surface wave that is generated on the FBHs and the incident compression wave can have a constructive interference that raises the amplitude of the smaller FBH. A phenomenon that the classic theory fails to predict is the generation of diffracted waves at the corners of the FBH. Fig. 2 depicts the various diffraction phenomena that occur for a plane compressional wave at perpendicular incidence on the

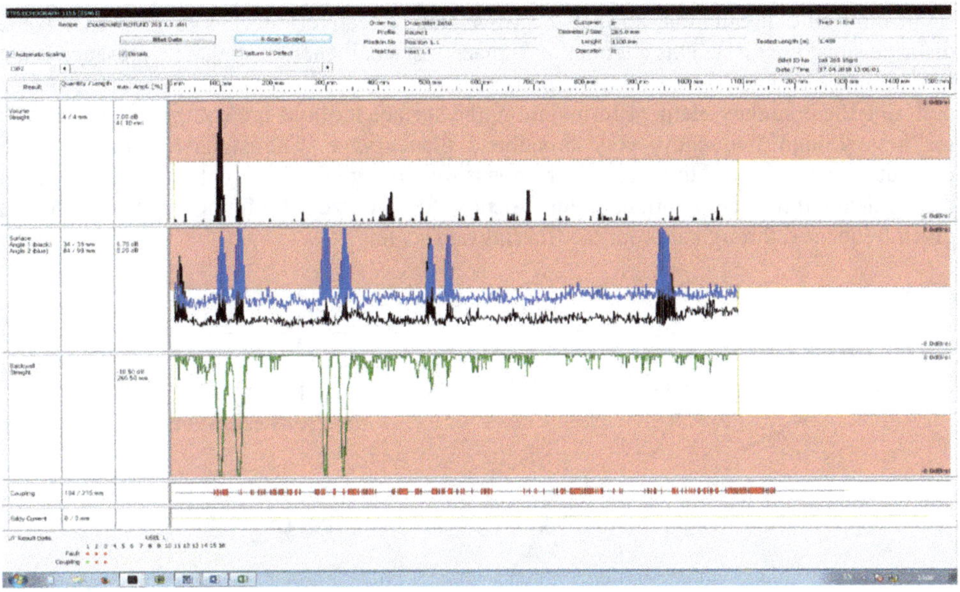

Figure 1: The difference between the signals of Φ 2.0 mm FBH and Φ 1.2 mm FBH.

FBH. It is seen that in addition to a reflected compressional wave, diffracted compressional and shear waves are generated, along with surface waves that propagate both down the bore of the FBH and across the top. Attention is directed to the surface waves propagating across the top of the FBH. Upon reaching the opposing corner of the FBH surface, the surface wave undergoes a second diffraction, during which a small amplitude diffracted compressional wave emerges from the FBH corner. Part of this secondary diffracted wave travels up to the transducer, slightly behind the primary compressional wave reflection from the FBH surface, as depicted in Fig. 3, and is received as a small signal trailing the main reflection, as depicted in Fig. 4. The time delay between these two waves is given by the product of the FBH diameter and the surface wave velocity.

If the time delay between these two signals is sufficiently small, an interaction could take place that would enhance or reduce the total signal amplitude through a constructive or destructive interference. Such an interaction might be the underlying cause of the deviation from the area-amplitude relation seen in experiments when looking at small reflectors.

The challenge in determining the significance of the surface wave interaction is quantitatively determining the amplitudes of the diffracted signals. For this purpose, a computer model was employed that used a boundary element method (BEM) formulation to solve the equations governing the surface wave diffraction phenomena on the FBH. The boundary element formulation uses a high-frequency computational ansatz based on an asymptotic analysis of the diffraction problem. Rather than using the asymptotic solution outright, this method uses the asymptotic result as a starting point, then seeks to find corrections to the asymptotic solution to obtain an exact numerical solution. The boundary elements are, therefore, used to compute corrections to the asymptotic solution, rather than represent the entire solution. Consequently, extremely large problems can be treated with practical computational efficiency. Boundary elements are prescribed over the top and sides of the FBH. The FBH is assumed infinitely long, and the medium is prescribed to have a small ultrasonic attenuation, so that the wave field effectively decays to zero after some distance along the FBH. This attenuation is made just small enough so that its presence is not noticed in the computed transducer response signals.

Using the boundary element formulation, the surface motions are computed on the FBH for a very high frequency, very broadband plane wave. The signals from one such computation can be used to predict the response for any signal with a centre frequency within the bandpass of the computation, or equivalently, for any size FBH for an incident pulse of a given frequency, through appropriate filtering and scaling.

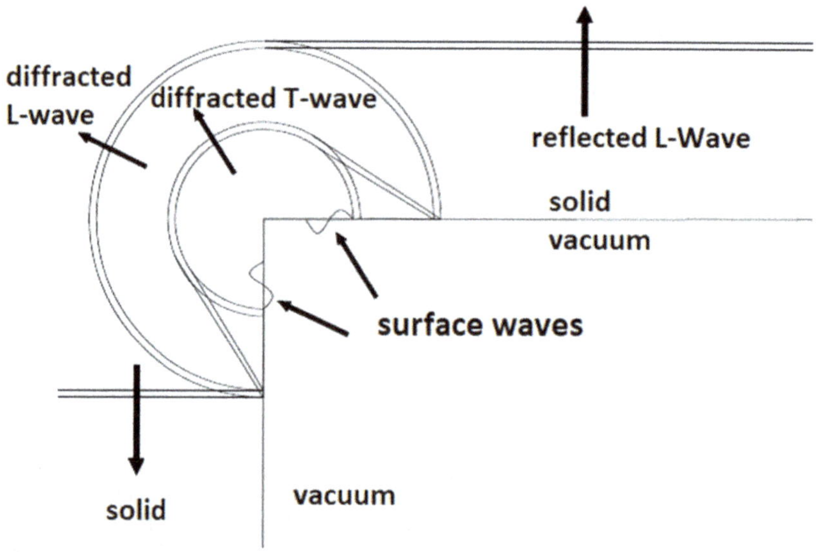

Figure 2: Wave modes generated when a compressional L-wave is incident on an FBH [3].

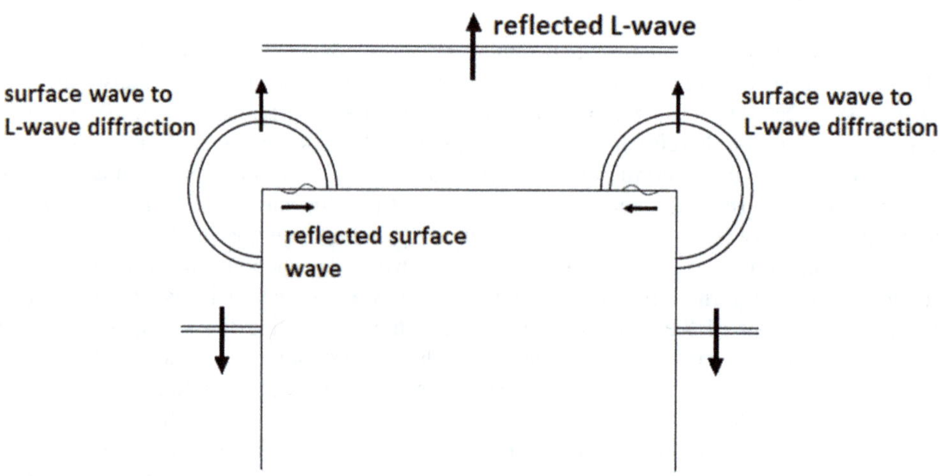

Figure 3: Refracted L-wave following the reflected L-wave [3].

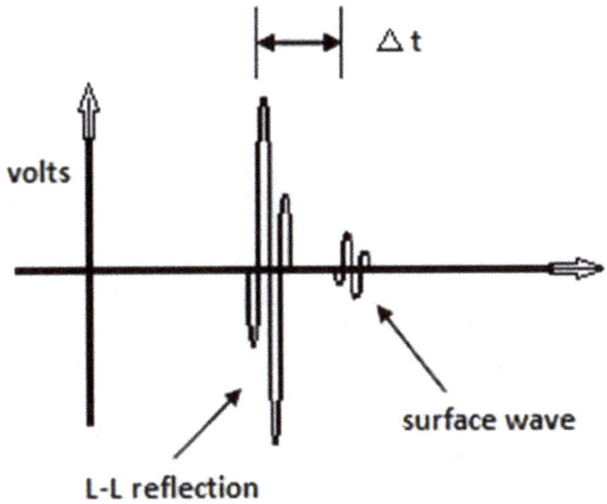

Figure 4: Two signals returned to the transducer [3].

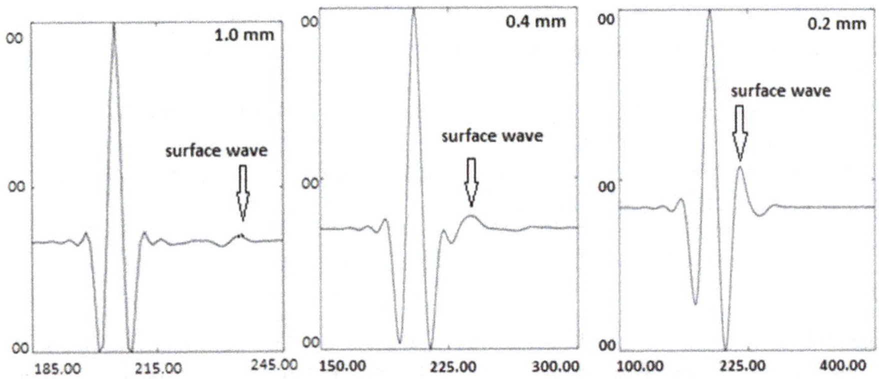

Figure 5: Interaction of the diffracted surface wave and the reflected L-wave for three FBH sizes [3].

Fig. 5 compares signals for Φ 1.0 mm, Φ 0.4 mm, and Φ 0.2 mm FBHs. It is seen that as the hole becomes smaller, the time delay between the two received signal components decreases. In the case of the Φ 0.2 mm FBH, the two signals are overlapping. The interaction between the two signals will depend on signal bandwidth: the narrower the bandwidth, the more significant the interaction.

3 SIMULATION SOFTWARE FOR NDT

In the simulation used for ultrasonic examination of the planar defects, two classical scattering models have been used: Kirchhoff approximation, to simulate the reflection, and Geometrical Theory of Diffraction, to simulate the diffraction. Recently, from the combination of these two theories, the Physical Theory of Diffraction (PTD) has appeared, which retains the advantages of both.

It is important to know the boundaries of such simulation tool to be able to assess the simulation results presented in a technical justification. Parameter studies can be used for review of important parameters, in order to find out limit values as well as which parameters are most important for the inspection system. In addition, optimization of defect content for the manufacture of test blocks can be done.

3.1 Principle of the Kirchhoff and Geometrical Theory of Diffraction model

Every one of the classical ultrasonic inspection methods, (pulse echo, tandem or Time of Flight Diffraction) perform the detection of the planar defects by interpreting their specular or diffraction echoes. The modern Non-Destructive Testing modelling is useful to assessing detection capability and to elaborate the procedures for inspections.

It assumes that the Kirchhoff scattered field can be decomposed in an approximate manner in two parts: a geometrical field which includes the specular reflected field and a contribution arising from the flaw edges corresponding to the edge's diffraction field. The contribution of this diffraction field at the observation point x is characterized by same form as the GTD field but a different edge diffraction coefficient (depending on the α incidence and β observation directions and polarizations):

$$U^{KA(Diff)}(x) = D_{\alpha\beta}^{KA}(x)\frac{e^{ikr}}{\sqrt{kr}}. \tag{2}$$

Note that this coefficient defines the directivity of edge diffraction contribution according Kirchhoff approximation.

The physical theory of diffraction (PTD) consists in correcting the Kirchhoff edge diffraction field by that modelled by GTD. This correction leads to add a corrective term to the KA scattered field (without far-field approximation). This corrective term is the difference of wave amplitudes diffracted by the edge, by GTD and KA.

$$U^{PTD}(x) = U^{KA}(x) + \left[D_{\alpha\beta}^{GTD}(x) - D_{\alpha\beta}^{KA}(x)\right]\frac{e^{ikr}}{\sqrt{kr}}. \tag{3}$$

The PTD field is the sum of the Kirchhoff field and a GTD modified field in which the GTD coefficient has been replaced by the difference between GTD and Kirchhoff edge diffraction coefficients. At the specular observation direction, the Kirchhoff field (without far-field approximation) is finite leading to an effective prediction of specular reflection. But the KA diffraction coefficient KA, $D_{\alpha\beta}^{KA}(x)$ for edge diffraction contribution (previously obtained from a far field approximation of the Kirchhoff field) diverges and has the same singularity as the GTD edge diffraction coefficient GTD, $D_{\alpha\beta}^{GTD}$(x). When making the difference of the two coefficients, their singularities cancel each other and the diffraction coefficients difference $D_{\alpha\beta}^{GTD}$(x)-$D_{\alpha\beta}^{KA}(x)$ is finite

$$U^{PTD}\,(x) \approx U^{KA}(x). \tag{4}$$

When the observation direction is far from to the specular direction, edge diffraction effects are predominant compared to reflection phenomena, the Kirchhoff field is equal to the Kirchhoff edge diffraction contribution and so cancels it so that the Kirchhoff and GTD model leads to similar results than the GTD model [4]

$$U^{KA}(x) \approx D_{\alpha\beta}^{KA}(x)\frac{e^{ikr}}{\sqrt{kr}} \text{ and } U^{PTD}(x) \approx D_{\alpha\beta}^{GTD}(x)\frac{e^{ikr}}{\sqrt{kr}} = U^{GTD}(x). \tag{5}$$

Flaws which can be modelled thanks to Kirchhoff and GTD are the same than with the GTD model: planar flaws (rectangular, semi-elliptical or CAD contour planar flaws), multi-facetted flaw and branched flaw [4].

Fig. 6 shows where Kirchhoff and GTD are applicable in relation to the orientation of the defect, with α = 0 representing a vertical flaw. This figure effectively illustrates what is described in para 3.1. The Kirchhoff approximation in CIVA gives rise to a tip response that is positioned correctly in terms of time of flight, but the amplitude may be inaccurate, with the error increasing with departure of the scatter direction from specular [5].

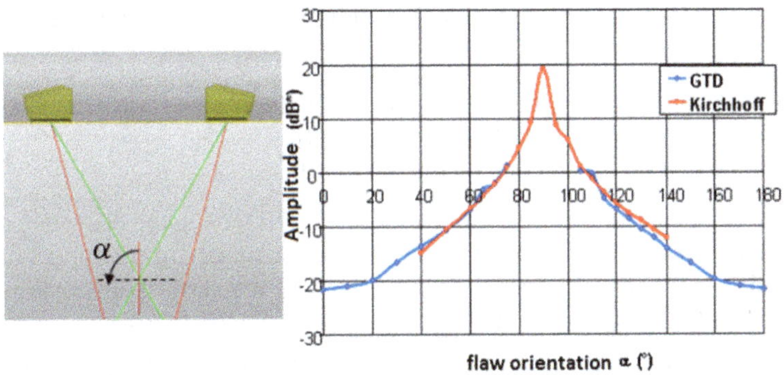

Figure 6: Applicability of Kirchhoff and GTD models.

3.2 Model capability to be assessed

The model verification activities are divided into five different partial phases. Each phase is divided into a number of different tasks with specific purposes (see Table 1).

Table 1: Phases of the verification activities.

Phase 1	Response prediction for simple/smooth defects in simple materials and probe modeling
Phase 2	Geometry handling with model
Phase 3	Complex materials – austenitic welds, inconels, dissimilar metal welds
Phase 4	Rough defects in simple materials
Phase 5	Rough defects in complex materials

The difference between simple/smooth defects and rough defects stated in Table 1 is that simple/smooth defects are typically artificial defects or an ideal fatigue crack. Rough defects are the type of defects that are typically service induced, with a clear morphology, following grain structure or other irregularities. By simple materials means carbon steel or stainless parent material that shows isotropic behaviour. Complex materials show anisotropic behaviour with significant influence on the sound beam giving effects such as large scattering, beam deflection and increased noise. The noise caused by the material structure is modelled as a separate layer which is super positioned on top of the defect response simulation, meaning that the defect response is not affected by the noise. If a noise simulation is used, it must be used together with additional attenuation modelling as mentioned above or else the result will be a non-conservative signal to noise ratio for any give indication [5].

4 CIVA 11 APPLICATIONS IN THE ULTRASONIC EVALUATION OF THE EXAMINATION STANDARDS

4.1 Determination of the FBH's ultrasonic responses

The goal of this virtual determination is to measure the difference (dB) between the signals obtained from Flat Bottom Holes of different diameters positioned at the same depth into the examined material. The obtained values will be compared with the theoretical difference, calculated according to the Kirchhoff Approximation.

The scanning of three reference blocks with the followings dimensions has been performed:

- Flat faced reference block, 111 x 111 mm square dimension (Fig. 7),
- Round reference block, 111 mm diameter (Fig. 8),
- Round reference block, 265 mm diameter (Fig. 9).

Each reference block is provided with one pair of FBH having the diameters of Φ1.2 and Φ 2.0 mm, first determination, and Φ 0.8 mm and Φ1.2 mm, second determination. All holes are positioned with the flat surface at 50 mm below the entry surface of the ultrasonic beam.

The ultrasonic transducers used for these simulations are the following:

STS 20 P5 – immersion transducer, non-focused, 20 mm crystal diameter, 5MHz central frequency.

STS 20 P5 L125 – immersion transducer, 125 mm focal distance in water, 20 mm crystal diameter, 5MHz central frequency.

STS 20 P5 L200 – immersion transducer, 200 mm focal distance in water, 20 mm crystal diameter, 5MHz central frequency.

Sound path in water is set 100 mm for all situation. Sound speed in material and specific attenuation are identical for all determinations (see Figs. 7–9).

Each reference block was scanned successively with each of the three transducers, respectively 18 measurement for the pair Φ 2.0-Φ 1.2 mm FBH and similarly for the pair Φ 1.2 – Φ 0.8 mm FBH. The results of the test are presented in Tables 2 and 3 (see below).

These figures present, in the left side, the echo-dynamic registration of the echoes obtained by scanning each of the two FBH. In the same time, the amplitude of the echoes and difference between signals from the holes can be direct read. In the right side of the pictures is represented the examination technique related to the reference block used and the holes positions.

5 RESULTS

5.1 Comparison between signal's amplitude

We observe that, for the FBH pair of Φ 1.2 – Φ 2.0 mm, the amplitude differences obtained with CIVA 11 are in the range of 8.6÷8.8 dB, very close of the theoretical value of 8.87 dB (see Table 2). Similarly for the FBH pair of Φ 0.8-Φ 1.2 mm, where the difference determined by the program are in the range of 6.8÷7.0 dB and the calculated value is 7.04 dB (see Table 3).

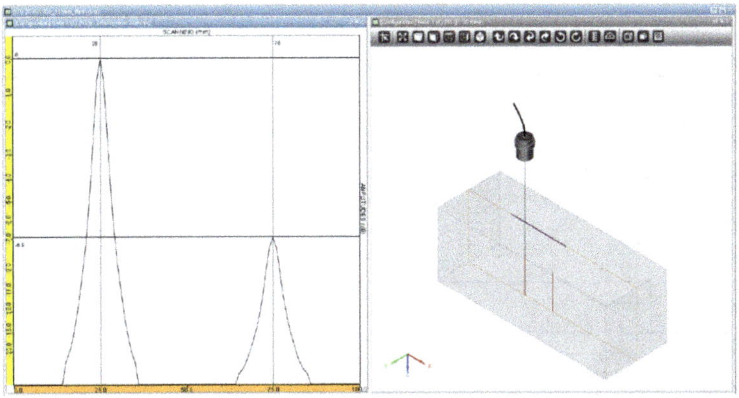

Figure 7: Scanning of the flat-faced 111 x 111 mm billet.

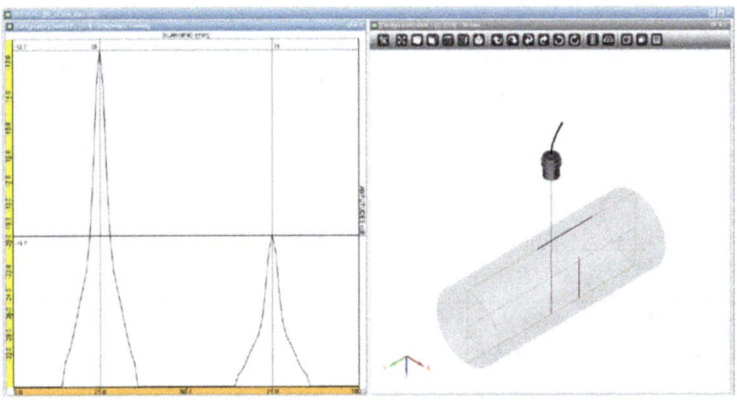

Figure 8: Scanning of the Φ 111 mm billet.

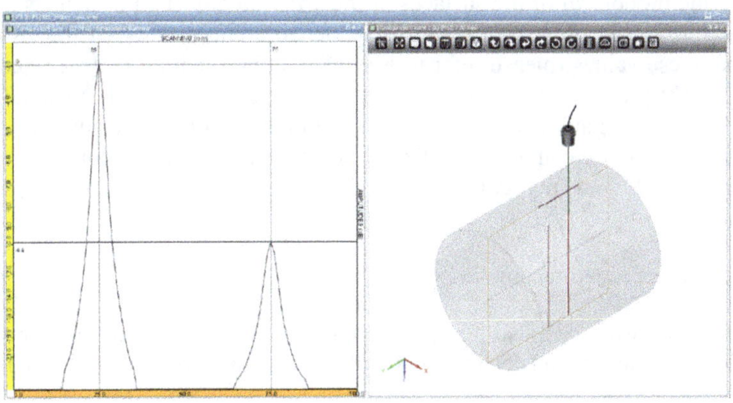

Figure 9: Scanning of the Φ 265 mm billet.

Table 2: Comparison of difference between signals of FBHs Φ 1.2 and Φ 2.0 mm.

Test specimen	Test flaw	STS 20 P5		STS 20 P5 L125		STS 20 P5 L200	
billet 111x111 mm	FBH Ø 1.2 mm	-12.6 dB	Δ=8.7 dB	-9.4 dB	Δ=8.7 dB	-8.6 dB	Δ=8.6 dB
	FBH Ø 2.0 mm	-3.9 dB		-0.7 dB		0 dB	
Bar Ø 111 mm	FBH Ø 1.2 mm	-30.7 dB	Δ=8.8 dB	-16.8 dB	Δ=8.8 dB	-21.3 dB	Δ=8.7 dB
	FBH Ø 2.0 mm	-21.9 dB		-8 dB		-12.6 dB	
Bar Ø 265 mm	FBH Ø 1.2 mm	-22.6 dB	Δ=8.8 dB	-11.6 dB	Δ=8.7 dB	-14.1 dB	Δ=8.7 dB
	FBH Ø 2.0 mm	-13.8 dB		-2.9 dB		-5.4 dB	

Table 3: Comparison of difference between signals of FBHs Φ 0.8 and Φ 1.2 mm.

Test specimen	Test flaw	TS 20 WB4		STS 20 P5 L125		STS 20 P5 L200	
billet 111x111 mm	FBH Ø 0.8 mm	-7.2 dB	Δ= 7 dB	-7.7 dB	Δ= 7 dB	-6.9 dB	Δ= 6.9 dB
	FBH Ø 1.2 mm	-0.2 dB		-0.7 dB		0 dB	
Bar Ø 111 mm	FBH Ø 0.8 mm	-21.9 dB	Δ= 7 dB	-15.2 dB	Δ= 7 dB	-19.7 dB	Δ= 7 dB
	FBH Ø 1.2 mm	-14.9 dB		-8.2 dB		-12.7 dB	
Bar Ø 265 mm	FBH Ø 0.8 mm	-14.5 dB	Δ= 6.8 dB	-9.9 dB	Δ= 6.9 dB	-12.3 dB	Δ= 6.8 dB
	FBH Ø 1.2 mm	-7.7 dB		-3 dB		-5.5 dB	

The resulting values (noted with Δ in Table 2 and Table 3) match the theoretical values obtained for the proportion of the surfaces of the artificial defects. For defects smaller than the wavelength, it was proven in reality that the differences between the two holes are different that these values (please refer to Fig. 1). As observed, these differences are more pronounced as the product ka >>1 (k = wave number and a = size of the artificial defect).

This proves the limitations of the CIVA 11 simulation software for the defects close to the specular area and geometrical reflex whereas the software applies the Kirchhoff Approximation and does not consider the diffraction phenomenon that appear at the edge of FBH.

5.2 Optimization of the ultrasonic inspection

The simulation software can also be used for the selection of the ultrasonic transducers in relation with the examination technique for ultrasonic inspection of immersed plates and forged bars.

Table 2 and Table 3 also provide the data relevant for the efficiency of each type of transducer at a certain depth for a specific radius of the entry surface.

The transducer with the best ultrasonic behaviour for this scenario can be decided by comparing the observed amplification reserve with the value initially selected for each of the 3 transducers.

For an example, if Table 2, row 1 – billet 111 x 111 mm and Table 2, row 2 – BAR Φ111 mm are selected:

Row 1, FBH Φ 1.2 mm represents the inspection of a defect located at a depth of 50 mm below the entry plane surface. The best result is provided by the transducer STS 20 P5 L 200 (-8,6 dB) due to optimal Pulse Volume for a defect located at a 50 mm depth related to its focal distance in water and the entry plane surface. The gain differences comparing with the other transducers are relatively small: 0.8 dB versus STS 20 P5 L 125 and 3.0 dB versus STS 20 P5 (unfocused).

The situation is significantly different when the Φ 111 mm billet is inspected. The most efficient transducer is STS 20 P5 L 125 with a focal length in water of 125 mm. Due to the radius of the entry surface, the focal distance in the material is increased (see para 6.) and this transducer presents now the focal distance in the area of depth of the artificial defect (50 mm). The gain difference is increased significantly comparing with the other 2 transducers: 4.5 dB versus the transducer focalized with the focal distance of 125 mm and 13.9 dB versus the unfocalized transducer. This is caused by the defocusing of the immersion transducers at incidence area of the ultrasonic beam with the cylindrical surface of the billet and due to the modification of the Pulse Volume in the area of the reference defect.

6 CONCLUSION

The accurate prediction of absolute noise levels requires detailed knowledge of the metal microstructure which enters the model calculations through certain frequency-dependent factors known as "backscatter coefficients" or "Figures-of-Merit". Unfortunately, the largest discrepancy between the software simulations and experiments is noise, or rather signal to noise ratio. Defects responses are often evaluated in relation to the surrounding noise levels rather than an arbitrary reference target, such as a notch or SDH.

The issues discussed above means that it is clear that it is not possible to simulate a complete inspection or validate an inspection procedure by simulations with CIVA at the current time. The conclusion is that simulations using CIVA can be used when specific problems or technical solutions must be solved or developed, e.g. the influence of the surface curvature over the Pulse Volume or the correct choice of the inspection configuration.

The inspection's optimization is achieved today both by minimizing the noise of the material and by adopting the inspection techniques capable to highlight and perform a correct evaluation of the discontinuities smaller than one wavelength, according to the highest quality standards.

ACKNOWLEDGEMENTS

This work was supported by S.C. ZIROM-S. A – Giurgiu, Romania and was performed by the NDT Consulting Company – DIAC SERVICII srl and AROEND, Romania.

REFERENCES

[1] Auld, B.A., *Acoustic Fields and Waves in Solids, vol. II*, Krieger Publishing Company: Malaber, 1990.

[2] Krautkrämer, J., Fehlergrößenermittlung mit Ultraschall. *Archiv für Eisenhüttenwesen*, **30**, pp. 693–703, 1959.

[3] Margetan, F.J.L. et al., Inspection Development for Titanium Forgings, Air Traffic Organization Operations Planning Office of Aviation Research and Development Washington, DC, 20591, May 2007.
[4] Darmon, M., Dorval, V. & Kamta Djakou, A., A system model for ultrasonic NDT based on the Physical Theory of Diffraction (PTD). *Ultrasonics,* **64**, pp. 115–127, 2016.
[5] Holmer, G., Daniels, W. & Zettervall, T., *Evaluation of the Simulation Software CIVA for Qualification Purpose,* SQC Swedish Qualification Centre, AMEC Foster Wheeler: Birchwood, 2017.

AUTOMATIC DETECTION OF LOITERING BEHAVIOUR USING SPATIOTEMPORAL IMAGE PROCESSING

YUTA EBIHARA, TERUOMI KATORI & TAKASHI IZUMI
Nihon University, Japan

ABSTRACT

In this paper, the authors propose a method for detecting loitering behaviour automatically from security camera images acquired in a corridor or passage, and the authors examine the performance of the proposed method. Image sensors (security cameras) are widely used for crime prevention. In this study, for educational settings, the authors developed a system for automatically detecting loitering behaviour where a student is worried about whether he or she is permitted to enter a laboratory on his/her first visit. Using the results, staff in the laboratory can approach them and appropriately guide the student during his or her visit. The purpose of this study is to detect loitering behaviour including fuzzy actions. Detecting loitering behaviour involves the ethical issue of ensuring that the captured images do not infringe an individual's privacy. In addition, there are a number of technical problems: What is a unique characteristic value indicating the target behaviour?; the method should not require much computational power; and it should be possible to explain the reason for the judgment result. In this study, to ensure privacy, the authors avoid using original images, for example, images in which the face or body of an individual can be recognized, and instead the authors use spatiotemporal images. General image processing is highly complex and requires computers using high-performance CPUs and a lot of memory. However, usual video capturing and behaviour recognition are expected to involve lower complexity. Spatiotemporal image processing can solve the technical problems mentioned above, for example, decreasing the computational complexity and maintaining high computational performance. In addition, as a measurement characteristic value, the authors adopt a simple staying time only, and the authors classify the behaviour into only two categories: "loitering behaviour" or "not loitering".

Keywords: spatiotemporal image processing, loitering behaviour, staying time.

1 INTRODUCTION

In an educational setting, when new students visit a laboratory for the first time, sometimes they hesitate to knock on the door or to enter the room, and they loiter in the corridor or passage. They usually wait to be found by somebody or may even go back home, which is very unfortunate. Therefore, the authors investigated how to automatically detect such loitering behaviour using image processing with an image sensor (camera) in front of a door [1], and when loitering is detected, the student is called into the room.

Recently, there have been significant advances in image processing technology, and there are many examples of successful applications involving moving images. However, these examples mainly recognize simple actions [2], or sign language and gestures [3]. From a security point of view, there are also many examples, but these systems cannot be developed without the verification of individuals. One example is tracking passengers in railway stations [4], [5], but the aim of those studies is crime prevention and avoiding trouble between passengers. In other examples, Bird et al. [6] describe how to detect drug dealers loitering at bus stops, using RGB intensity matching. Min et al. [7] describe loitering detection using characteristic point tracking.

In this paper, the authors describe how to automatically detect loitering behaviour with a security camera installed in a corridor or passage. Generally, image processing has high computational complexity, and the computer executing it needs a high-performance CPU and a lot of memory. However, video capturing and behaviour recognition processing require low

WIT Transactions on Engineering Sciences, Vol 125, © 2019 WIT Press
www.witpress.com, ISSN 1743-3533 (on-line)
doi:10.2495/CMEM190131

complexity. Therefore, the authors adopt spatiotemporal image processing to realize high performance. Real-time person tracking by spatiotemporal image processing has also been described by Yann and Bouthemy [8]. First, for behaviour recognition, the authors classify the behaviour into only two categories, either "loitering" or "other behaviour", and the characteristic value is only the staying time.

2 LOITERING BEHAVIOUR

For behaviour recognition, the target action needs to be expressed by a characteristic value. However, there are cases in which similar actions have different characteristic values. Loitering behaviour involves not only staying in one place for a long time, but also different characteristic values. Our ultimate goal is to express and discriminate many behaviours using characteristic values. As a first step, loitering behaviour is detected.

In this study, the loitering behaviour does not include actions related to dementia; the behaviour is limited to whether a person goes into a room or not, or hesitates in front of the door for a long time. For detecting this behaviour, the used sensors need to measure the time the person stays in one place and loiters in a spatial area, using moving images.

The authors use an image sensor to detect loitering behaviour on the basis of the staying time and area. The authors use spatiotemporal image processing because the authors place priority on reducing the computation time, and the authors use the staying time only. In the future, higher performance loitering detection can be expected by expanding the loitering area.

3 DETECTION OF LOITERING BEHAVIOUR IN FRONT OF A DOOR

A spatiotemporal image is formed of multiple scanning lines obtained from the same specific location in multiple images. Fig. 1 shows the concept of a spatiotemporal image.

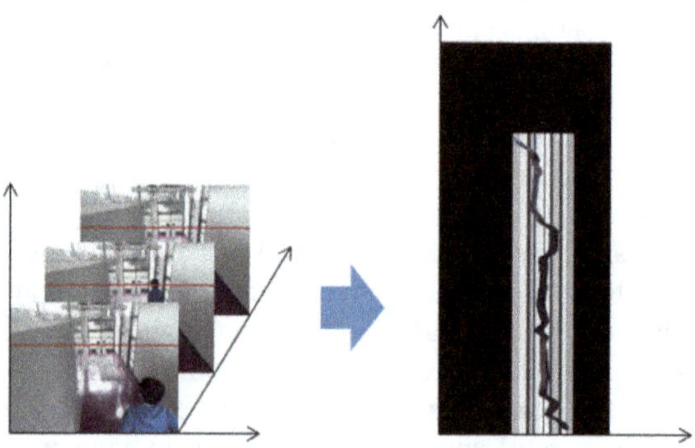

(a) (b)

Figure 1: Concept of spatiotemporal image. (a) Original image; and (b) Spatiotemporal image.

In a spatiotemporal image, only the variation of the scanning lines is detected. A significant advantage of this approach is the ability to perform processing at high speed, because the whole frame is not processed.

3.1 Overview of loitering behaviour detection

Fig. 2 shows a block diagram of loitering behaviour detection.

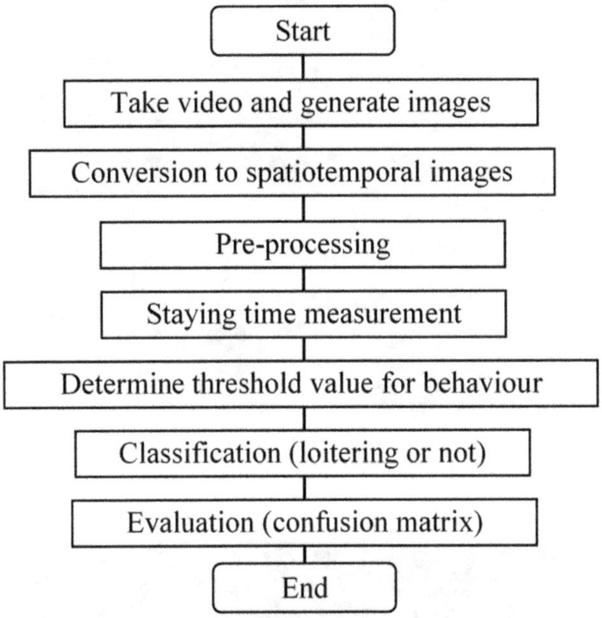

Figure 2: Block diagram of loitering behaviour detection.

A video of the scene in a corridor is captured using an image sensor, and the video is read in a personal computer. In the future, the read video will be processed in real time, but for now, the video is temporarily stored in memory, and detection processing is performed off-line.

The specific location of the scanning line in the images is determined, and a spatiotemporal image and its background image are generated. These images are subjected to pre-processing steps including background difference processing, binarization and noise reduction processing.

All scenes are manually judged to be "loitering" or "not loitering", and the scene data are divided into two groups.

For loitering behaviour detection, many sample scenes are needed. Half of these data are provided for learning (to determine a threshold value), and a threshold staying time is determined based on the relation between time and loitering. The other half of the data are unknown data, and these data are judged as to whether they involve loitering by using the threshold staying time, and loitering results are evaluated.

3.2 Spatiotemporal image generation

In this study, the scanning line direction in the images is the horizontal direction at the doorknob, because a person holds the doorknob when coming into the room. There are some windows at the side of the corridor, and since outside lighting has a bad influence on the detection, the image areas that are not required are masked manually to reduce the influence of outside lighting.

Fig. 3 shows an example of a captured picture, and Fig. 4 shows an example of a spatiotemporal image generated from a video that includes Fig. 3.

Figure 3: An example of a captured picture influenced by outside light.

Figure 4: Example of generated spatiotemporal image.

In Fig. 3, a person's reflection appears on the panel shown by the red ellipse, and this is the influence of the outside lighting. Therefore, Fig. 4 is generated using masking processing, where the areas at both sides are blacked out. The spatiotemporal image is generated only during the time a person walks; therefore, the upper area in Fig. 4 is also blacked out. To process images in real time, usually the authors need to capture a video, and the background spatiotemporal image continues to appear in the upper area in Fig. 4 when a person does not walk there.

3.3 Pre-processing

The generated spatiotemporal image is pre-processed by using background difference processing, binarization and noise reduction in this order. Fig. 5 shows an example of a background difference image, Fig. 6 shows an example of a binarized image, and Fig. 7 shows the result of noise reduction.

(a) (b) (c)

Figure 5: Background difference processing. (a) Original spatiotemporal image; (b) Background spatiotemporal image; and (c) Difference image.

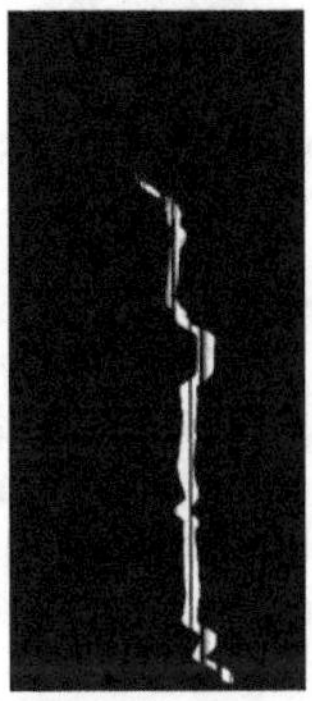

Figure 6: A binary image.

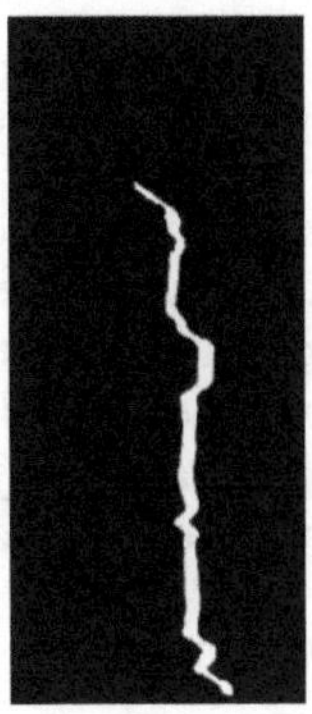

Figure 7: A noise-reduced image.

A moving object is detected by background difference processing and binarization processing. Then the image is subjected to noise reduction processing. Thus, the precision of staying time detection can be improved, but excessive noise reduction results in a shorter staying time.

3.4 Staying time measurement

In a noise-reduced spatiotemporal image, the number of lines in the same label between the first and last time are counted after labelling processing. The staying time is calculated from the measured number of frames divided by the frame rate.

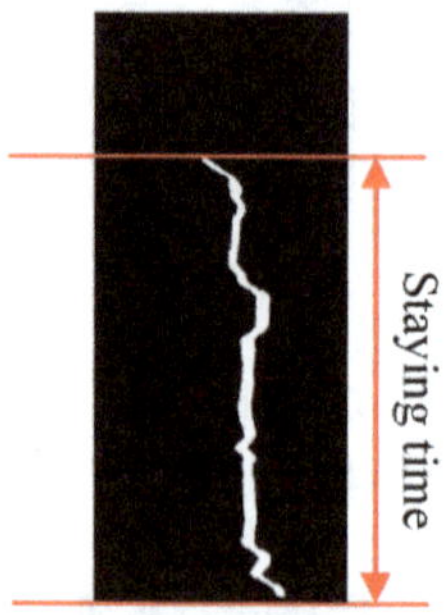

Figure 8: Staying time measurement.

3.5 Loitering behaviour detection in front of a door using staying time

The action of a person moving in front of a door is judged as "loitering" or "not loitering" by using the measured staying time. The staying time in loitering behaviour is longer than the staying time when a person passes the door, and in the characteristic space (staying time zone), a longer time means loitering, and a shorter one means passing, as shown in Fig. 9.

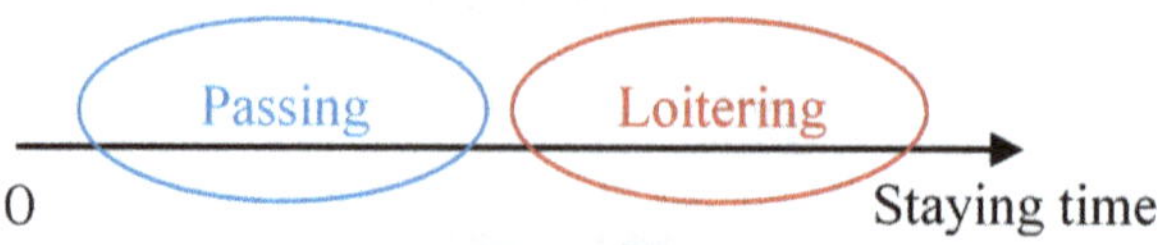

Figure 9: Distribution of staying time and loitering behaviour.

Therefore, a threshold value that can be used to classify the behaviour as "loitering" or "other" is very important. The threshold value is an intermediate value between the minimum loitering time and the maximum passing time in the determination group.

Fig. 10 shows the distribution, which was clearly divided into two distinct categories, without any overlap. The staying time threshold value for loitering was 27.3 s (=819.5 frames)

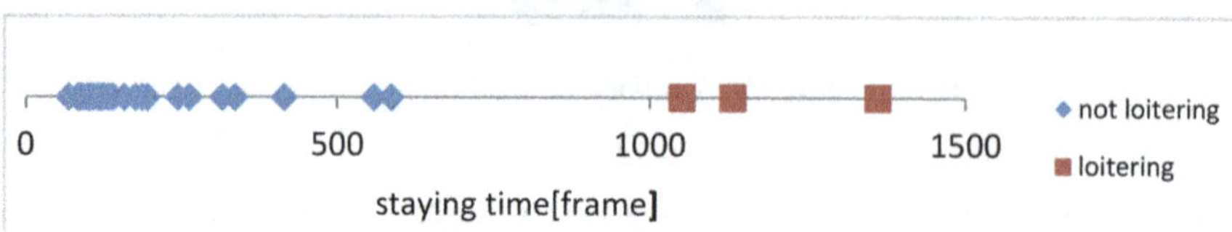

Figure 10: Loitering behaviour on staying time distribution.

$$\begin{cases} t \geq th & \text{loitering,} \\ t < th & \text{others,} \end{cases} \quad (1)$$

where t: staying time; th: threshold value.

3.6 Evaluation of loitering behaviour detection

The loitering behaviour detection was evaluated using another dataset which is used for determining a threshold value. The evaluation value is the accuracy (right rate).

4 RESULTS AND DISCUSSION

A camera was set over a corridor in our campus. There were windows at the side of the corridor, and this environment was thus easily influenced by the outside lighting. The number of captured video scenes was 58. The authors divided the scenes into two groups: one group (33 persons in 29 scenes) was used for the threshold value determination, and the other group (35 persons in 29 scenes) was used for evaluation of loitering detection.

Table 1 shows the confusion matrix for the results.

Table 1: Confusion matrix (number of persons).

		Proposed method		
		Loitering	Not loitering	Non-detection
Manual	Loitering	2	2	0
	Not loitering	0	21	4

In Table 1, four persons were not detected for not-loitering; however, the purpose of this study is loitering detection, the non-detection is not a big problem. However, when someone passes, blocks are connected in the spatiotemporal image. In this case, the authors measure the whole length, the measured staying time is shorter, and this time causes an error detection. The reason for this is an excessively large threshold value determination.

Fig. 11 shows an example result in which the authors can detect loitering behaviour. This sample is judged correctly because the staying time was over the threshold value.

Fig. 12 shows loitering behaviour detected with both manual and automatic (presented) methods. This sample was also judged correctly.

Fig. 13 shows two persons (one is loitering, the other one is passing) in the same scene. This result was misjudged as loitering by one person only, because their traces crossed in the spatiotemporal image. In this case, the trace has one label only; the label is judged as one person's behaviour, and other person's behaviour is not detected. Time measurement used the label length. The person staying for a short time is not detected. This scene was judged as loitering by only one person.

When the traces of two people cross, the traces form an "X" shape in the spatiotemporal image. In Fig. 13(b), a block forms an "x". Therefore, the centre lines are extracted for Fig. 13(b), and two people crossing are detected. Fig. 13(c) shows centre line extraction for Fig. 13(b). Because the lines intersect, it is understood that this label includes two people. The correct staying time was measured due to each person's track.

There were no samples for two people who did not cross. The track has other labels in the spatiotemporal image, and thus the labels could be judged as correct behaviour.

In Fig. 14(a), there is a walking person only, but in Fig. 14(b), the track is divided into two blocks. Therefore, with the automatic (presented) method, this result was misjudged as one person loitering and one person not loitering.

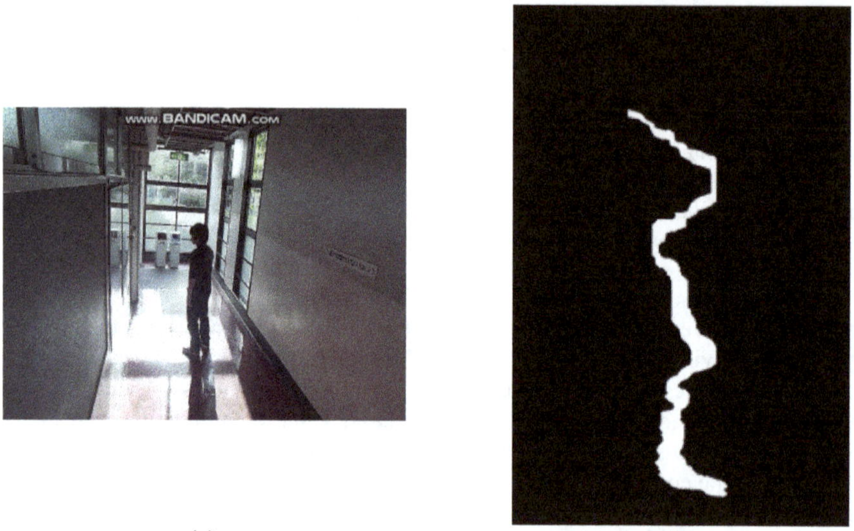

(a) (b)

Figure 11: Loitering for loitering. (a) Original image; and (b) Spatiotemporal image.

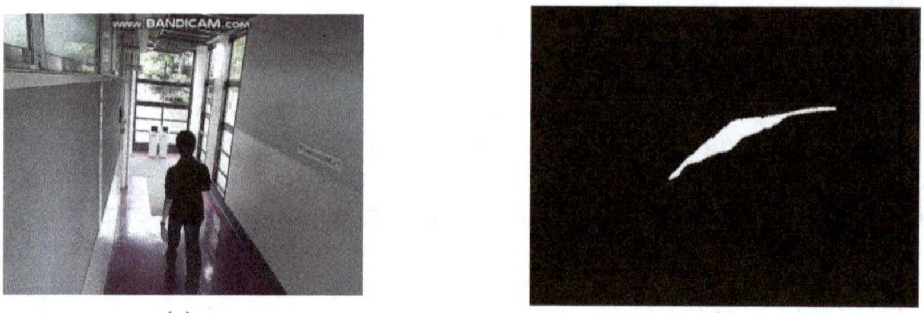

(a) (b)

Figure 12: Non-loitering behaviour detected with both manual and automatic (presented) methods. (a) Original image; and (b) Spatiotemporal image.

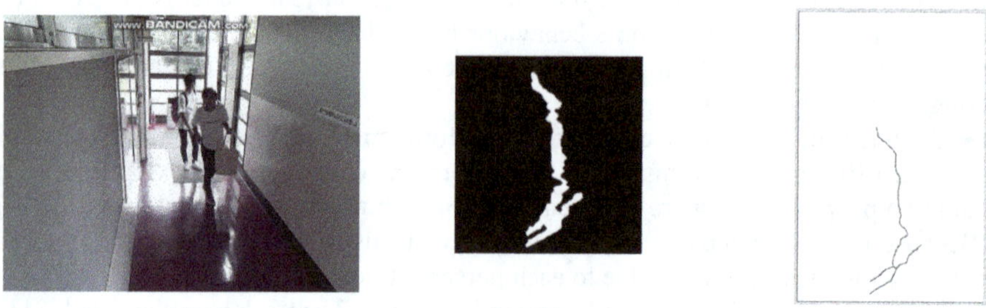

(a) (b) (c)

Figure 13: One person passing and one person loitering, but detected result is judged as one person loitering. (a) Original image; (b) Spatiotemporal image; and (c) Centre lines.

(a) (b)

Figure 14: Image judged as one person loitering and one person not loitering, in the case of only one person loitering. (a) Original image; and (b) Spatiotemporal image.

In this method, when the moving person does not cross the scanning line, the person is divided into two blocks. This problem can be solved by comparison with the RGB intensity distribution for two blocks, and the two blocks are judged as being the same label.

Fig. 15 shows an image of a scene in which a loitering person was misjudged as not loitering (passing).

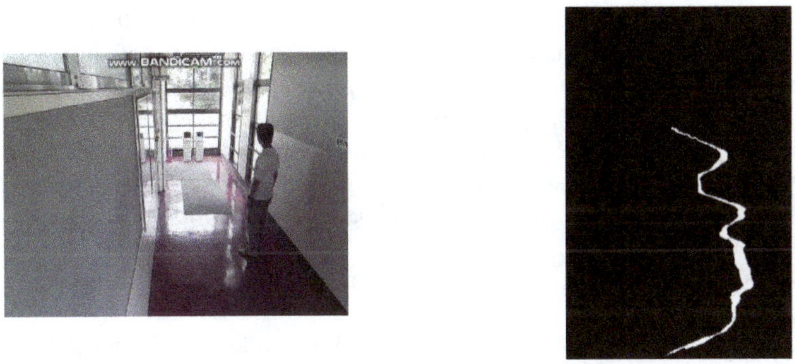

(a) (b)

Figure 15: Loitering misjudged as not loitering. (a) Original image; and (b) Spatiotemporal image.

The reason for this misjudgement is determination of the threshold value. The learning scene group includes a passing person who stays for a long time.

Fig. 16 shows the staying time distribution used for evaluation.

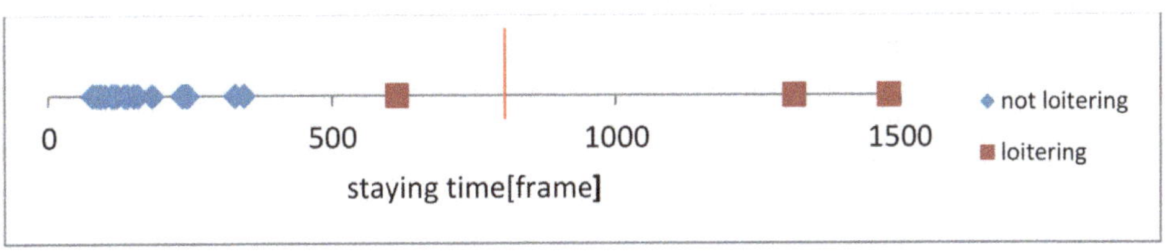

Figure 16: Staying time distribution for evaluation.

There is a "loitering" point in the not loitering zone. The distribution depends on the number of scenes. To determine appropriate threshold values, many scenes are needed.

5 CONCLUSION

The authors proposed a method of detecting loitering behaviour automatically by image processing, to positively guide a person to enter a room in an educational establishment. Image processing has high computational complexity, and the complexity is decreased by using spatiotemporal image processing. The proposed method greatly improves the processing speed.

This loitering detection method used only the staying time as a characteristic value. Behaviour was classified as "loitering" and "not loitering" only, and the authors showed that it was possible to detect the correct behaviour with a simple process.

In our future research, the authors will add other characteristic values, for example, the location and area of a person. In addition, the authors will expand the method to track any number of people by assigning different labels to each track, the authors will increase the number of applicable scenes (by capturing more sample videos), employ machine learning, and so on.

REFERENCES

[1] Tateno, J., Katori, T. & Izumi, T., Wandering behavior detection in front of the door using time-spatial image. *I.E.E. of Japan, 2018*, 3-040, pp. 61–62, 2018. (In Japanese.)

[2] Yamamoto, M., Measurement, recognition and surveillance of human gesture. *Information Processing Society of Japan, Computer Vision and Image Media (CVIM) 2005,* (18(2004-CVIM-148)), pp. 59–66, 2005.

[3] Tanigaki, N., Shimada, N. & Shirai, Y., Extraction of hand and finger regions, and Recognition of Sign Language words. *The Institute of Electronics, Information and Communication Engineers (IEICE), Well-being Information Technology (WIT)*, **101**(264), pp. 37–42, 2001.

[4] Ukai, M., Abnormal statement detection system using image processing. *Railway and Electrical Engineering*, **19**(8), pp. 9–13, 2008.

[5] Zin, T., Tin, P., Toriu, T. & Hama, H., A Markov random walk model for loitering people detection. *IEEE, 2010 Sixth International Conference on Intelligent Information Hiding and Multimedia Signal Processing*, pp. 680–683, 2010.

[6] Bird, N., Masoud, O., Papanikolopoulos, N. & Isaacs, A., Detection of loitering individuals in public transportation areas. *IEEE Transactions on Intelligent Transportation Systems*, **6**(2), pp. 167–177, 2005.

[7] Min, C., Zhang, J., Xu, H., Miao, Z. & Liu, F., A method of video loitering detection based on dynamic programming. *IEEE, 2012 Symposium on Photonics and Optoelectronics,* 2012.

[8] Yann, R. & Bouthemy, P., Real-time tracking of moving persons by exploiting spatio-temporal image slices. *IEEE Transactions on Pattern Analysis and Machine Intelligence*, **22**(8), pp. 797–808, 2000.

FUNDAMENTAL STUDY OF AN ADAPTIVE EVACUATION GUIDE SIGN BY USING A DISCRETE ELEMENT METHOD-BASED SIMULATION

HIROSUKE HORII[1], AKIHITO ITO[2], NOBUTAKA TSUJIUCHI[2], TAKAHIRO FUKUDA[2] & YUSUKE KASAHARA[2]
[1]Kokushikan University, Japan
[2]Doshisha University, Japan

ABSTRACT

In this study, an adaptive evacuation guide sign is proposed, and a fundamental evaluation is performed by using a discrete element method (DEM)-based simulation. When a disaster occurs in a large, closed space that has many emergency exits, such as a building or an underground area, having an evacuation route guide to an emergency exit that avoids intensive congestion of evacuees is important. The adaptive evacuation guide sign monitors the situations of the disaster and the evacuation which changes frequently and controls the direction of evacuation. A simulation model is constructed by using a DEM-based simulation which introduces pseudo-spring and pseudo-radius to interactions among discrete elements for representing evacuees' behaviour. The pseudo-spring and the pseudo-radius are representing the personal space of evacuees and they affect interaction among evacuees during evacuation. At a fundamental evaluation, a limited area that has six emergency exits with three adaptive evacuation guide signs is modelled for evaluating efficiency of the guide sign.

Keywords: crowd behaviour simulation, evacuation simulation, discrete element method.

1 INTRODUCTION

When constructing a large architecture such as a building or an underground area is planned, evacuation plan is pre-examined to respond to disaster by investigating efficiency of an evacuation route or safety of a floor plan. But pre-examined evacuation plan is not enough to respond to the disaster, since the situations on evacuation route are changing dynamically such as blocking by obstacles or congesting by concentration of evacuees. In this study, we propose an adaptive evacuation guide sign which monitors situations of the disaster and the evacuation situations and controls the direction of evacuation dynamically.

A numerical simulation is applied to evaluate proposed system, since examination at actual large architecture is difficult. Investigation of occurrence mechanisms of panic and congestion at evacuation studied by Helbing et al. was one of the early studies of numerical simulation for analysing evacuation behaviour [1]. Since then, various computational approaches such as cellular automata or the lattice gas model have been investigated and applied to planning and evaluating actual evacuation plans such as evacuating at high-rise building [2] or optimizing a vessel layout [3].

Though the discrete element method (DEM) [4] has been mainly used in the field of geotechnical engineering or structural engineering, it has been extended for analysing human crowd behaviour by introducing human behavioural factors so far [5]. In this study, a DEM-based simulation is applied since the features of the DEM is appropriate for our investigation. The features of the DEM are: (a) solving and tracking position of each discrete element and (b) enabling quantitative evaluation of force transfer among discrete elements.

Firstly, simulation model by the DEM and introduced human behavioural factors are described. Then the system and the algorithm of proposed adaptive evacuation guide sign are described. Finally computational results are discussed.

WIT Transactions on Engineering Sciences, Vol 125, © 2019 WIT Press
www.witpress.com, ISSN 1743-3533 (on-line)
doi:10.2495/CMEM190141

2 SIMULATION MODEL

Evacuation simulation model is constructed by using an extended DEM which introduced a pseudo-spring and a pseudo-radius. The pseudo-spring and the pseudo-radius represent personal space of each evacuee.

2.1 Discrete element method

In order to grasp complicated situation at evacuation quantitatively, a DEM-based simulation is employed. The DEM represents each evacuee as a discrete element, and then behaviour of each evacuee is observed as a movement of the element. Force transfer among evacuees in contact is represented by spring and dashpot at contact point shown in Fig. 1. Position of each element at every moment is solved and tracked by an independent equation of motion for each element. The equation of motion of p-th element which constitutes a crowd of evacuees is represented as follows:

$$m_p \ddot{x}_p(t) = f_p^x(t), \tag{1}$$

$$m_p \ddot{y}_p(t) = f_p^y(t). \tag{2}$$

Here, m_p, x_p and y_p represent mass, displacement of x and y direction of the p-th element. f_p^x and f_p^y represent resultant force of x and y direction affecting to the p-th element respectively. The resultant force consists of resistant force of elastic spring of elastic coefficient *k*, resistant force of viscous spring of damping coefficient *c* and individual's thrust force. The individual's thrust force is defined as force which receives until reaching a target walking velocity.

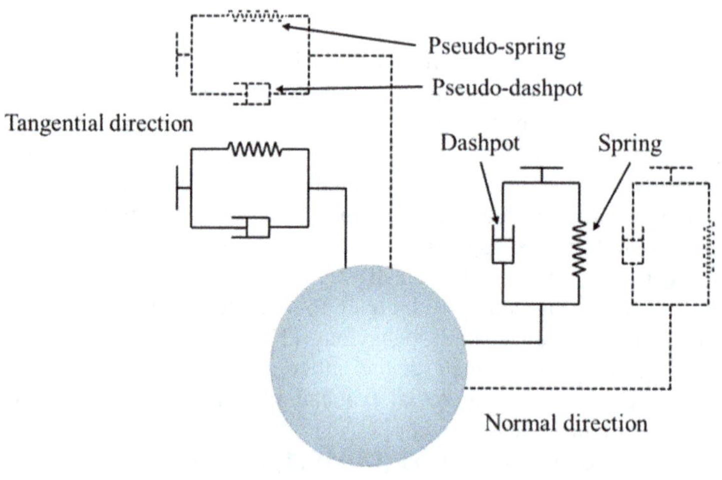

Figure 1: Force transfer among evacuees in a DEM.

2.2 Introducing of pseudo-spring and pseudo-radius

In this study, a pseudo-spring and a pseudo-radius shown in Fig. 2 are introduced for representing personal space which is the distance that a person likes to keep among other people. Before physical contact, a person will try to avoid an obstacle object and keep a certain distance psychologically. The pseudo-radius represents the psychological distance and is set to 0.8 m in this study. That is the spring represents physical reaction and the pseudo-spring represents psychological reaction. A viewing angle represents a person's viewing range and an expression area represents the activation area of the psychological avoidance.

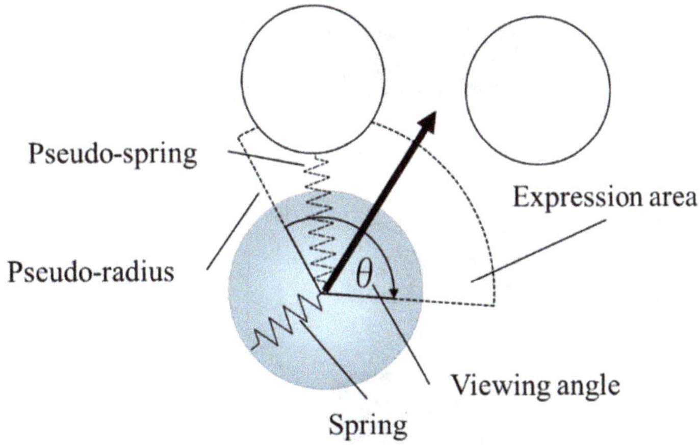

Figure 2: Pseudo-spring and pseudo-radius.

3 ADAPTIVE EVACUATION GUIDE SIGN

3.1 Adaptive evacuation guide algorithm

In this study, we propose an adaptive evacuation guide sign to evacuate smoothly. Though standard evacuation guide sign guides an evacuation direction according to a certain predetermined plan and the evacuation direction is set static, situations on the evacuation route are changing dynamically, such as blocking by obstacles or congesting by concentration of evacuees. The adaptive evacuation guide sign switches the evacuation direction dynamically with observing the situation. Fig. 3 shows an example of switching the evacuation direction. Firstly, the adaptive guide sign indicates a predetermined direction to evacuees as shown in Fig. 3(a). When congestion occurs at an emergency exit as shown in Fig. 3(b), the sign switches to an appropriate direction, the opposite direction here.

The adaptive evacuation guide sign proposing in this study observes the situations and reflects on switching of direction indication. The algorithm sets a certain threshold for the number of evacuees on evacuation routes and switches the directions to prevent the congestions when exceeding the threshold.

3.2 Recognition area of adaptive evacuation guide sign

In this study, recognition area of the adaptive evacuation guide sign is set on crossroads as shown in Fig. 4. Evacuees start evacuation at initial position and when they enter the recognition area, they recognize the guide sign and move to the indicated direction. The recognize rate of the guide sign is set to 100% here and the evacuees obey the direction.

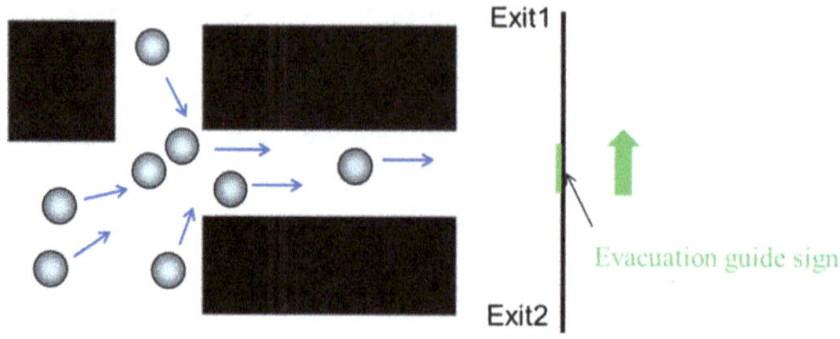

(a)

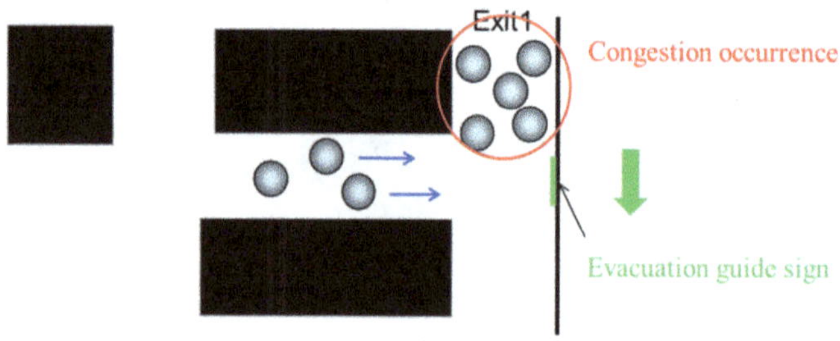

(b)

Figure 3: An example of the switching of evacuation direction indication. (a) Indication of predetermined direction at initial condition; and (b) Switching to the opposite direction when congestion occurred.

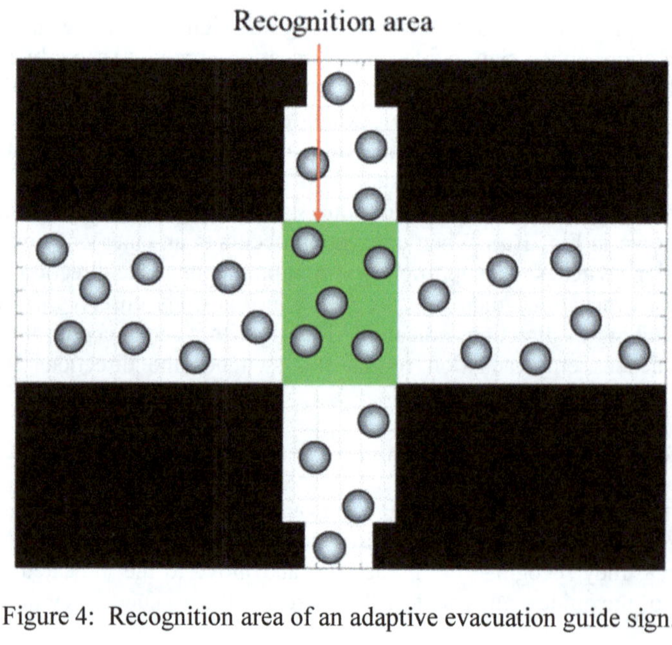

Figure 4: Recognition area of an adaptive evacuation guide sign.

3.3 Investigations of criteria for detecting congestion for switching evacuation direction

For implementing the algorithm described the above, we investigate the threshold of the number of evacuees on evacuation routes as a criterion for detecting congestion for switching the evacuation direction. The investigations are carried out by using a crossroad as shown in Fig. 5. Range of the main passageway of the horizontal direction is 13 m distance and 3 m width. Range of the sub-passageway to exits is 10 m distance and 2.5 m width, and the exits are 1.3 m width. 75 evacuees are set at initial position on the passageway. The investigations are carried out by changing the threshold in the range of 1 to 30 and comparing with a fixed direction case as a standard evacuation guide sign. The other parameters are shown in Table 1.

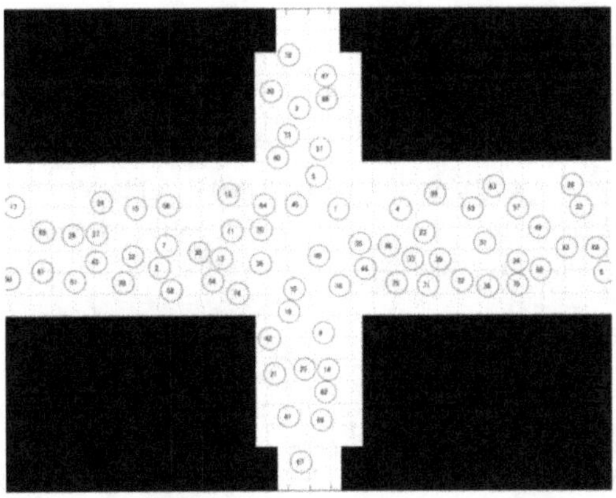

Figure 5: Schematic of passageway for investigation.

Table 1: Parameters of simulation.

Radius of particle (evacuee): r(m)	2.15e-1
Pseudo-radius: Ah(m)	8.00e-1
Mass: m(kg)	6.0e1
Moving velocity: v(m/s)	0.8-1.8
Individual's thrust force: Fg(N)	1.74e2
Time step: Δt(s)	1.0e-2
Viewing angle: θ(deg.)	100
Spring constant (Normal, Tangential): k^u, k^v(N/m)	9.18e3, 4.09e2
Damping constant (Normal, Tangential): c^u, c^v(Ns/m)	1.10e3, 2.47e2
Pseudo-spring constant (Normal, Tangential): k'^u, k'^v(N/m)	4.50e2, 2.71e0
Pseudo-damping constant (Normal, Tangential): c'^u, c'^v(Ns/m)	3.39e0, 2.63e1

Ten trials of each condition are carried out by changing the initial position of the evacuees. Fig. 6 shows the mean and the standard deviation of completion time of evacuation at 10 trials of each condition. The mean and the standard deviation at standard guide sign are shown on the left side. According to the computational results, 16 of the thresholds is the most efficient value for preventing congestion and shorten the evacuation time. Thus we employ 16 of the thresholds for evacuation simulation at below section.

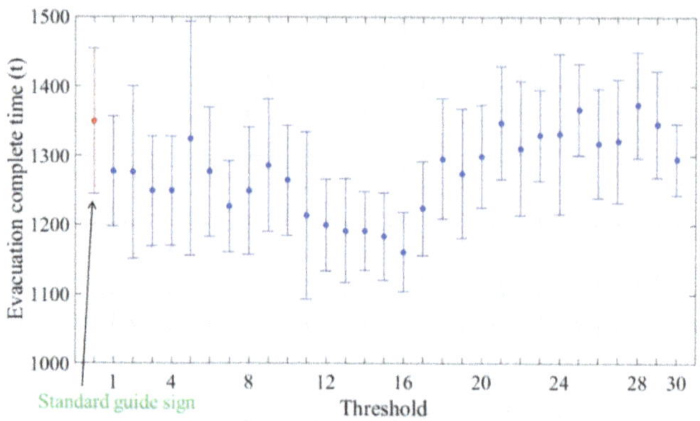

Figure 6: Mean and standard deviation of completion time of evacuation.

4 EVACUATION SIMULATION

In this section, evacuation simulations by using the adaptive evacuation guide sign described the above are carried out for investigating extension to massively evacuation simulation.

4.1 Simulation conditions

Schematic of passageway of the simulation is shown in Fig. 7. Main passageway of 28 m distance and 3 m width is set along with horizontal direction. Three sub-passageways of 10 m distance and 2.5 m width are set crossing to the main passageway and connecting to exits of 1.3 m width. Initial positions of 250 evacuees are set on the passageway at random as shown in Fig. 8. The other parameters are set as same as the above section on Table 1. The evacuation simulations are carried out with two guide policies. The two guide policies are (A) fixed evacuation direction of the standard guide sign and (B) variable evacuation direction of the adaptive guide sign proposed in this study. In policy A, each evacuee goes to the nearest exit. In policy B, each evacuee obeys the direction indicated by the adaptive guide sign. If the number of evacuees near an exit is less than the threshold, then the adaptive guide sign indicates both directions and every evacuee goes to the nearest exit respectively. And if the number exceeds the threshold, then the guide sign indicates to the opposite direction or more vacant direction.

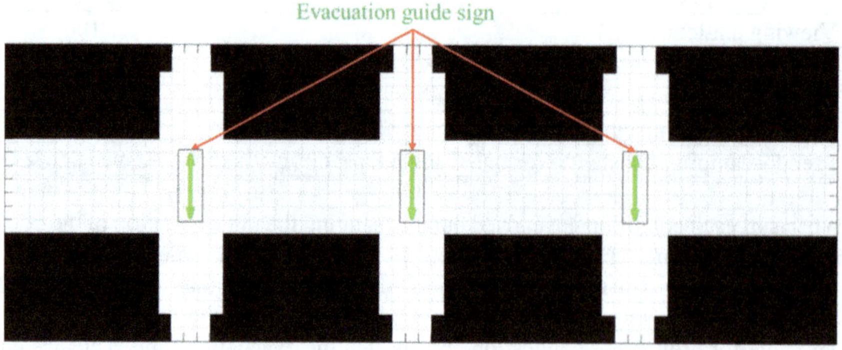

Figure 7: Schematic of passageway of evacuation simulation.

WIT Transactions on Engineering Sciences, Vol 125, © 2019 WIT Press
www.witpress.com, ISSN 1743-3533 (on-line)

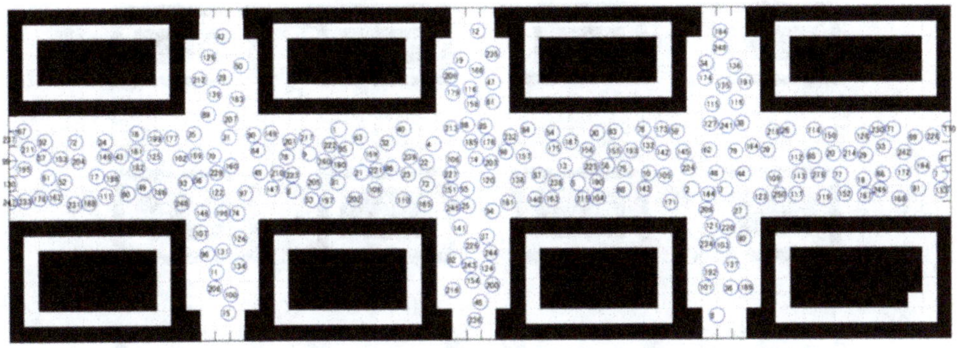

Figure 8: Initial position of evacuees.

4.2 Results and discussion

Completion time of evacuation at each exit is shown in Table 2 and Fig. 9 shows an evacuation situation at time of t = 6 sec. When we focus on evacuees on places circled in red, congestions were occurred at sub-passageways near exits by biasing to one side at policy A as shown in Fig. 9(a). On the other hand, the biases were disappeared at policy B as shown in Fig. 9(b), and the completion time of evacuation was shortened. However a problem has remained at the policy B. Though congestion was disappeared at centre sub-passageway, congestions were still occurred at both left and right sub-passageways. In this case, each evacuation guide sign observes the situation around itself and assigns evacuees to upper and lower exits in charge of guide independently. But if the guide signs share their observed overall situations and guide to another vacant exits in cooperate such as evacuating from left and right sub-passageways to centre sub-passageway, optimum evacuation guide will be realized. And further investigation and improvement of this point is necessary for applying this system in massively area.

Table 2: Completion time of evacuation at each exit.

<table>
<tr><th rowspan="2">Exit name</th><th colspan="2">Policy A</th><th colspan="2">Policy B</th></tr>
<tr><th>Completion time (sec.)</th><th>Difference (sec.)</th><th>Completion time (sec.)</th><th>Difference (sec.)</th></tr>
<tr><td>Exit 1 (upper left)</td><td>10.83</td><td rowspan="2">3.54</td><td>12.12</td><td rowspan="2">0.25</td></tr>
<tr><td>Exit 2 (lower left)</td><td>14.37</td><td>11.87</td></tr>
<tr><td>Exit 3 (upper centre)</td><td>7.31</td><td rowspan="2">3.52</td><td>9.36</td><td rowspan="2">0.7</td></tr>
<tr><td>Exit 4 (lower centre)</td><td>10.83</td><td>10.06</td></tr>
<tr><td>Exit 5 (upper right)</td><td>10.33</td><td rowspan="2">2.44</td><td>11.21</td><td rowspan="2">0.36</td></tr>
<tr><td>Exit 6 (lower right)</td><td>12.77</td><td>11.57</td></tr>
</table>

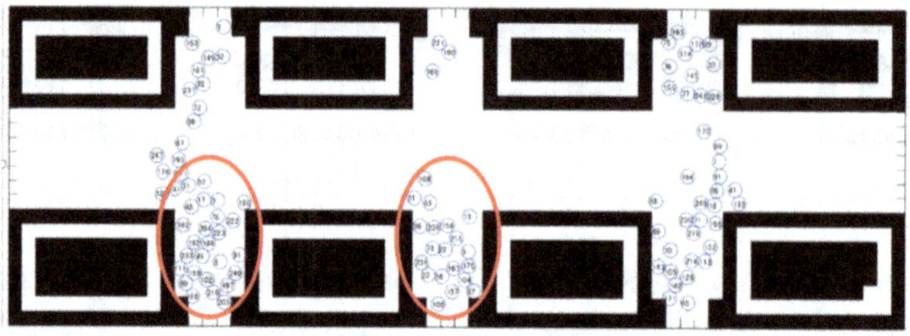

(a)

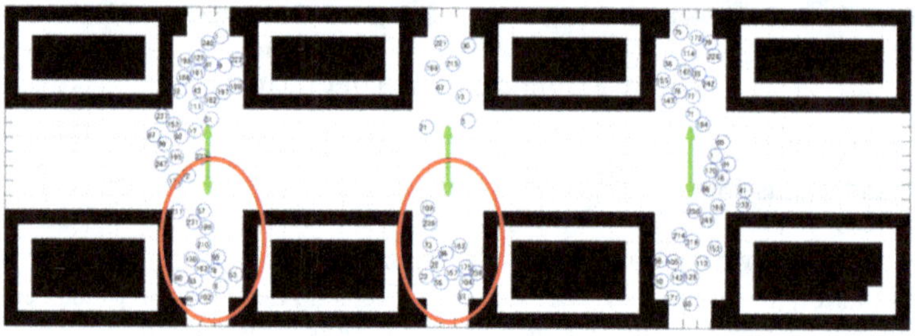

(b)

Figure 9: Evacuation situation at time of t = 6 sec. (a) Policy A; and (b) Policy B.

5 CONCLUDING SUMMARY

In this study, an adaptive evacuation guide sign was proposed, and a fundamental evaluation was performed by using a DEM-based simulation. Firstly, the number of evacuees on evacuation routes as a criterion for detecting congestion for switching the evacuation direction was investigated. And then, a fundamental investigation was carried out by using a limited area model. As a result, although the effectiveness of the adaptive evacuation guide sign was shown, in addition to that, the necessity of cooperation among them was understood at applying to massively structures in future.

ACKNOWLEDGEMENT

This work was partly supported by a grant from the Suzuki Foundation.

REFERENCES

[1] Helbing, D., Farkas, I. & Vicsek, T., Simulating dynamical features of escape panic. *Nature*, **407**, pp. 487–490, 2000.

[2] Kinsey, M.J., Galea, E.R. & Lawrence, P.J., Investigating the use of elevators for high-rise building evacuation through computer simulation. *Proceedings of Human Behaviour in Fire Symposium 2009*, pp. 85–96, 2009.

[3] Deere, S.J., Galea, E.R. & Lawrence, P.J., Optimising vessel layout using human factors simulation. *Proceedings of Pedestrian and Evacuation Dynamics 2008*, pp 597–603, 2010.

[4] Cundall, P.A. & Strack, O.D.L., A discrete numerical model for granular assembles. *Geotechnique*, **29**, pp. 47–65, 1979.

[5] Lim, S.W.H. & Lim, E.W.C., Human Evacuation Modeling. *Discrete Event Simulations - Development and Applications*, IntechOpen, pp. 135–142, 2012.

EFFECT OF CARBON NANOTUBE GRAFTING ONTO CARBON FIBRE ON THE CONDUCTIVITY OF CARBON FIBRE-REINFORCED THERMOPLASTIC LAMINATES

KAZUTO TANAKA, KEN UZUMASA & TSUTAO KATAYAMA
Department of Biomedical Engineering, Doshisha University, Japan

ABSTRACT

Carbon fibre reinforced thermoplastics (CFRTP) can be expected to be used as a structural material for aircraft and automobiles, due to not only mechanical properties such as high specific strength and high specific rigidity, but also for their high recyclability and short moulding time. When lightning hits directly upon an airplane body made from CFRTP, the lightning current does not spread isotropically, which is different from a conductor, causing exterior damage and malfunction of the internal electronic equipment. In recent years, carbon nanotubes (CNTs) were expected to improve the mechanical properties and conductivity of composite materials, due to their excellent mechanical properties, conductivity and high aspect ratio. In this study, we evaluated the effect of CNT grafting onto carbon fibres, on the conductivity of CFRTP. CNT-grafted carbon fibres showed lower resistance values in the *X* and *Y* directions compared to carbon fibre alone. Although the resistance of CFRTP laminates using CNT-grafted carbon fibre in the *X* direction had the same value of CFRTP laminates using control as-received carbon fibre, we found that the resistance of CFRTP laminates using CNT-grafted carbon fibre in both the *Y* and *Z* directions had lower values.

Keywords: carbon fibre reinforced thermoplastics, carbon nanotube, carbon nanotube grafted carbon fibre, conductivity, material properties, reinforced thermoplastics, resistance, resistivity.

1 INTRODUCTION

In recent years, weight reduction of automobiles and aircraft is required, in order to follow regulations on the reduction of carbon dioxide emissions. Carbon fibre reinforced plastics (CFRPs) are expected to be of use for structural parts, due to their mechanical properties that include high stiffness, high strength and low density. Within CFRPs, the carbon fibre reinforced thermoplastics (CFRTP) are especially expected to be applied to mass-produced products in automobiles and aircraft, from the point of view that CFRTP uses thermoplastics which are superior in both recyclability and productivity [1]. Although CFRTP has superior mechanical properties, it should have anisotropy of having not only mechanical properties, but also the right amount of electrical conductivity [2]. When lightning hits an airplane body made from CFRTP directly, the lightning current does not spread isotropically, which is different from a conductor causing exterior damage and malfunction of the internal electronic equipment [3].

Carbon nanotubes (CNTs) are expected to improve the mechanical properties and the conductivity of composite materials, due to their excellent mechanical properties, conductivity and high aspect ratio. One of the issues in using CNT is the difficulty of CNT dispersion within the matrix of CFRTP. In recent years, CNT grafting methods onto carbon fibre have been developed [4]. By using CNT-grafted carbon fibre for reinforcement, the CNTs are automatically dispersed within the matrix after placing the reinforcement.

In this study, we evaluated the effect of CNT grafting onto carbon fibre, on the conductivity of spread carbon fibre and CFRTP laminates.

WIT Transactions on Engineering Sciences, Vol 125, © 2019 WIT Press
www.witpress.com, ISSN 1743-3533 (on-line)
doi:10.2495/CMEM190151

2 MATERIALS AND EXPERIMENTAL PROCEDURES

2.1 Materials and CNT grafting

Spread PAN-based carbon fibres of 24K with the width of 23 mm (Nippon Tokushu Fabric, hereafter referred to as the CF) were used in this study. Nickel (Ni) is used as the catalyst for carbon nanotubes in a chemical vapor deposition (CVD) process that carries it onto the surface of the study's carbon fibres: an electrolytic Ni plating method to make Ni-plated carbon fibre (Ni-CF). The plating bath components were Ni sulfate hexahydrate (240 g/L), Ni chloride hexahydrate (45 g/L) and boracic acid (30 g/L). The spread fibres were connected on one end of the electrode and a Ni plate was connected to the other end: they were then dipped in the plating bath, as shown in Fig. 1. The plating time was set for 15 s and the current was set for 0.3 A.

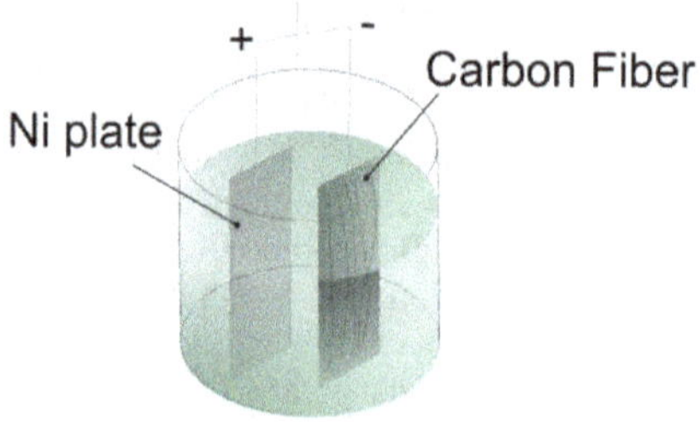

Figure 1: Schematic drawing of Nickel (Ni) plating method.

The CVD system (MPCVD-70, Microphase) shown in Fig. 2 was used for CNT grafting onto Ni-plated spread carbon fibres. The grafting temperature was set at 600°C, with a flow of ethanol as the carbon source to graft the CNTs, and Argon (Ar) gas was used as the carrier gas. To evaluate the effect of the amount of grafted CNTs on the material's electrical resistance, CNT-grafted carbon fibres were prepared for 10 min, 30 min and 60 min. In this study, these CNT-grafted carbon fibres are referred to as CNT_{10}-CF, CNT_{30}-CF and CNT_{60}-CF, respectively.

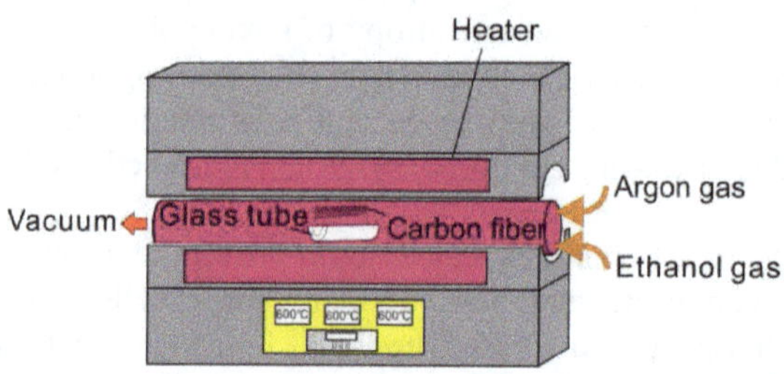

Figure 2: Schematic drawing of the chemical vapor deposition apparatus.

The morphology of the surface and a cross-section of spread carbon fibres were examined by scanning electron microscope (SEM: JSM-6390LT, JEOL) and field emission scanning electron microscope (FE-SEM: JSM-7500FD, JEOL). By using the fibre length and the binary image of the SEM image of CNT-CF, we calculated the apparent outer diameter of CNT-CF.

2.2 Moulding of CFRTP laminate

In order to clarify the effect of CNT on electrical resistance, we used the CF, CNT_{10}-CF, CNT_{30}-CF and CNT_{60}-CF for the reinforcement of CFRTP. Non-woven fabric (50 g/m^2, Prototype, Kuraray) manufactured from polyamide 6 (1015B, Ube Industries) was used for the matrix of CFRTP. CFRTP ($[0°]_{24}$ with the volume fraction (V_f) of 45%) was moulded by using a stripe-shaped mould, attached to a universal precision testing machine (Autograph, AG-250 kN, Shimadzu), as shown in Fig. 3. The moulding conditions were: 250°C for moulding temperature, 3 MPa for moulding pressure and 600 s for the press holding time. Hereafter, CFRTP moulded by using CF, CNT_{10}-CF, CNT_{30}-CF and CNT_{60}-CF as the reinforcing fibre are referred to as CFRTP, CNT_{10}-CFRTP, CNT_{30}-CFRTP and CNT_{60}-CFRTP, respectively. The cross sections of these laminates were polished using a cross section polisher (CP, SM-09010, JEOL Ltd.) at a voltage of 5.2 kV for a polishing time of 102 h and observed by using FE-SEM (SU8020, Hitachi High-Technologies Corporation).

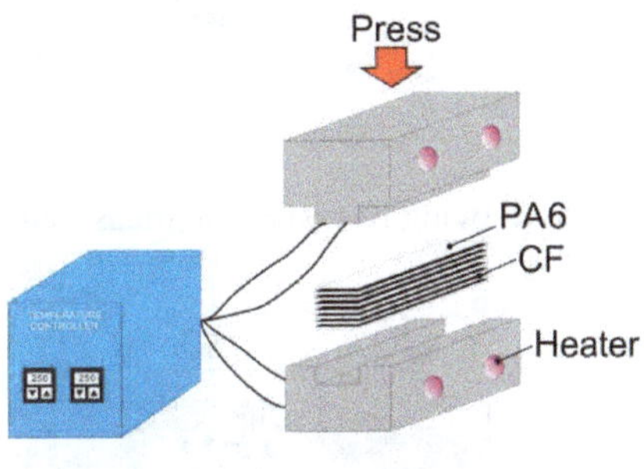

Figure 3: Mould of laminated plate.

2.3 Resistance measurement of spread carbon fibres

The electric resistance value of each spread carbon fibre in the fibre longitudinal direction (*X* direction) and transverse direction (*Y* direction) was measured by the direct current (DC) milli-ohm meter (GOM-802, GW Instek), as shown in Fig. 4. The spread carbon fibres are glued together at both ends with electroconductive resin material (Fujikura Kasei Company, D-550).

2.4 Resistance measurement of CFRTP

We measured the electric resistance value of CFRTP samples cut out from moulded CFRTP laminates with a digital multimeter (HIOKI, DT4282), as shown in Fig. 5. The measured surfaces were coated by electroconductive resin material. As shown in Fig. 6, we measured the electric resistance values in the fibres' longitudinal (*X*) direction, transverse (*Y*) direction, and for thickness (*Z*). Each volume resistivity was calculated by using eqn (1).

$$\rho = \frac{RA}{l}. \quad (1)$$

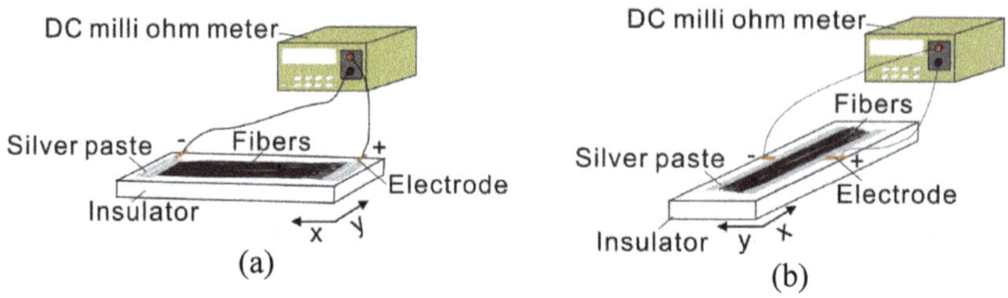

Figure 4: Schematic drawing of resistance measurement of spread carbon fibres. (a) *X* axis direction; (b) *Y* axis direction.

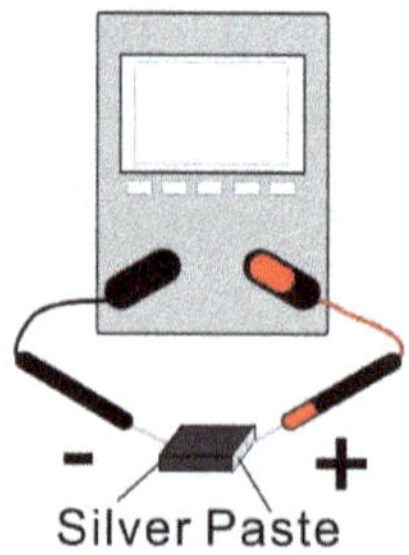

Figure 5: Schematic drawing of resistance measurement of CFRTP.

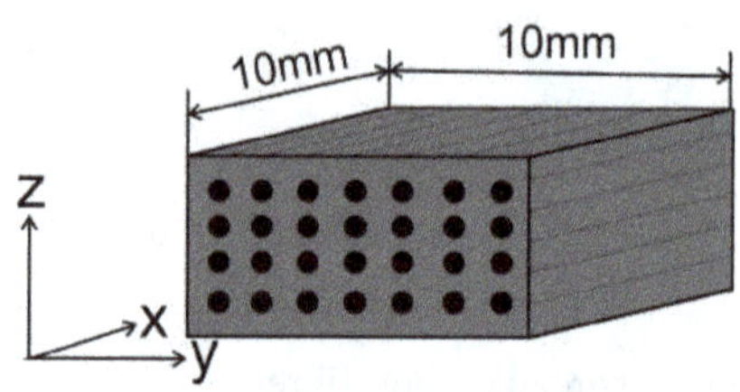

Figure 6: Sample dimension and direction.

3 RESULTS AND DISCUSSION

3.1 SEM observation of carbon fibre surface and cross section

Fig. 7 shows the surface of (a) CF, (b) Ni-CF, (c) CNT_{10}-CF, (d) CNT_{30}-CF and (e) CNT_{60}-CF; as observed by SEM. On the Ni-CF surface, the Ni particles are plated onto the carbon fibre surface [5]. On the surface of CNT_{10}-CF, CNT_{30}-CF and CNT_{60}-CF, the CNTs are grafted and longer CNTs are observed when the grafting duration is longer. A cross-section of each carbon fibre is show in Fig. 8. CNTs are grafted onto the carbon fibre surface and the CNTs grafted on the neighbouring carbon fibres are connected with each other. The CNT-CF with a longer grafting duration has more connected CNTs. The apparent outer diameter of each carbon fibre is shown in Fig. 9. The CNT-CF with longer grafting duration has larger apparent outer diameters.

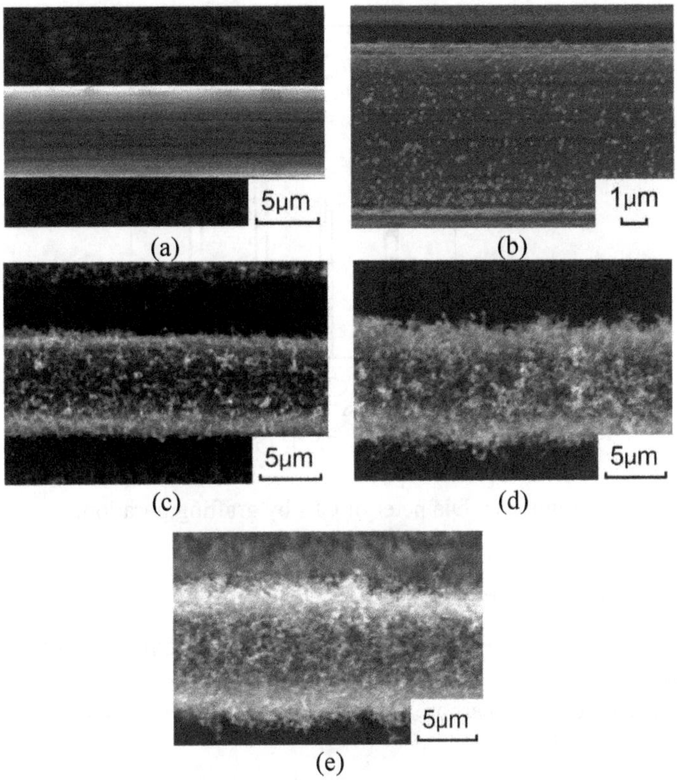

Figure 7: SEM images of each type of CF. (a) Control CF; (b) Ni-CF; (c) CNT_{10}-CF; (d) CNT_{30}-CF; and (e) CNT_{60}-CF. (c)–(e) are our samples, grafted for the different experimental times.

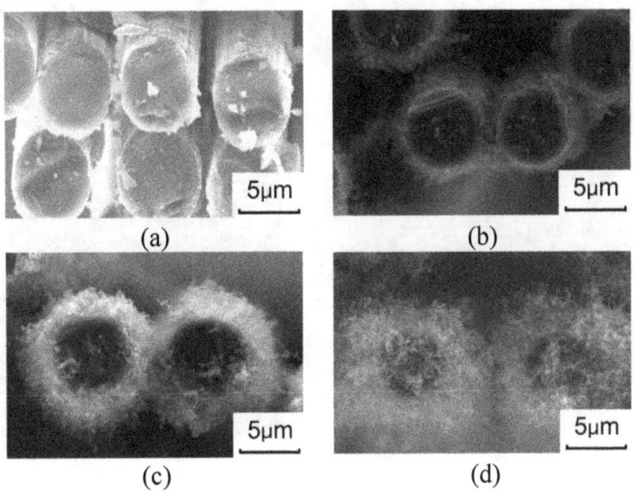

Figure 8: Cross-sections of the experimental CFs. (a) Control CF; (b) CNT_{10}-CF; (c) CNT_{30}-CF; and (d) CNT_{60}-CF. 10 to 60-minute samples.

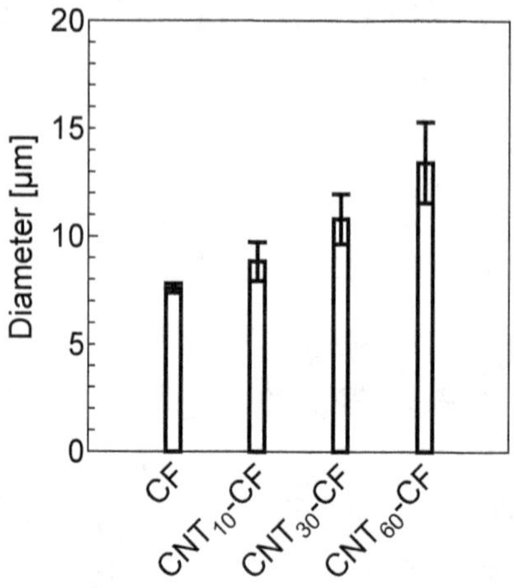

Figure 9: Diameter of CFs by grafting duration.

3.2 SEM observation of cross-section of CFRTP

The cross-sections of CFRTP, CNF_{10}-CFRTP, CNF_{30}-CFRTP and CNF_{60}-CFRTP, which were polished using a cross-section polisher, are shown in Fig. 10. In all laminates, resin was impregnated in between the carbon fibres and the CNTs: no void was observed by the resolution with FE-SEM.

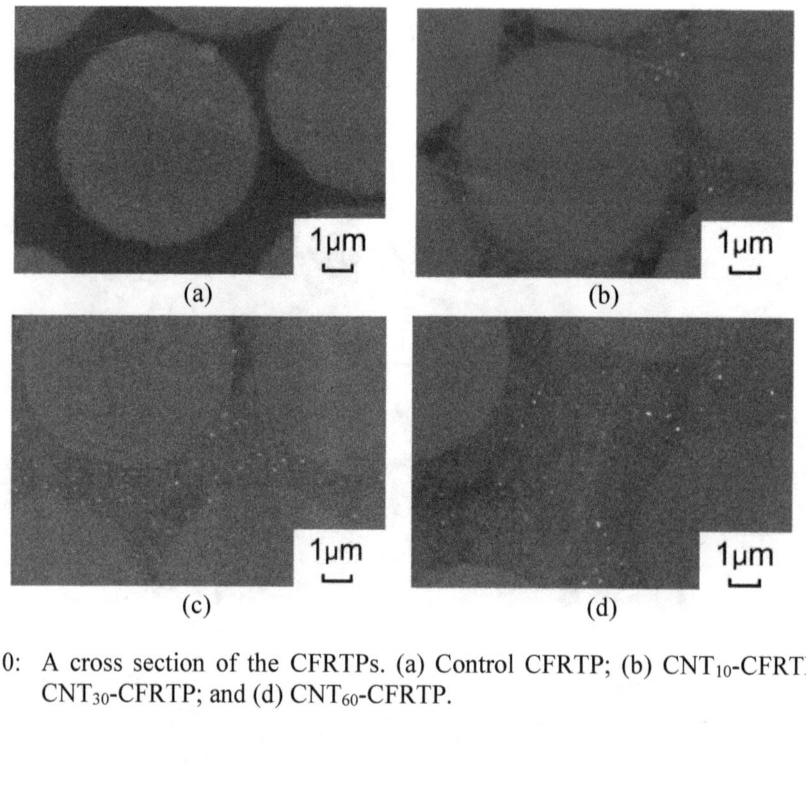

Figure 10: A cross section of the CFRTPs. (a) Control CFRTP; (b) CNT_{10}-CFRTP; (c) CNT_{30}-CFRTP; and (d) CNT_{60}-CFRTP.

Fig. 11 shows the carbon-emphasized FE-SEM images. CNTs grafted onto neighbouring carbon fibres are connected with each other. The white particles in Fig. 11(a) are plated Ni, which exists on both the carbon fibre surface and the tip of the CNTs, as was shown in our previous study [6].

Fig. 12 shows the cross-section in between the 0° layers. A resin rich layer was observed between the 0° layers.

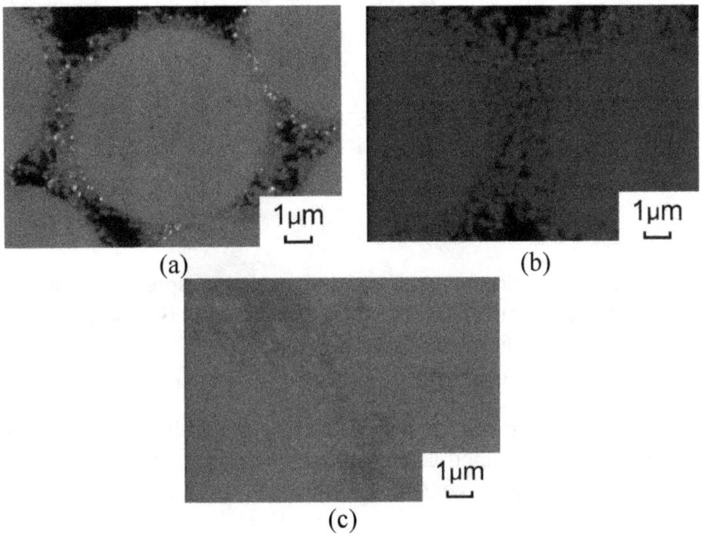

Figure 11: The cross-sections of experimental CNT-CFRTPs. (a) CNT_{10}-CFRTP; (b) CNT_{30}-CFRTP; and (c) CNT_{60}-CFRTP.

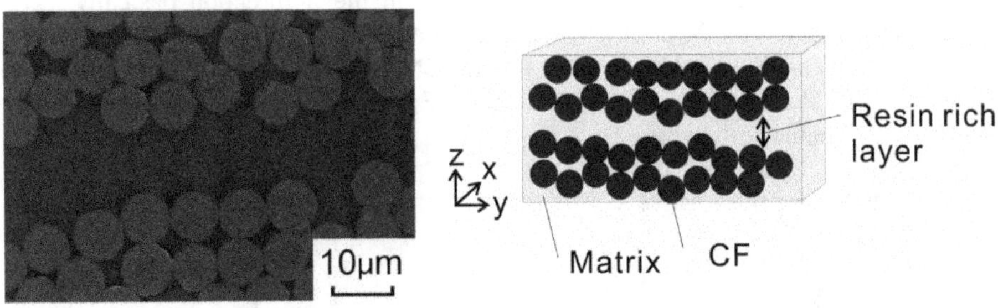

Figure 12: Resin-rich layer in between 0° layers.

3.3 Electrical resistance of spread carbon fibres

Fig. 13 shows the electric resistance value of each spread carbon fibre in the fibres' longitudinal direction (*X* direction) and transverse direction (*Y* direction). We found that the electric resistance value in the *X* direction of CNT_{10}-CF, CNT_{30}-CF and CNT_{60}-CF is lower than that of CF. There is no big difference in the electric resistance value in the *X* direction for CNT_{10}-CF, CNT_{30}-CF and CNT_{60}-CF. In contrast, the electric resistance value in the *Y* direction of CNT_{10}-CF, CNT_{30}-CF and CNT_{60}-CF is lower than that of CF.

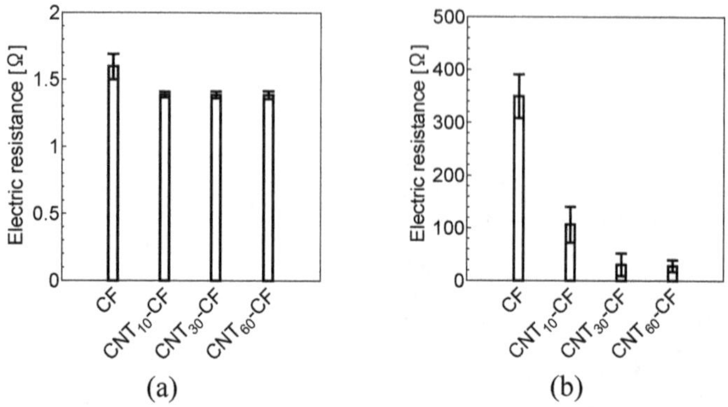

Figure 13: Electrical resistance in the *X* and *Y* direction of spread carbon fibre. (a) Spread carbon fibre measured in the *X* direction; and (b) Spread carbon fibre measured for resistance in the *Y* direction.

The CNT-CF with longer grafting durations tend to show a lower electric resistance value. As shown in Fig. 11, the connected CNTs on neighbouring carbon fibres are considered to contribute towards the making of conductive paths and in giving lower electrical resistance values [7], [8].

3.4 Volume resistivity of CFRTP

The volume resistivities of CFRTP are shown in Fig. 14. The volume resistivities of CFRTP, CNT_{10}-CFRTP, CNT_{30}-CFRTP and CNT_{60}-CFRTP in the *X* direction basically have no significant differences. On the contrary, we found that the volume resistivities in the *Y* and *Z* directions of CNT_{10}-CFRTP, CNT_{30}-CFRTP and CNT_{60}-CFRTP were lower than those of our control CFRTP. The CFRTP that used CNT-CF with a longer grafting duration, tended to show lower volume resistivity.

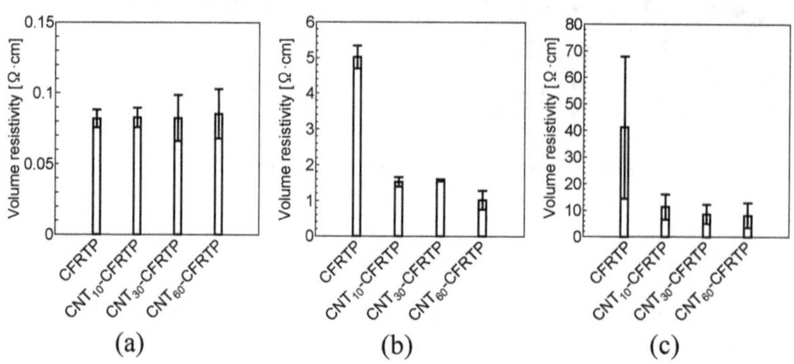

Figure 14: Resistivity of CFRTP. (a) Control and sample measurements in the *X* direction; (b) Measurements in the *Y* direction; and (c) Measurements in the *Z* direction.

Fig. 15 shows the summarized graph of volume resistivity in the X and Y directions. The difference of volume resistivity in the X and Y directions becomes smaller with the larger diameter of carbon fibre by CNT grafting, as shown in Fig. 9. Grafting CNTs onto carbon fibres can contribute to the reduction of the anisotropy of resistivity of the CFRTP laminates.

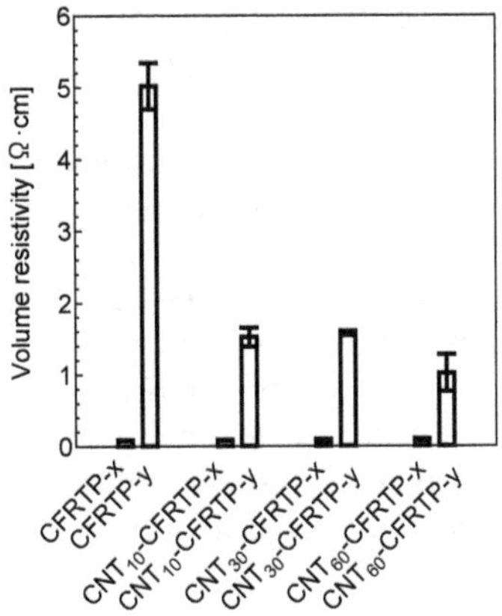

Figure 15: Volume resistivity in the X and Y direction.

4 CONCLUSIONS

The effect of CNT grafting onto carbon fibres on the conductivity of the spread carbon fibres and CFRTP was evaluated. The investigation yielded the following conclusions:

1. The CNT-grafted carbon fibre shows lower resistance values in the X and Y directions, compared to the control as-received carbon fibre.
2. Although the resistance of CFRTP laminates using CNT-grafted carbon fibre in the X direction has the same value of control CFRTP laminates using as-received carbon fibre, the resistance of CFRTP laminates using CNT-grafted carbon fibre in the Y and Z directions have lower values. Thus, grafting CNTs onto carbon fibre can contribute to reducing the anisotropy of resistivity of CFRTP laminates.

ACKNOWLEDGEMENTS

This work was partially supported by KAKENHI (Japan Society for the Promotion of Science), Grant-in-Aid for Scientific Research (B) number 26289011; and also, a research project on "Research and Development Centre for Advanced Composite Materials" through Doshisha University and MEXT (Ministry of Education, Culture, Sports, Science and Technology, Japan), a Supported Program for the Strategic Research Foundation at Private Universities, provided for 2013–2017, project S1311036.

REFERENCES

[1] Marsh, G., Reinforced thermoplastics: The next wave? *Reinforced Plastics*, **58**(4), pp. 24–28, 2014.

[2] Todoroki, A., Suzuki, H., Kobayashi, H., Nakamura, H. & Shimamura, Y., Evaluation of orthotropic electrical resistance for delamination detection of CFRP by electrical potential method. *Transactions of the JSME*, **64**(622), pp. 1654–1659, 1998.

[3] Yamamoto, K., New lightning protection technologies for airplanes using composite materials. *IEEJ Transactions on Power and Energy*, **133**(9), pp. 690–693, 2013.

[4] Tanaka, K., Okumura, Y., Katayama, T. & Morita, Y., Effect of carbon nanotubes deposition form on carbon fiber and polyamide resin interfacial strength. *Journal of the Society of Materials Science*, **65**(8), pp. 586–591, 2016.

[5] Tanaka, K., Uzumasa, K. & Katayama, T., Effect of CNT grafting on carbon fibers on impact properties of CFRTP laminate. *Key Engineering Materials*, **774**, pp. 410–415, 2018.

[6] Tanaka, K., Nishikawa, T., Aoto, K. & Katayama, T., Effect of carbon nanotube deposition time to the surface of carbon fibres on flexural strength of resistance welded carbon fibre reinforced thermoplastics using carbon nanotube grafted carbon fibre as heating element. *Journal of Composites Science*, **3**(1), p. 9, 2019.

[7] Takase, H., Mikata, Y., Matsuda, S. & Murakami, A., Dispersion of carbon nanotubes in a polymer matrix by a twin-screw extruder. *Seikei-Kakou*, **14**(2), pp. 126–131, 2002.

[8] Kumagai, T., Suzuki, W., Yamada, E., Noguchi, T., Iinou, S. & Endo, M., Properties of multiwall carbon nanotubes/polystyrene-block-polybutadiene-block-polystyrene composites. *Journal of Japan Rubber Association*, **83**(8), pp. 258–263, 2010.

SPECIAL SESSION

COMPUTER MODELLING APPLICATIONS FOR CIVIL ENGINEERING

ORGANISED BY
Y. VILLACAMPA

3D MODELLING OF DUNE ECOSYSTEMS USING PHOTOGRAMMETRY FROM REMOTELY PILOTED AIR SYSTEMS SURVEYS

JOSE IGNACIO PAGÁN, ISABEL LÓPEZ, LUIS BAÑÓN & LUIS ARAGONÉS
Department of Civil Engineering, University of Alicante, Spain

ABSTRACT

Dune ecosystems play a key role in coastal dynamics, where a constant exchange of sediments between the sand dune and the beach exists. Thus, it is essential to measure with high precision its movements and monitor its evolution. Classical topographical surveys – even those based on real-time kinematic GPS surveys – are suitable for obtaining profiles with high accuracy, but their cost, in terms of time and personnel, make them inappropriate for large areas. Moreover, these techniques do not precisely measure volumetric changes in high-variable environments such as sand dunes. LiDAR is an active remote sensing method that reflects laser pulses off the land and records topographic positions, but LiDAR scanners are still an expensive and evolving technology. The use of Remotely Piloted Air Systems (RPAS) aims to solve this problem. RPAS map sand dunes using photogrammetric methods, such as Structure-from-Motion (SfM), to produce three-dimensional models from two-dimensional images. This methodology has the advantage of needing only a digital camera and is, therefore, cost effective. The objective of this study is to demonstrate the use of RPAS to map and model the sand dunes of Guardamar del Segura (Spain). The survey covered an extension of 226,000 m^2. The results allow us to obtain a point cloud with a density of 30 points per sq m. A 3D digital elevation model of the dune has been created from the dense point cloud, with a vertical accuracy of 0.2 m Root Mean Square Error. An orthophoto was also created with a spatial resolution of 5 cm. The main conclusion is that the use of RPAS and SfM is an adequate technique to periodically monitor dune ecosystems with high accuracy and cost savings.
Keywords: RPAS, dune, photogrammetry, DSM, SfM, cloud point.

1 INTRODUCTION

The accurate knowledge of the dune-beach ecosystem is essential, since it is a dynamic area where the interaction between terrestrial and marine ecosystems takes place. Moreover, in these areas there is a constant exchange of sediments with the sea. These natural interactions can be altered by human action, modifying coastal dynamics, causing alterations in beaches and dunes, and affecting their stability [1].

Coastal dunes are difficult landforms to study due to the complex interaction between topography, vegetation and the aeolian processes that move the sand throughout the system. To monitor the changes, it is necessary to repeatedly measure the topography over time. Collecting precise data using classical topographical surveys is both labour-intensive and time-consuming, even using modern techniques such as real time kinematic GPS surveys. Furthermore, volumetric measurements become difficult to obtain due to the low resolution of the data. Light detection and ranging (LiDAR) systems deployed on low-level aircraft, have proven to give accurate horizontal and vertical measurements [2]. However, their cost usually relies on government-sponsored flights, affecting the temporal coverage of the available data.

In recent years, the emergence of affordable unmanned aerial vehicles (RPAS, acronym for Remotely Piloted Aircraft System) for civilian use, together with the development of computational techniques based on images, such as Structure from Motion (SfM), has dramatically increased the use of photogrammetry based on RPAS to produce high-resolution

WIT Transactions on Engineering Sciences, Vol 125, © 2019 WIT Press
www.witpress.com, ISSN 1743-3533 (on-line)
doi:10.2495/CMEM190161

digital surface models (DSM) for the study of different surface processes [3]. SfM uses several digital images and automatically solves the geometry of the scene and camera position and orientation by means of an iterative bundle adjustment procedure [4]. SfM, along with multi-view stereo (MVS) algorithms, enables the generation of dense 3D point clouds from a set of digital photos. The use of Ground Control Points (GCPs) with known coordinates enables the projection of the photogrammetric results to a system of absolute coordinates [5]. SfM can be applied to the geomorphological analysis of landforms and their evolution by means of 3D models [6].

Once the data have been obtained, it is necessary to manipulate them in order to analyse spatial characteristics, detect changes and to represent them cartographically. Geographic Information System (GIS) technology allows such analysis [7]. RPAS-based photogrammetry has several advantages in comparison with the methods mentioned before. It can replace the conventional surveys carried out in current coastal monitoring programs, with a considerable reduction in data acquisition costs while maintaining the quality aerial imagery data and increasing the topographic accuracy [6]–[7].

This study is focused on the beaches of Guardamar del Segura, Alicante (Spain), where the continued regression of the shoreline [8] has led to the destruction of buildings located within the Maritime and Terrestrial Public Domain (DPMT) and eroded the dune system. A campaign of flights with a RPAS has been conducted to map and model the sand dunes and the beach. The main aim of this work is to illustrate the procedure of field data collection and the SfM-based photogrammetric modelling performance.

2 FIELD CAMPAIGN AND DATA ACQUISITION

The field campaign and data acquisition occurred on June 2017. The area of study encompasses the coastal area that extends from the north of the town of Guardamar del Segura (Alicante, Spain) to the Segura River mouth. The area of interest is the stretch between the shoreline and the inland road, covering an extension of 226,000 m^2. It forms a rectangle with a length of 1,500 m and an width of 150 m (Fig. 1).

2.1 Image acquisition

The data used in this study were obtained using a DJI Phantom 4 Advanced quadcopter. The RPAS is equipped with a 1" CMOS sensor camera (20 Mpixel of resolution). Its focal length ranges from 8.8 mm to 24 mm. The 5870 mAh battery allows a flight time of aproximately 30 minutes, so the flight planning was developed to cover in 3 passes the entire study area at 60 m above sea level. The latitudinal overlap was 60% and the longitudinal overlap was 75%. The spatial resolution of the images at this altitude is 2 cm per pixel. 1,200 images were recorded. The meteorological conditions were fair, the wind speed was weak and regular, so the flights occurred without any obstacles.

2.2 GNSS survey

GCPs are required for the georeferencing. 150 GCPs were acquired using a Leica Viva GS16 GPS reciever, eqquiped with Real Time Kinematic (RTK) technology used to enhance the precision of position data derived from the GNSS (Fig. 1). A topographic GNSS measurements with 2069 points was also surveyed to compare and contrast the data acquired by RPAS. In order to evaluate vertical and horizontal errors, a total of 531 complementary GNSS measurements were carried out on stable areas of the study site (road and wooden walkways).

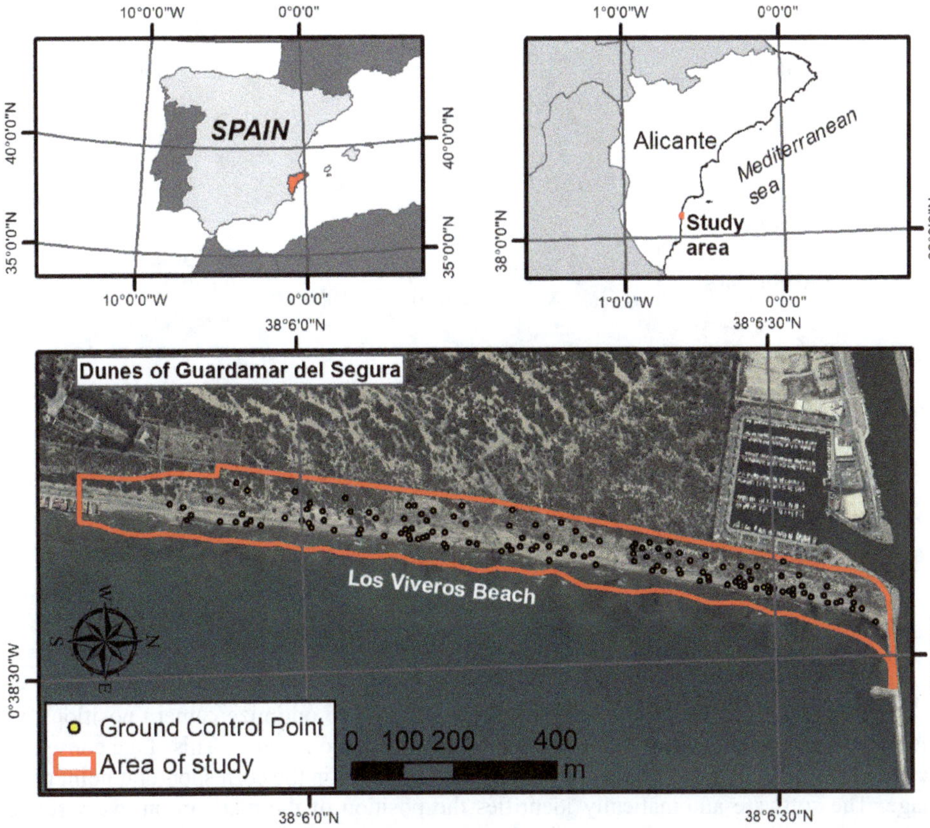

Figure 1: Location of the study area.

3 SFM-BASED PHOTOGRAMMETRIC PROCESS

The photogrammetric process based on SfM algorithm is required for 3D coastal topography reconstruction from 2D RPAS images. The SfM algorithm allows reconstruction of a 3D scene geometry by solving the geometry of the scene, camera positions and orientation, extracted from a set of multiple overlapping images [4]. Recently, several photogrammetric software packages based on the SfM method that bring knowledge from computer vision and traditional photogrammetry together, have been developed and are more dedicated to the processing of RPAS images. One commercial package, Agisoft Photoscan [9], was chosen for this study due to its essentially automatic process, based on multi-view 3D reconstruction technology, and its suitability for RPAS image processing. The general workflow for the RPAS images processing is divided into 8 different steps (Fig. 2).

The first step consists of loading the photos from the RPAS survey and its camera positions, saved in the EXIF metadata of the images. This information contains the coordinates of the GNSS signal received by the RPAS and its orientation. In the second stage, Align photos, PhotoScan finds matching points between overlapping images, estimates the camera position for each photo and builds the sparse point cloud model.

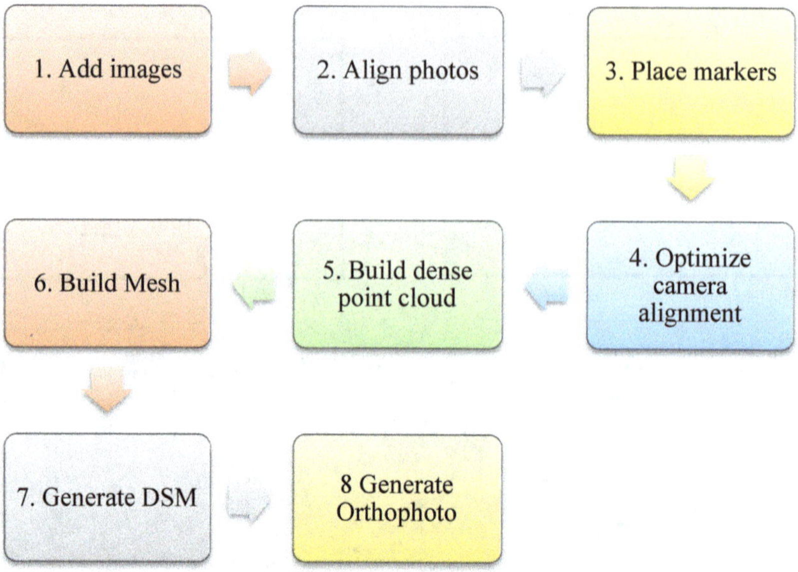

Figure 2: Worflow followed for SfM modelling.

The third stage is placing markers. Markers are used to optimize camera positions and orientation data, which allows for better model reconstruction results. This stage consists of locating the GCPs visible on each photo and place a marker in the corresponding point of the image. The software automatically identifies the position of the markers on every related photo but checking the marker location and refining its position is necessary to provide maximum accuracy. This step is the most time-consuming. Finally, marker coordinates are imported from a file containing the accurate coordinates of the GCPs from the GNSS survey.

The next step is to optimize camera alignment. The position of the images is recalculated considering the established GCPs, so that the result will already be perfectly georeferenced in its XYZ coordinates (Fig. 3).

In stage five the software calculates the depth information for each camera to be combined into a single dense point cloud. After the dense point cloud has been reconstructed it is possible to generate a polygonal mesh model based on the dense cloud data (Fig. 4). Finally, the result output can be exported as a raster DSM and as an orthophoto with a spatial resolution of 5 cm/pixel, both in the desired coordinate system in a GeoTIFF format.

The SfM technique demands high computing capability for processing and viewing. Thus, the area of study was divided into 3 zones coinciding with the 3 flights performed. These zones were processed independently with the aim of not overloading the capacity of processing of the computer used, which had the following characteristics:

- Windows 10 64-bit.
- Intel Core i7-2670QM @ 2.2 GHz 4 cores 8 threads 6 MB cache.
- 8 Gb RAM.
- 250GB SSD disk Samsung 850 EVO.

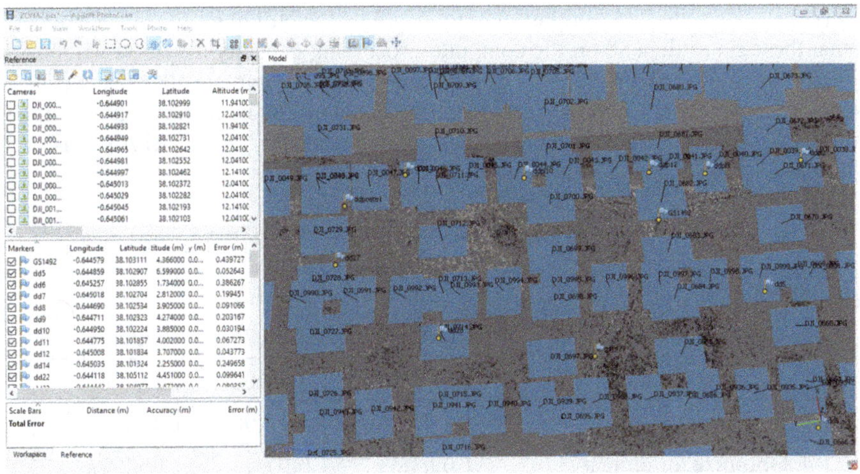

Figure 3: Snapshot with the position of the images and the markers of the GCP identified on the photos. The panel on the left shows the coordinates of the images and the GCPs as well as the errors of the georeferencing process.

Figure 4: 3D model view of the dune-beach with the dense cloud point coloured in RGB.

4 RESULTS

The dune-beach system of Guardamar del Segura has been revealed as an excellent test site for the methodology proposed in this research. The complex dune ecosystem and the adjacent beach have been surveyed using a small, commercial RPAS. Although the area of interest extends over 226,000 m^2, the actual survey covered 286,100 m^2.

The performance of each stage of the SfM process is shown in Table 1. The most time-consuming step is placing markers, since it requires the human visual identification of each GCP on all the images. Due to the high number of images and GCP surveyed (8–12 GCP visible per image), to complete this step an average of 5 h was required. Besides that, aligning photos is the second time-consuming step, followed by the building of the dense point cloud. Note that these times depend on the computer processing capacity available, so they can be improved using more powerful PCs. The total time spent to complete all the steps of the SfM methodology was 33 h. However, considering the commitment of time and manpower to conducting traditional field surveys, SfM photogrammetry from RPAS has a better performance.

Table 1: Performance of the SfM process.

Stage	Zone 1	Zone 2	Zone 3
1 Add photos	0:00:30	0:00:26	0:00:46
2 Align photos	3:13:00	3:05:00	3:08:00
3. Place markers	4:45:00	5:15:00	4:20:00
4. Optimize alignment	0:01:59	0:01:14	0:02:28
5. Build dense point cloud	2:27:00	2:49:00	1:44:00
6. Build mesh	0:27:50	0:39:53	0:22:02
7. Generate DSM	0:05:12	0:08:22	0:04:48
8. Generate orthophoto	0:04:29	0:06:22	0:03:59
Total	11:05:00	12:05:17	9:46:03

The results from each zone are shown in Table 2. After processing the zones independently, the dense cloud point, DSM (Fig. 5) and orthophoto (Fig. 6) were merged into one single dataset for the whole area of interest. A total of 8,335,766 points were obtained from the dense point cloud generated, modelling the surface with a density of 30 points per sq m. A 3D digital surface model of the dune and the beach was created from the dense point cloud (Fig. 5), with an average horizontal accuracy of 0.08 m and a vertical accuracy of 0.2 m RMSE Z. The orthophoto generated from the images captured from the RPAS and the DSM had a spatial resolution of 2.25 cm per pixel, identifying even the footsteps on the sand dune (Fig. 6(d)).

Table 2: Results of the SfM process.

Parameter	Zone 1	Zone 2	Zone 3
Number of images	622	689	548
Flying altitude	61 m	59 m	59 m
Coverage area	98,600 m^2	79,500 m^2	108,000 m^2
Number of GCP	62	50	42
RMSE XY	0.098 m	0.063 m	0.069 m
RMSE Z	0.238 m	0.231 m	0.229 m
Dense Cloud Points	2,792,354	2,306,957	3,236,455
DSM Point density	29.6 points per m^2	32.7 points per m^2	31.8 points per m^2
Ground resolution	2.3 cm/pix	2.19 cm/pix	2.21 cm/pix

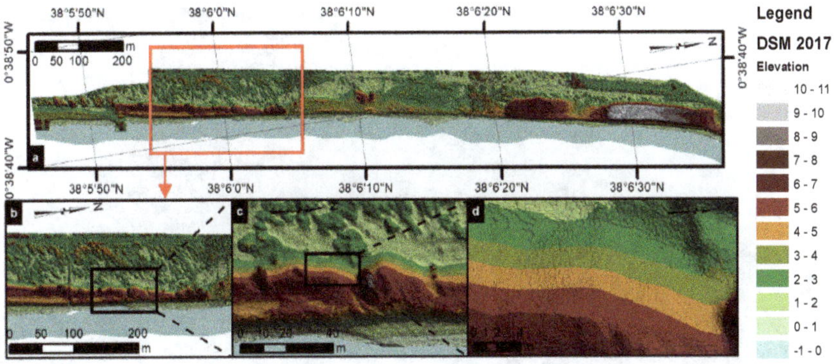

Figure 5: (a) DSM of the area of study; (b) Detail at 1:10000 scale; (c) Detail at 1:2500 scale; and (d) Detail at 1:500 scale.

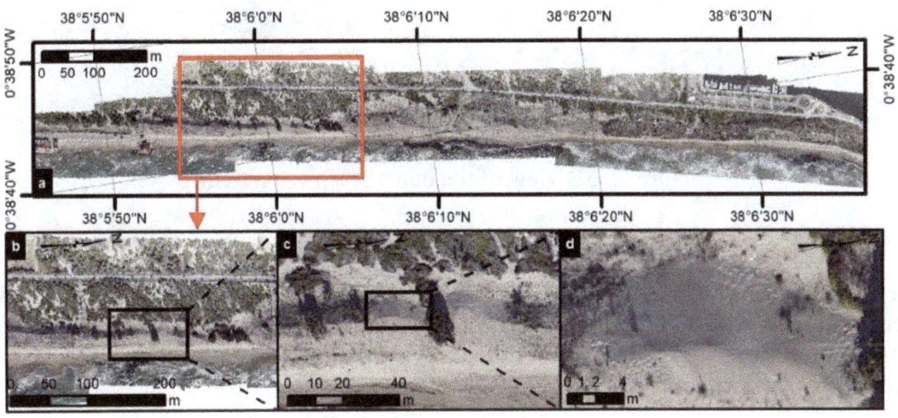

Figure 6: (a) Orthophoto of the area of study; (b) Detail at 1:10000 scale; (c) Detail at 1:2500 scale; and (d) Detail at 1:500 scale.

The dense cloud point can be exported to other software, like Cloud Compare or GIS environments, for better visualization or to compare it with other 3D points clouds available, such as LiDAR datasets. In addition, the DSM and orthophoto can be also exported for the same purpose, incrementing the possibilities of analysis (Fig. 7). For instance, seasonal sediment balance between dune and beach, or analysis of the slopes or dune erosion can be performed.

However, there are some drawbacks to this approach. Firstly, flights regulations can be restrictive on the areas to survey, especially near to populated areas. Secondly, the SfM technique generates a dense point cloud of the surface that can be converted in a Digital Surface Model (DSM), but height can come from the top of buildings, tree canopy, powerlines and other features. However, a bare-earth elevation model is particularly useful in some engineering fields, such as in hydrology, soils and land use analysis. Although there are some techniques to avoid this problem, they usually require the use of Near Infrared (NIR) sensors to better detect and filter vegetation [10], but unfortunately commercial RPAS are not equipped with them.

Figure 7: 3D view of the dune-beach ecosystem.

5 CONCLUSION

The advantages of RPAS together with SfM methodology are clear: a fast, reliable and highly accurate technique that allows to model complex and changing environments such as dune ecosystems with a scarce inversion. Moreover, analysing the evolution of these systems regularly and in detail are proved as possible in this research.

REFERENCES

[1] Aragonés, L., Pagán, J.I., López M.P. & García-Barba, J., The impacts of Segura River (Spain) channelization on the coastal seabed. *Science of The Total Environment,* **543**, pp. 493–504, 2016.

[2] Woolard, J.W. & Colby, J.D., Spatial characterization, resolution, and volumetric change of coastal dunes using airborne LIDAR: Cape Hatteras, North Carolina. *Geomorphology,* **48**(1–3), pp. 269–287, 2002.

[3] Colomina, I. & Molina, P., Unmanned aerial systems for photogrammetry and remote sensing: A review. *Journal of Photogrammetry Remote Sensing,* **92**, pp. 79–97, 2014.

[4] Snavely, N., Seitz, S.M. & Szeliski, R., Modeling the World from Internet Photo Collections. *International Journal of Computer Vision,* **80**(2), pp. 189–210, 2008.

[5] Riquelme, A., Del Soldato, M., Tomás, R., Cano, M., Jordá Bordehore, L. & Moretti, S., Digital landform reconstruction using old and recent open access digital aerial photos. *Geomorphology,* **329**, pp. 206–223, 2019.

[6] Westoby, M.J., Brasington, J., Glasser, N.F., Hambrey, M.J. & Reynolds, J.M., "Structure-from-Motion" photogrammetry: A low-cost, effective tool for geoscience applications. *Geomorphology,* **179**, pp. 300–314, 2012.

[7] Andrews, B.D., Gares, P.A. & Colby, J.D., Techniques for GIS modeling of coastal dunes. *Geomorphology,* **48**(1–3), pp. 289–308, 2002.

[8] Pagán, J.I., López, I., Aragonés, L. & García-Barba, J., The effects of the anthropic actions on the sandy beaches of Guardamar del Segura, Spain. *Science of The Total Environment,* **601–602**, pp. 1364–1377, 2017.

[9] Agisoft, Agisoft PhotoScan Professional Edition, 2016.

[10] Kyratzis, A.C., Skarlatos, D.P., Menexes, G.C., Vamvakousis, V.F. & Katsiotis, A., Assessment of vegetation indices derived by UAV imagery for durum wheat phenotyping under a water limited and heat stressed Mediterranean environment. *Plant Science*, **8**(1114), 2017.

COMPRESSIVE STRENGTH CLASSIFICATION OF LIGHTWEIGHT AGGREGATE CONCRETE USING A SUPPORT VECTOR MACHINE MODEL

ANTONIO JOSÉ TENZA-ABRIL[1], ROSANA SATORRE-CUERDA[2], PATRICIA COMPAÑ-ROSIQUE[2], FRANCISCO JOSÉ NAVARRO-GONZÁLEZ[3] & YOLANDA VILLACAMPA[3]
[1]Department of Civil Engineering, University of Alicante, Spain
[2]Department of Computer Science and Artificial Intelligence, University of Alicante, Spain
[3]Department of Applied Mathematics, University of Alicante, Spain

ABSTRACT

Lightweight aggregates (LWA) are used to produce low-density concretes required for building applications. Lightweight aggregate concrete (LWAC) is a multi-purpose material for construction, which offers technical, economical and environment benefits, and it is produced by replacing the normal-weight aggregates with LWA, depending upon the requirements of density and strength. LWAC is a complex composite material, and a model of its compressive strength must be highly nonlinear because it is very sensitive to its ingredients, so modelling its behaviour is a difficult task. Many studies have tried to develop accurate and effective predictive models for LWAC compressive strength. In this study, a support vector machine (SVM) learning algorithm is used to propose a model to classify the compressive strength of a wide range of LWAC. A dataset of 241 different LWACs were used for classifying the compressive strength into six different classes (from low-strength to high-strength) using different variables – quantity of cement, water and LWA in the dosage and density of the LWAC produced. The results show that increasing the variables means the model becomes more accurate up to approximately an 80% rate of success. The SVM model proved to be a significant tool to classify the compressive strength of LWAC contributing to engineers avoiding costly experimental trial tests.
Keywords: lightweight concrete, aggregate, vibration, classification, modelling.

1 INTRODUCTION

Due to the increasing energy consumption, the requirements for higher building energy efficiency have been widely considered [1], [2]. Insulation materials, such as, cement-based porous materials, including aerated concrete, foamed concrete and lightweight aggregate concrete (LWAC), exhibit superior fire resistance and more desirable insulation properties than the conventional concrete materials [3], [4].

LWAC is an important and versatile material for use in modern construction [5]. The application of lightweight concrete is not only used as an insulation material, it has other actual engineering applications such as large-span bridges, high-rise buildings, tall concrete wind towers [6] and offshore oil platforms because of the lower density, higher strength/weight ratio and better durability properties [7].

Research has been conducted worldwide on a large number of natural and artificial lightweight aggregates used in the manufacture of mortar and concrete. The absorption caused by the lightweight aggregate is mainly responsible for the difficulty during the production of lightweight aggregate concrete (LWAC) in practical situations [8]. LWA are generally porous materials, which tend to absorb water in aqueous [9] and viscous medium [10]. The mix design of LWAC is more complex than a conventional normal-weight concrete because there are more design parameters – such as absorbed water during the mixing of concrete and proportioning of different aggregate sizes. Changing the LWA used in the concrete varies other properties of LWAC like compressive strength and density which is not predictable for all types of lightweight aggregates. For this reason, many different techniques

WIT Transactions on Engineering Sciences, Vol 125, © 2019 WIT Press
www.witpress.com, ISSN 1743-3533 (on-line)
doi:10.2495/CMEM190171

have been used to provide better mixtures design using different types of approaches [11], such as genetic algorithms [5], artificial neural networks [12], [13] or finite elements [14].

There are also difficulties in the placement of the LWAC. During the mixing of LWAC, due to the different densities of its components, LWAC is susceptible to segregation of the aggregates [15]. In fact, during the vibration of the concrete, lightweight aggregates tend to float. Segregated concretes are susceptible to an increased risk of cracking due to the separation of the aggregates from the rest of the mixture reducing their durability [13]. There are several methods to measure the segregation [16], which can be detected by ultrasonic pulse velocity [13] and can be measured through image analysis [17] because the manufacture process can affect it [18]. There are numerical models for describing it [15] and also neural networks have been applied to analyse the phenomenon [13]. The supplementary materials can also affect the segregation [19].

The most important required property of LWC is its compressive strength and its density. The compressive strength of LWC has great influence on its structural performance and is significantly affected by the composition —materials, the dosage of chemical and mineral admixtures, types of aggregate used, packing density and water to binder ratio (W/B) [20].

Support-vector machines (SVMs) are a supervised learning model with associated learning algorithms developed by Cortes and Vapnik [21] used for classification and regression analysis. The technique appeared from the theory of statistical learning postulated by Vapnik and Chervonenkis [22], [23]. SVMs were introduced in 1992 and became popular when they obtained better results than neural networks in handwriting recognition using pixels as input data [21]. The technique transforms the input space into another space of higher dimension, in which the problem can be solved by means of an optimal hyperplane (of maximum margin). It performs extremely complex data transformations and then finds out how to separate the data according to the labels or defined outputs (Fig. 1). The technique has been used to solve different problems in engineering [24]–[26], used in normal-weight concrete [27], [28] as well as in LWAC [29].

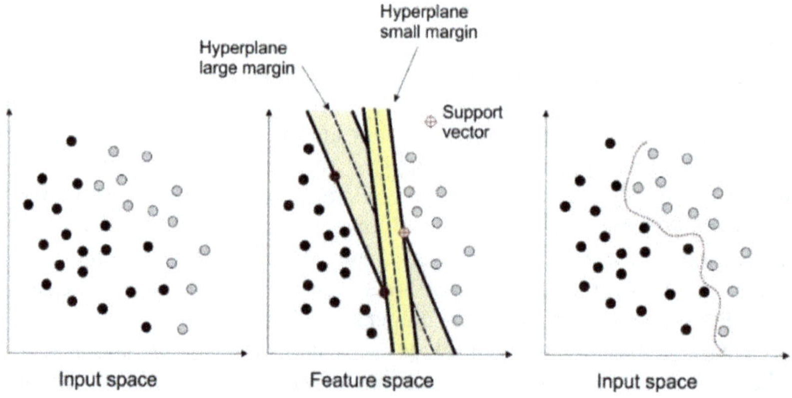

Figure 1: Concept of support vector machine classification. (Source: Figure adapted from Hariri-Ardebili and Pourkamali-Anaraki [24].)

The original proposal using SVM is to classify the points into two categories, but it is also extendable to n categories. In this study the classification of the compressive strength of LWAC using SVM is investigated, the points were classified into 6 different categories corresponding to compressive strength classification.

The main objective using SVM is to take as an input a set of points (parameters of the LWAC design from the literature) in which each of them belongs to one of several possible categories (compressive strength category). An algorithm based on SVM builds a model capable of predicting to which category a new point belongs (using Matlab®).

2 MATERIALS AND METHODOLOGY

2.1 Experimental database

In this research, 241 groups of test data of LWAC from the literature were collected [30]–[38]. The concretes studied used different types of LWA: Argex®, Arlita Leca®, natural pumice stone, Slate®, Rthyolite®, oil palm shell and recycled LWA.

2.1.1 Input variables: Mixture proportions

The goal is to design a classifier from the training data that allows us to predict the classification of compressive strength for a set of materials properties and some quantities for the dosage. The inputs used were:

- LWA dry density: Depending upon the source of material, the dry density of the aggregate from the literature ranges from 573 to 2650 kg/m^3.
- Amount of water: The water used to manufacture the concrete used as data, ranges from 85 to 287 litres per cubic meter. In this study, only the effective water is considered because the water absorption of the LWA varies depending on the aggregate pore structure [39] and the value differs if the test is carried out in water or in the mass of the concrete [10].
- Amount of cement: The LWACs of the study were manufactured with a range of 150 to 640 kg in one cubic meter.
- Lightweight aggregate: from 0 to 863 kg per cubic meter (from normal-weight concrete to LWAC).

As seen in Fig. 2, the LWAC compressive strength does not present a linear trend because the LWAC used in this research had a wide range of results and input parameters (Fig. 2).

2.1.2 Output variable: Compressive strength

The compressive strength is a key issue but it is a random variable that can vary over the time, even in a single structure [40]. To enhance professional practice an accurate and appropriate classification of compressive concrete strength is required [41]. Literature reveals gaps and overlapping in compressive concrete strength classifications. In this study the classification proposed by Sojobi et al. [41] is used because it avoids gaps, vague classification and it is easily understood by the professionals (Table 1).

As can be observed in Fig. 2, the maximum class achieved for lightweight concrete used in the database is a high-strength (81.6 MPa) and the minimum is the low compressive strength (0.6 MPa).

2.1.3 Supporting Vector Machine (SVM)

As explained in the introductory section, a Support Vector Machine (SVM) is a discriminating classifier formally defined by a separation hyperplane. In other words, given the training data label (supervised learning), the algorithm produces an optimal hyperplane that categorizes the new examples. In the case of a two-dimensional space, this hyperplane is a line that divides space into two zones where the samples are located to one side or to the other of the line depending on their class. This is extendable to more dimensions.

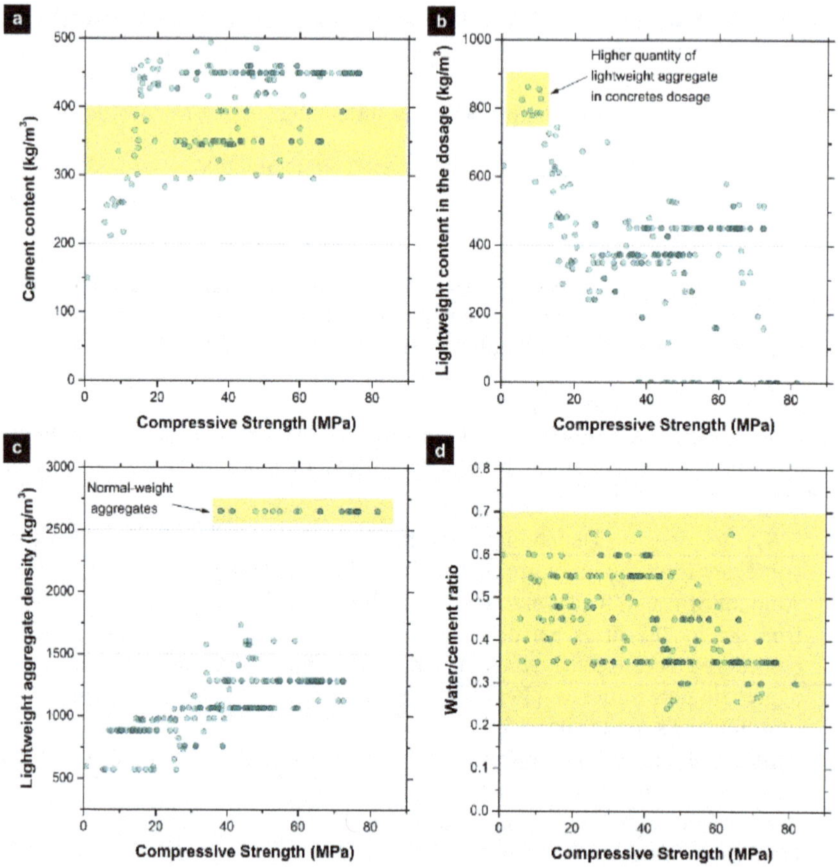

Figure 2: Variation of the properties used for SVM. Compressive strength of the concretes vs (a) Cement content; (b) LWA content; (c) LWA density; (d) W/c ratio.

Table 1: Classification used in the study according to Sojobi et al. [41].

Class	Classification of concrete according to its compressive strength		
	Type of concrete	**Subclasses**	**Compressive strength (MPa)**
1	Low-strength		0–19
2	Normal-strength		20–39
3	Medium-strength	Low	40–49
4		Medium	50–59
5		High	60–69
6	High-strength	Low	70–84
7		Medium	85–99
8		High	100–119
9	Ultra-high strength	Low	120–200
10		Medium	200–400
11		High	400–600
12		Very high	600–1000

The support vector machine is a generalization of a simple and intuitive classifier called the maximal margin classifier, which was introduced in Section 1. Though it is elegant and simple, this classifier cannot be applied to most data sets, since it requires that the classes be separable by a linear boundary. SVM is intended for the binary classification setting in which there are two classes, it was later extended to handle more than two classes.

To understand the classifier maximal margin, it is necessary to introduce the concept of an optimal separating hyperplane. In a *p*-dimensional space, a hyperplane is a flat affine subspace of dimension *p*-1. It is simple to visualize a hyperplane in two dimensions. However, as the number of dimensions increases, it becomes more difficult to visualize. Fig. 3 shows an example of two-dimensional classification using the online application provided by LIBSVM [42].

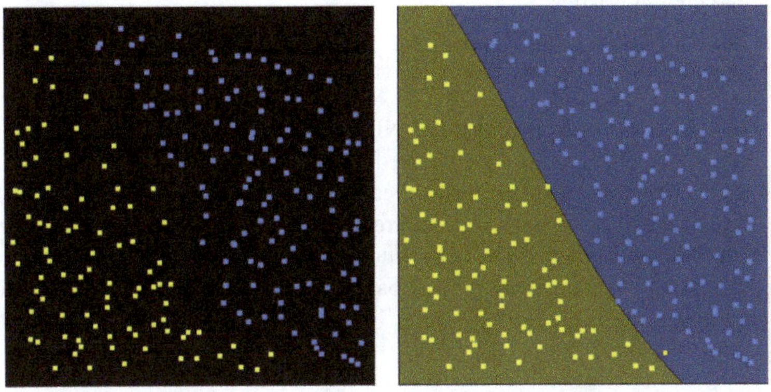

(a) (b)

Figure 3: Two-dimensional classification. (a) Two classes of observations are shown in blue and yellow; (b) An SVM applied to the samples on the left image.

As mentioned above, the support vector classifier is a natural approach for classification in the two-class setting if the boundary between the two classes is linear [43]. The linear support vector classifier can be represented by

$$f(x) = \beta_O + \sum_{i=1}^{n} \alpha_i \langle x, x_i \rangle, \tag{1}$$

where there are n parameters α_i, i=1,…, n one per training observation and $\langle x, x_i \rangle$ is the inner product of two r-vectors, in this case, x and x_i. To estimate the parameters $\alpha_i,\dots,\alpha_n$ and β_O, it is necessary to evaluate the $\binom{n}{2}$ inner products $\langle x_i, x_{i'} \rangle$ between all pairs of training observations. Thus, to evaluate the function f(x) it is required to compute the inner product between the new point x and each of the training points x_i. To summarize, the function is replaced with a generalization of the inner product of the form

$$K(x_i, x_{i'}), \tag{2}$$

where *K* is some function that we will refer to as a kernel. A kernel is a function that quantifies the similarity of two observations.

Some different types of kernel used in SVM algorithms are:

- The linear kernel. It uses Pearson correlation to quantify the similarity of two pairs of observations (eqn (3))

$$K(x_i, x_{i'}) = \sum_{j=1}^{p} x_{ij} \cdot x_{i'j}. \quad (3)$$

- Polynomial kernel. It is possible to replace each instance of eqn (3) with eqn (4), where d is the polynomial degree and $d > 1$. This kernel leads to a more flexible decision boundary. This is equivalent to adapting an SVM in a higher-dimensional space involving polynomials of degree d rather than in the original characteristic space

$$K(x_i, x_{i'}) = \left(1 + \sum_{j=1}^{p} x_{ij} \cdot x_{i'j}\right)^d. \quad (4)$$

- Radial kernel. This kernel also matches samples in a higher-dimensional space but it can handle scenarios where the relationship between classes and attributes is not linear. This kernel takes the form given by eqn 5, where γ is a positive constant in a space of infinite dimensional characteristics

$$K(x_i, x_{i'}) = exp\left(-\gamma \cdot \sum_{j=1}^{p} \left(x_{ij} - x_{i'j}\right)^2\right). \quad (5)$$

3 RESULTS AND DISCUSSION

The prediction system is based on a standard C-parameterized margin, C-SVC [21] with Radial kernel [22]. Initially, the features are scaled to (0–1) range before passing them to the SVM, preserving the many values that are already 0 and maintaining matrix scarcity. With the scaled features, a model selection is conducted to boost the performance of the SVM by picking up the best set of parameters. As two parameters have to be selected (C regularization parameter and amplitude), a 2D grid-search is performed, with parameter values varying exponentially in powers of 2 (eqn (6))

$$C \in \{2^i \colon i \in [-7,7] \cap \mathbb{Z}\}, \gamma \in \{2^j \colon j \in [-7,7] \cap \mathbb{Z}\}. \quad (6)$$

As previously mentioned in Section 2.1, the dataset was composed of 241 elements classified into 6 different types (according to Table 1). There are 12 possible classifications for compressive strength according to Table 1 but the database used for the research had a range of results from 0.6 MPa (Class 1) to 81.6 MPa (Class 6). The independent variables used were the input parameters presented in Section 2.1.1.

For each pair of parameters, a leave-one-out Cross-Validation training gives its accuracy estimation (success percentage in Table 2). Finally, the 2 parameters corresponding to the best accuracy estimation are selected. For the training tests, 85% (209 samples) of the samples were used, reserving the remaining 15% for the test (32 samples).

Table 2: Success percentage according the number of independent variables.

	Independent variables					
	2 variables			3 variables		All
Cement content	×	×		×		×
Water content	×		×	×	×	×
LWA content		×	×	×	×	×
LWA density					×	×
Success (%)	35.27	50.62	60.17	65.56	71.78	78.13

The success rate was calculated as the number of successes in the classification by comparing the actual classification and the classification predicted by the model.

As can be seen in Table 2, when the input variables increase, higher success in the classification model was obtained. This trend is a common behaviour in concrete compressive strength modelling. A better knowledge of the parameters of the concrete dosage allows one to reduce the uncertainty in the prediction and to adjust better to the real result [44]. As can be seen in Table 2, with two independent variables – where the amount of water and cement in the concrete dosage are known (water to cement ratio shown in Fig. 2(d) – the model does not allow obtaining a high success percentage (35.27%). This fact is contrary to expectation, because w/c is one of the most important factors when dosing a concrete (only very low a/c ratios allow us to obtain concretes with very high resistance [45]). However, when the quantity of water and the quantity of LWA in the manufacture of concrete is known, it is possible to estimate with higher success percentage the LWAC strength class (60.17%). With 3 independent variables (Table 2), it is observed again that the quantity of cement in the LWAC dosage is not a decisive factor in the prediction of compressive strength classification. Considering the quantity of cement, the success rate for predicting the compressive strength class is 65.56% and considering the LWA density is reached the 71.78% of success. Because of it, in LWAC, the LWA density has more influence in the prediction success than the quantity of cement used in the dosage when a prediction of compressive strength class is required.

The best model had a success of 78.125%, that is, 25 success in the classification of 32 in total used for the test (Fig. 4 shows the prediction vs real classification of the 32 data from the dataset used for the test). As can be seen in Fig. 4, there was a total success in the prediction of LWAC classified as category 1 (low compressive strength) and 6 (ultra-high compressive strength low). The worst classification can be seen in classification 5 (real compressive strength classification) because, as can be seen in Fig. 4, the model classified results as model 2 to 5. Further research is needed for increasing the success predicting the classification in the middle range of results (class 5) where lower success was observed due to the high variations in the dosage parameters that result in lower accurate prediction.

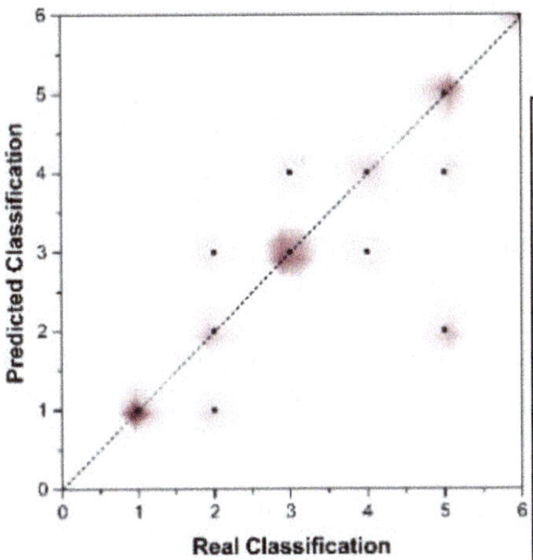

(a)

Predicted classification	Real classification 1	2	3	4	5	6
1	5 15.6%	0 0.0%	0 0.0%	0 0.0%	0 0.0%	0 0.0%
2	1 3.1%	2 6.3%	0 0.0%	0 0.0%	2 6.3%	0 0.0%
3	0 0.0%	1 3.1%	7 21.9%	1 3.1%	0 0.0%	0 0.0%
4	0 0.0%	0 0.0%	1 3.1%	2 6.3%	1 3.1%	0 0.0%
5	0 0.0%	0 0.0%	0 0.0%	0 0.0%	4 12.5%	0 0.0%
6	0 0.0%	0 0.0%	0 0.0%	0 0.0%	0 0.0%	5 15.6%

(b)

Figure 4: Prediction vs real classification of the data. (a) Real and predicted classification; (b) Success number and percentage in the classification using the SVM model with four parameters as input variables and 32 results used for test. Green colour in the diagonal indicates success and red indicates errors.

4 CONCLUSIONS

In this study, a prediction model of LWAC compressive strength was established by SVM. 241 sample data collected from experimental tests were used to develop the SVM model to predict compressive strength. The input variables selected to develop the model were cement, water, LWA content and LWA dry density. The predicted compressive strengths from the developed SVM model matched well and can be potentially used to predict the compressive strength. However, there are ranges of compressive strength where the model does not present accurate results due to the variation of the LWAC properties.

ACKNOWLEDGEMENT

This research was supported by the University of Alicante (GRE13-03) and (VIGROB-256).

REFERENCES

[1] Rodriguez-Ubinas, E., Rodriguez, S., Voss, K. & Todorovic, M.S., Energy efficiency evaluation of zero energy houses. *Energy and Buildings*, 2014.

[2] Korjenic, A., Petránek, V., Zach, J. & Hroudová, J., Development and performance evaluation of natural thermal-insulation materials composed of renewable resources. *Energy and Buildings*, 2011.

[3] Xie, Y., Li, J., Lu, Z., Jiang, J. & Niu, Y., Preparation and properties of ultra-lightweight EPS concrete based on pre-saturated bentonite. *Construction and Building Materials*, **215**, 2019.

[4] Solak, A.M. & Pereiro-Barceló, J., Cost-optimal analysis of concrete solutions for single-family nZEBs applying an Open BIM Workflow. Presented at *SynerCrete'18: Interdisciplinary Approaches for Cement-based Materials and Structural Concrete: Synergizing Expertise and Bridging Scales of Space and Time*, 2018.

[5] Jafari, S. & Mahini, S.S., Lightweight concrete design using gene expression programing. *Construction and Building Materials*, **139**, pp. 93–100, 2017.

[6] Jin, Q. & Li, V.C., Development of lightweight engineered cementitious composite for durability enhancement of tall concrete wind towers. *Cement and Concrete Composites*, **96**, pp. 87–94, 2019.

[7] Pla, C., Tenza-Abril, A.J., Valdes-Abellan, J. & Benavente, D., Influence of microstructure on fluid transport and mechanical properties in structural concrete produced with lightweight clay aggregates. *Construction and Building Materials*, **171**, 2018.

[8] Nadesan, M.S. & Dinakar, P., Mix design and properties of fly ash waste lightweight aggregates in structural lightweight concrete. *Case Studies in Construction Materials*, **7**, 2017.

[9] Fernández-Fanjul, A., Tenza-Abril, A.J. & Baeza-Brotons, F., A new methodology for determining particle density and absorption of lightweight, normal-weight and heavy weight aggregates in aqueous medium. *Construction and Building Materials*, **146**, pp. 630–643, 2017.

[10] Fernández-Fanjul, A., Tenza-Abril, A.J. & Baeza-Brotons, F., A new methodology for determining water absorption of lightweight, normal-weight and heavyweight aggregates in a viscous medium. *Construction and Building Materials*, **165**, 2018.

[11] Fernández-Fanjul, A. & Tenza-Abril, A.J., Méthode Fanjul: Dosage pondéral des bétons légers et lourds. *Annales du Bâtiment et des Travaux Publics*, **5**, pp. 32–50, 2012.

[12] Kewalramani, M.A. & Gupta, R., Concrete compressive strength prediction using ultrasonic pulse velocity through artificial neural networks. *Automation in Construction*, 2006.

[13] Tenza-Abril, A.J., Villacampa, Y., Solak, A.M. & Baeza-Brotons, F., Prediction and sensitivity analysis of compressive strength in segregated lightweight concrete based on artificial neural network using ultrasonic pulse velocity. *Construction and Building Materials*, **189**, pp. 1173–1183, 2018.

[14] Chung, S.Y., Elrahman, M.A., Stephan, D. & Kamm, P.H., The influence of different concrete additions on the properties of lightweight concrete evaluated using experimental and numerical approaches. *Construction and Building Materials*, **189**, 2018.

[15] Tenza-Abril, A.J., Villacampa, Y., Baeza-Brotons, F., Navarro-González, J.F. & Solak, A.M., Numerical model for describing the segregation phenomenon in lightweight concrete using density sections. *International Journal of Computational Methods and Experimental Measurements*, **6**, pp. 726–736, 2017.

[16] Solak, A.M., Tenza-Abril, A.J., Baeza-Brotons, F., Garcia-Vera, V.E. & Lanzón, M., Experimental procedures for determining the level of segregation of lightweight aggregate concretes. Presented at *SynerCrete'18: Interdisciplinary Approaches for Cement-based Materials and Structural Concrete: Synergizing Expertise and Bridging Scales of Space and Time*, 2018.

[17] Solak, A.M., El fenómeno de la segregación en hormigones ligeros. Análisis mediante procesamiento de imágenes y estudio ultrasónico. *Departamento de Ingeniería Civil*, 2017.

[18] Solak, A.M., Tenza-Abril, A.J., Baeza-Brotons, F., García-Vera, V.E. & Lanzón, M., New insights on the segregation due to manufacture conditions of lightweight aggregate concretes. Presented at *SynerCrete'18: Interdisciplinary Approaches for Cement-based Materials and Structural Concrete: Synergizing Expertise and Bridging Scales of Space and Time*, 2018.

[19] Solak, A.M., Tenza-Abril, A.J., Saval, J.M. & García-Vera, V.E., Effects of multiple supplementary cementitious materials on workability and segregation resistance of lightweight aggregate concrete. *Sustainability*, 2018.

[20] Ting, T.Z.H., Rahman, M.E., Lau, H.H. & Ting, M.Z.Y., Recent development and perspective of lightweight aggregates based self-compacting concrete. *Construction and Building Materials*, **201**, pp. 763–777, 2019.

[21] Cortes, C. & Vapnik, V., Support-vector networks. *Machine Learning*, 1995.

[22] Vapnik, V.N., An overview of statistical learning theory. *IEEE Transactions on Neural Networks*, 1999.

[23] Gaenssler, P. & Stute, W., Vapnik-Chervonenkis-theory BT. *Seminar on Empirical Processes*, eds P. Gaenssler & W. Stute, Birkhäuser: Basel, pp. 87–109, 1987.

[24] Hariri-Ardebili, M.A. & Pourkamali-Anaraki, F., Support vector machine based reliability analysis of concrete dams. *Soil Dynamics and Earthquake Engineering*, **104**, 2018.

[25] Liu, J., Yan, K.Z., Zhao, X. & Hu, Y., Prediction of autogenous shrinkage of concretes by support vector machine. *International Journal of Pavement Research and Technology*, 2016.

[26] Koo, B., La, S., Cho, N.-W. & Yu, Y., Using support vector machines to classify building elements for checking the semantic integrity of building information models. *Automation in Construction*, **98**, pp. 183–194, 2019.

[27] Reuter, U., Sultan, A. & Reischl, D.S., A comparative study of machine learning approaches for modeling concrete failure surfaces. *Advances in Engineering Software*, **116**, pp. 67–79, 2018.

[28] Sonebi, M., Cevik, A., Grünewald, S. & Walraven, J., Modelling the fresh properties of self-compacting concrete using support vector machine approach. *Construction and Building Materials*, **106**, pp. 55–64, 2016.

[29] Abd, A.M. & Abd, S.M., Modelling the strength of lightweight foamed concrete using support vector machine (SVM). *Case Studies in Construction Materials*, **6**, pp. 8–15, 2017.

[30] Bogas, J.A., Gomes, M.G. & Real, S., Capillary absorption of structural lightweight aggregate concrete. *Materials and Structures*, **48**, pp. 2869–2883, 2015.

[31] Videla, C. & López, M., Efecto de la resistencia intrínseca del árido ligero en la resistencia a compresión y rigidez del hormigón ligero. *Materiales de Construcción*, **52**, 2002.

[32] Bogas, J.A. & Gomes, A., Compressive behavior and failure modes of structural lightweight aggregate concrete – Characterization and strength prediction. *Materials and Design*, **46**, pp. 832–841, 2013.

[33] Bogas, J.A., Gomes, M.G. & Gomes, A., Compressive strength evaluation of structural lightweight concrete by non-destructive ultrasonic pulse velocity method. *Ultrasonics*, **53**, pp. 962–972, 2013.

[34] Youm, K.S., Moon, J., Cho, J.Y. & Kim, J.J., Experimental study on strength and durability of lightweight aggregate concrete containing silica fume. *Construction and Building Materials*, **114**, 2016.

[35] Choi, Y.W., Kim, Y.J., Shin, H.C. & Moon, H.Y., An experimental research on the fluidity and mechanical properties of high-strength lightweight self-compacting concrete. *Cement and Concrete Research*, **36**, 2006.

[36] Mo, K.H., Alengaram, U.J. & Jumaat, M.Z., Experimental investigation on the properties of lightweight concrete containing waste oil palm shell aggregate. *Procedia Engineering*, **125**, 2015.

[37] Nepomuceno, M.C.S. & Silva, P.D., Experimental evaluation of cement mortars with phase change material incorporated via lightweight expanded clay aggregate. *Construction and Building Materials*, **63**, 2014.

[38] Bogas, J.A., de Brito, J. & Cabaco, J., Long-term behaviour of concrete produced with recycled lightweight expanded clay aggregate concrete. *Construction and Building Materials*, **65**, pp. 470–479, 2014.

[39] Domagała, L., The effect of lightweight aggregate water absorption on the reduction of water-cement ratio in fresh concrete. *Procedia Engineering*, **108**, pp. 206–213, 2015.

[40] Croce, P., Marsili, F., Klawonn, F., Formichi, P. & Landi, F., Evaluation of statistical parameters of concrete strength from secondary experimental test data. *Construction and Building Materials*, **163**, pp. 343–359, 2018.

[41] Sojobi, A.O., Aladegboye, O.J. & Awolusi, T.F., Green interlocking paving units. *Construction and Building Materials*, **173**, pp. 600–614, 2018.

[42] Chang, C., Lin, C. & Tieleman, T., LIBSVM : A library for support vector machines. *ACM Transactions on Intelligent Systems and Technology (TIST)*, 2008.

[43] James, G., Witten, D., Hastie, T. & Tibshirani, R., Moving beyond linearity. *An Introduction to Statistical Learning*, Springer: New York, 2013.

[44] Ni, H.-G. & Wang, J.-Z., Prediction of compressive strength of concrete by neural networks. *Cement and Concrete Research*, **30**, pp. 1245–1250, 2000.
[45] Yeh, I.C., Modeling of strength of high-performance concrete using artificial neural networks. *Cement and Concrete Research*, **28**, 1998.

MODIFIED SOIL TESTS FOR SCOUR ANALYSIS ON OFFSHORE WINDFARM FOUNDATIONS

JORGE SORIANO-VICEDO[1], JORGE LUENGO FRADES[1],
JAVIER GARCÍA-BARBA[1] & VICENTE NEGRO-VALDECANTOS[2]
[1]University of Alicante, Spain
[2]Technical University of Madrid, Spain

ABSTRACT

The analysis of the scour processes following the pile driving procedures in offshore wind platforms is one of the major problems associated with this new form of clean energy generation. At present there are no scaled studies carried out analyzing the mechanical and deformational behavior of both the material of which the pile supporting the engine and the wind shovels is made (large steel hollow piles with a diameter of 8 m and a thickness of 15–20 cm) as well as the soil where the pile is driven. Usually these elements are installed on a very low granulometry sand displaced from the limits of dry beach–wet beach towards the offshore limits and where, due to this, it cannot return back to its natural state in its previous location. In addition, the few scaled studies carried out on dry soil similar to the existing one in the area of installation of the turbines has not considered a fundamental aspect such as the presence of seawater. This paper presents results obtained from tests to analyze the behavior of the base material on which the turbines are installed (degraded sands). These tests are the direct shear testing and a modification of the California bearing ratio (CBR) test to be carried out on completely submerged samples. The objective is to compare the results with those obtained in standard tests and analyze their suitability for field studies in offshore windfarms.

Keywords: offshore, wind platform, CBR, modified, soil, foundation.

1 INTRODUCTION

The objective of this study is to analyze the behavior of the degraded sands on which offshore wind farms are installed [1], [2], trying to avoid scouring and liquefaction processes [3]–[6]. In the works that have been carried out, the real state of the submerged soil where the foundation is built has not been considered. A material with similar characteristics to the real one has been assessed but under dry conditions, as has happened with the PISA project [1], where the trials of the foundation study for marine turbines were carried out onshore.

For this research, the following tests similar to "in situ" trials [7] were carried out: granulometric analysis, direct shear test and a compaction test such as the California bearing ratio (CBR) test.

2 MATERIALS AND METHODS

2.1 Granulometric analysis

The study has been carried out according to the UNE-EN 933-1 standard [8] "Ensayos para determinar las propiedades geométricas de los áridos (Tests to determine the geometric properties of aggregates)". The test was realized on two types of quarry granular samples, one from Algueña and another from Abanilla. The objective was to determine the basic granulometric characteristics to characterize the sands, such as the granulometric curves and the grain diameter (Table 1, Fig. 1).

WIT Transactions on Engineering Sciences, Vol 125, © 2019 WIT Press
www.witpress.com, ISSN 1743-3533 (on-line)
doi:10.2495/CMEM190181

Table 1: Grain size data.

Algueña				Abanilla		
Mesh width (mm)	% Partial retention	% Cumulative retention	% Pass	% Partial retention	% Cumulative retention	% Pass
4	0.00	0.00	100.00	0.00	0.00	100.00
2	0.22	0.22	99.78	0.45	0.45	99.55
1	32.99	33.20	66.80	36.95	37.40	62.60
0.5	26.79	59.99	40.01	22.54	59.95	40.05
0.315	12.70	72.70	27.30	11.47	71.42	28.58
0.25	6.63	79.33	20.67	5.21	76.63	23.37
0.2	4.35	83.68	16.32	4.15	80.77	19.23
0.16	4.87	88.54	11.46	4.83	85.61	14.39
0.125	4.09	92.64	7.36	3.80	89.40	10.60
0.09	4.09	96.73	3.27	5.19	94.60	5.40
0.063	3.19	99.91	0.09	4.76	99.35	0.65
Base	0.09	100.00	0.00	0.65	100.00	0.00
TOTAL	100.00	100.00		100.00	100.00	
D50 (mm)	0.686			0.721		

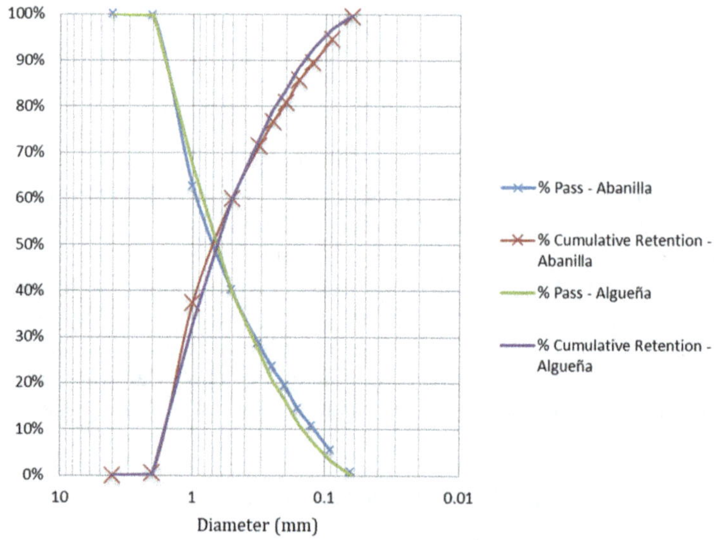

Figure 1: Distribution of grain size.

2.2 Direct shear test

The direct shear test was carried out according to the UNE 103 401 standard [9] "Determinación de los parámetros resistentes al esfuerzo cortante de una muestra de suelo en la caja de corte directo (Determination of the shear resistant parameters of a soil sample in the direct cutting box)" for each sand sample. The test is the consolidate-drained (CD) one, recommended for granular soils.

For the preparation of the sample, it has been chosen to pour the sand into the box, making sure that it falls from a height that prevents its compaction (Fig. 2). This option has been chosen to test the material in the conditions that would be found on the seabed, on which the material is deposited by its own weight. According to the standard, as in sandy soils, the consolidation occurs almost instantaneously; a speed of 1 mm/min was chosen so that the break lasts between 5 and 10 minutes.

Figs 3–6 show the graphs of the relationship of displacements, the break and the relationship of tensions, which determines the friction angle and the cohesion of each material (Figs 7 and 8).

2.3 CBR test

Two types of CBR tests were carried out, one as stated in the UNE 103 502 standard [10] "Método de ensayo para determinar en laboratorio el índice CBR de un suelo (Test to determine the CBR index of a soil in the laboratory)" and another with an important difference that is explained below. Both tests share the following aspects in the preparation of the samples:

Figure 2: Pouring the sand into the box.

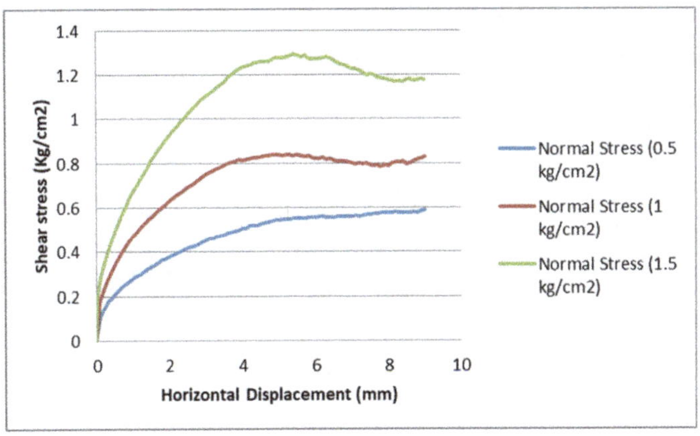

Figure 3: Breaking point on Abanilla sand.

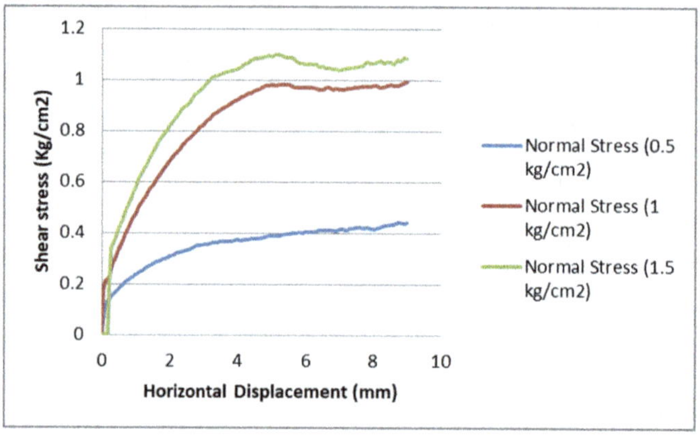

Figure 4: Breaking point on Algueña sand.

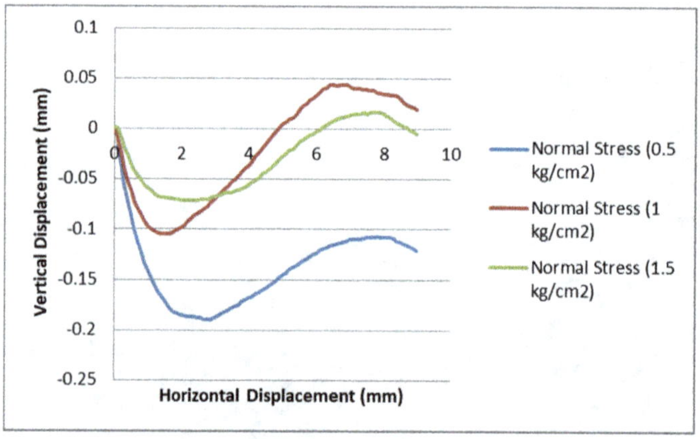

Figure 5: Displacements on Abanilla sand.

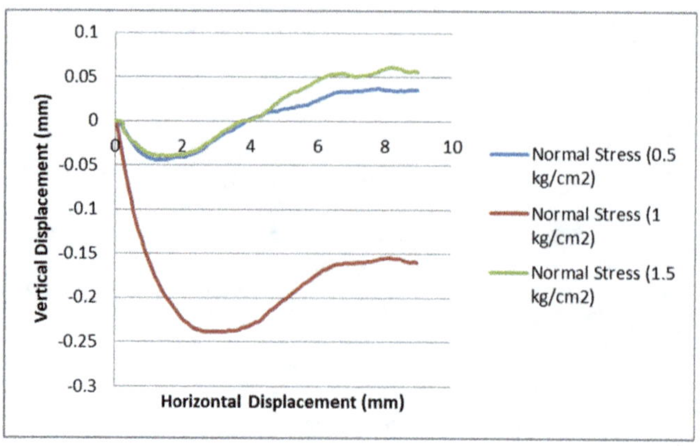

Figure 6: Displacements on Algueña sand.

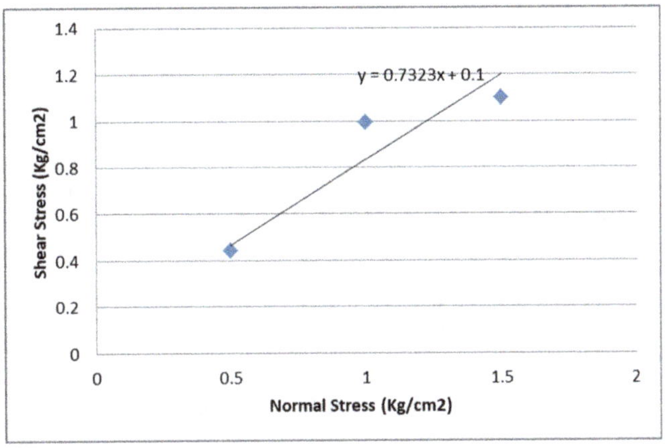

Figure 7: Stresses on Algueña sand (friction angle: 36.2°; cohesion ≈ 0 kg/cm²).

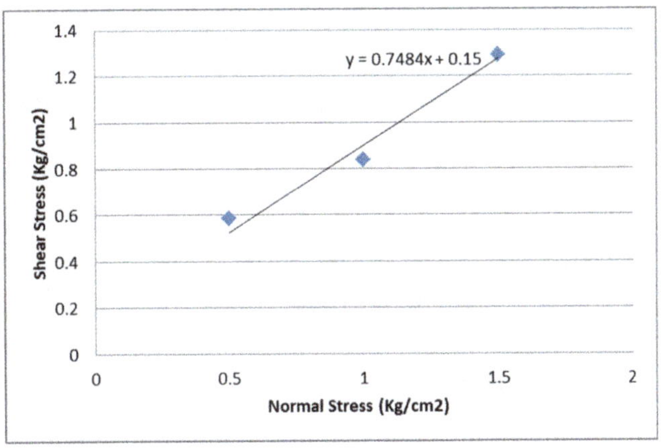

Figure 8: Stresses on Abanilla sand (friction angle: 36.8°; cohesion ≈ 0 kg/cm²).

- Compaction in three layers mediating 15 blows per layer with the 2.5 kg mace that falls freely from a height of 305 mm. The sample has only been compacted by 15 blows because it was intended to apply a corresponding compaction energy with 25% of the compaction energy of the normal proctor. This is done to simulate the conditions of the material on the seabed, where it is not compacted, but has been deposited over time.
- The samples are immersed in containers filled with water for 72 hours (3 days). This time period is enough to saturate the granular soil.

The objective of these two tests was to compare the behavior of soil in dry penetration and submerged situations.

2.3.1 CBR standard test

For this test, three samples were tested for each type of granular soil. Next, the obtained results up to reaching a certain penetration are represented in Figs 9 and 10.

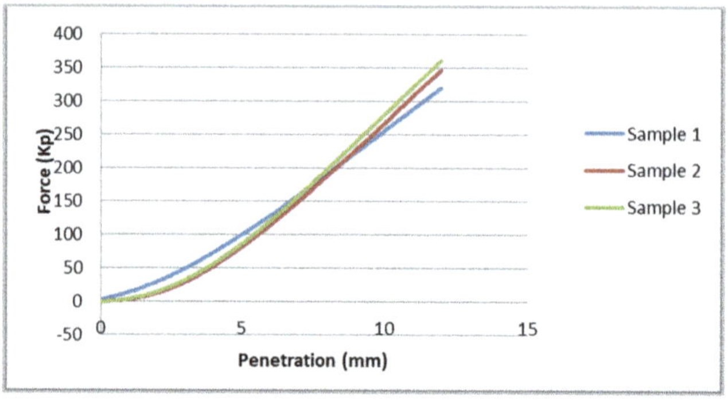

Figure 9: Standard CBR test on Algueña sand.

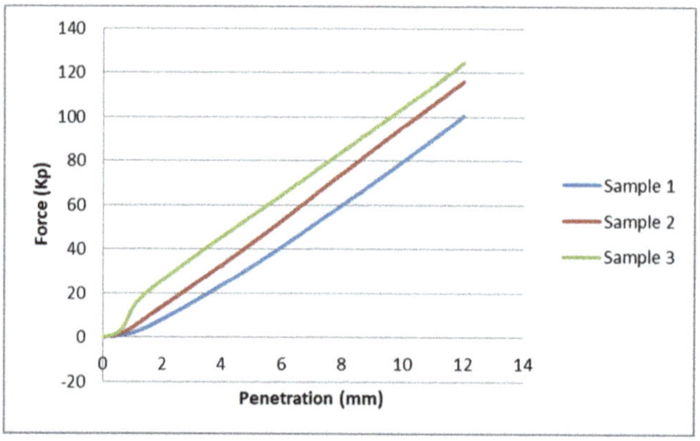

Figure 10: Standard CBR test on Abanilla sand.

Figure 11: Submerged penetration.

2.3.2 CBR modified test
It is the same methodology as that of the previous test, the only difference being that during penetration, the samples were immersed in a cubic container filled with water (Fig. 11). The obtained results up to reaching a certain penetration are represented in Figs 12 and 13.

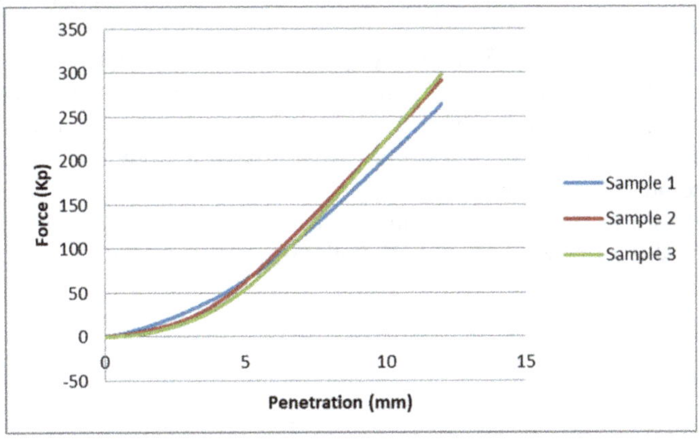

Figure 12: Submerged penetration for samples on Algueña sand.

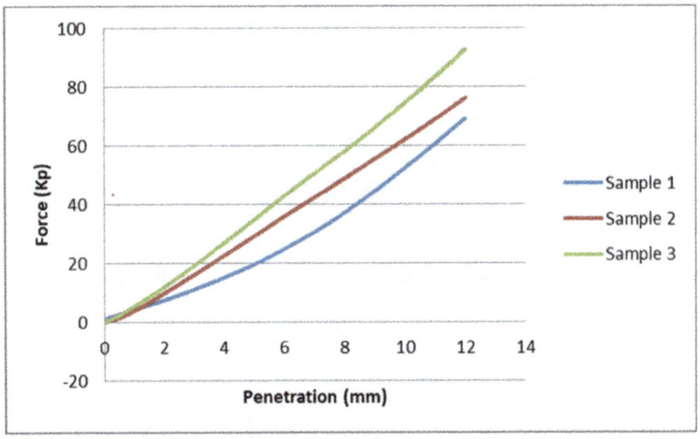

Figure 13: Submerged penetration for samples on Abanilla sand.

3 CONCLUSIONS

The objective of this research was to carry out tests on similar materials to existing ones in the foundation area for offshore wind platforms trying to avoid scour and liquefaction cases. In many cases, the tests carried out in the different research works, including the PISA report involving entities such as the Royal Geographical Society (London), British Petroleum or the Norwegian Institute NGI, have been carried out onshore and on unsaturated samples of soil. Therefore, knowing the behavior of the soil in a submerged situation is a fundamental aspect for evaluating the results obtained on the trials.

In this respect, a modified CBR test that has been carried out allows one to have an approximate idea of the mechanical and resistant behavior of the submerged foundation soil, obtaining the penetration of a piston in a sample of submerged material. The comparison of the results with those obtained in the normalized standard test, would allow one to obtain a relationship between dry and submerged materials, and the penetration values can be estimated without carrying out an "in situ" test on the seabed, but simply with a granulometry of the material and the normalized CBR test, a correction factor would be applied to obtain submerged real values.

The values in Figs 14 and 15 show a decrease in penetration force of 27% and 44%, respectively, to obtain the same value. Although the granulometries and direct cutting tests are similar, although with lower values in the direct cut test on the Algueña sand, there may be some relationship between this value and the one from the submerged penetration test that has been carried out (Figs 16 and 17).

Therefore, it appears that a relationship can be established between the actual submerged values for an offshore foundation and those achieved for a similar material under standard CBR test conditions.

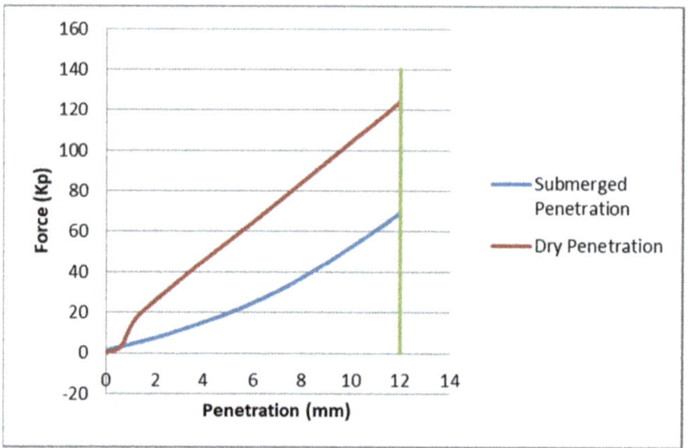

Figure 14: Comparative of penetration (Abanilla sand).

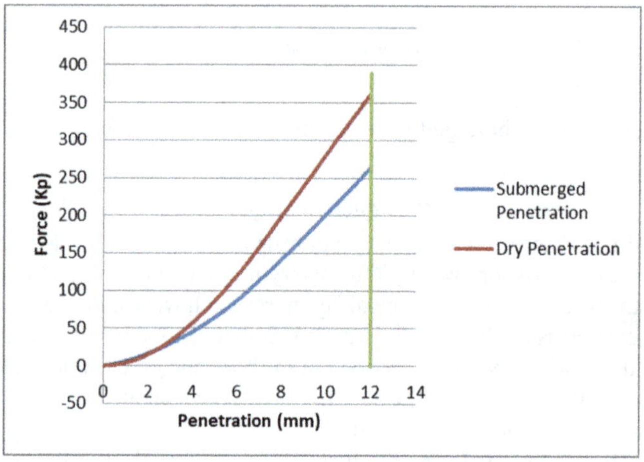

Figure 15: Comparative of penetration for sample 3 (Algueña sand).

Figure 16: Sample after dry penetration.

Figure 17: Sample after submerged penetration.

REFERENCES

[1] Byrne, B.W., McAdam, R., Burd, H.J. & Houlsby, G.T., *PISA: New Design Methods for Offshore Wind Turbines,* 2017.

[2] DNVGL-ST-0126, *Support Structure for Wind Turbines,* 2016.

[3] Lam, I.P.O. & Martin, G.R., Seismic Design of High-Way Bridge Foundations. US Department of Transportation, Report No. FHWA/RD-86/102, 1986.

[4] Papadimitriou, A.G. & Bouckovalas, G.D., Plasticity model for sand under small and large cyclic strains: A multiaxial formulation. *Soil Dyn. and Earthquake Engn.*, **22**(3), pp. 191–204, 2002.

[5] Reese, L., Cox, W.R. & Koop, F.D., Analysis of laterally loaded piles in sand. *Proceedings of the Offshore Technical Conference,* Houston, 1974.

[6] Whitehouse, R.J.S., Sutherland, J. & O'Brien, D., *Seabed Scour Assessment for Offshore Windfarm*, 2006.

[7] Kelly, R.B., Houlsby, G.T. & Byrne, B.W., A comparison of field and laboratory caisson tests in sand and clay. *Géotechnique*, **56**(9), pp. 617–626, 2006.

[8] UNE-EN 933-1 standard, Ensayos para determinar las propiedades geométricas de los áridos (Tests to determine the geometric properties of aggregates).
[9] UNE 103 401 standard, Determinación de los parámetros resistentes al esfuerzo cortante de una muestra de suelo en la caja de corte directo (Determination of the shear resistant parameters of a soil sample in the direct cutting box).
[10] UNE 103 502 standard, Método de ensayo para determinar en laboratorio el índice CBR de un suelo (Test to determine the CBR index of a soil in the laboratory).

NON-LINEAR NUMERICAL MODELS FOR PREDICTING THE BOND STRENGTH OF FIBRE-REINFORCED CONCRETE AT HIGH TEMPERATURES

FRANCISCO B. VARONA[1], YOLANDA VILLACAMPA[2], FRANCISCO JOSÉ NAVARRO-GONZÁLEZ[2], DAVID BRU[1] & FRANCISCO JAVIER BAEZA[1]
[1]Department of Civil Engineering, University of Alicante, Spain
[2]Department of Applied Mathematics, University of Alicante, Spain

ABSTRACT

The steel to concrete bond mechanism is critical to address the behaviour of reinforced concrete structural members. Although this mechanism can be compromised during a fire, it may be one of the least researched phenomena in concrete technology and is not addressed in the design codes and standards. In this work, we present a thorough review of the experimental data available on this topic, focusing on fibre-reinforced concrete. The data allow us to study the evolution of the bond strength as a function of three variables: the exposure temperature, the type of fibre and the volume fraction. A linear multiple regression is initially carried out, followed by a series of non-linear numerical models. These models are built using a methodology based on the finite element method combined with the formulation of the Galerkin method. The numerical models have been developed for different degrees of complexity. The error measurements obtained with the linear regression and the numerical models are compared in order to present a prediction model. The selected model is then validated for different values of the independent variables. This process supports the discussion of the influence that the independent variables have in the evolution of the bond strength between steel reinforcement and fibre-reinforced concretes exposed to high temperatures.

Keywords: fibre-reinforced concrete, high temperature, bond strength, non-linear numerical models.

1 INTRODUCTION

Reinforced concrete (RC) became a structural material during the second half of the 19th century. Reportedly, one of several reasons was its good behaviour exposed to the high temperatures in accidental fire situations, especially when compared to other construction materials of that age, e.g. wrought iron. Concrete exhibits better fire performance than structural steel and recent history provides some notable though dramatic examples for this assertion: the fire in Windsor Tower (Madrid, 2005) and the fire in Grenfell Tower (London, 2017).

The behaviour of concrete exposed to high temperature has been thoroughly researched during the 20th century. One of the earliest pieces of research dates from 1920 [1]. The effects of the exposure of concrete to elevated temperatures can be described through the following stages [2]: (i) drying of capillary water occurs between 20 and 100°C, with a slight reduction of the compressive strength; (ii) for temperatures up to 300°C, an increase of the compressive strength is reported for young and dry concretes; (iii) the dehydration of the Calcium-Silicate-Hydrate gel starts around 150–180°C and the dehydration of the portlandite takes place between 400 and 600°C; (iv) a notorious drop of the mechanical properties of concrete is observed in the interval 300–650°C; (v) in the case of quartzitic aggregates, the transformation of α-quartz into β-quartz takes place at 570°C and entails a volumetric increase which is thought to be responsible for the higher drop of mechanical properties in concretes made with this type of aggregates as compared to other types; (vi) the decarbonation of limestone aggregates occurs between 600 and 900°C; (vii) by 700°C the dehydration of the CSH gel is mostly complete; (viii) the rate of reduction of mechanical

WIT Transactions on Engineering Sciences, Vol 125, © 2019 WIT Press
www.witpress.com, ISSN 1743-3533 (on-line)
doi:10.2495/CMEM190191

properties slows down for temperatures higher than 650°C; (ix) melting of some types of aggregates takes place at around 1200°C and melting of the Portland cement paste is reported at around 1350°C.

Thermal spalling is one of the major concerns in concrete structures exposed to fire. This phenomenon can be described as the progressive separation of surface and outer cover of a concrete member when heated. This process can be either prolonged or sudden and violent. The latter is known as explosive spalling and is typical of dense or compact microstructures – e.g., high strength concrete and self-compacting concrete [3]–[6] – and is also likely to occur at high rates of heating [2]. In this regard, notable fire catastrophes can be mentioned, such as fire in the Eurotunnel of the English Channel in 1996, the Tauern tunnel in 1999 (Austria) and the Mont Blanc tunnel in 1999 (France). In these examples, the thermal exposure in a confined environment, and with a high combustible charge, caused a fast temperature rise, which led to severe concrete damage and explosive spalling. Thus, the structural section was rapidly reduced and inner layers became exposed, hence compromising the structural safety. Another cause of spalling is associated with the thermal gradients and the thermal incompatibility of the different components of the reinforced concrete elements: the cement paste, the aggregates, the steel rebars and tendons [7], [8], which have different coefficients of thermal expansion. Another cause of spalling is the pore pressure increase due to the water steam expelled by the dehydration of the CSH gel and the portlandite. And yet another cause suggested in [9] is the calcination of $CaCO_3$ (600–900°C) when calcareous aggregates are used; this chemical process expels CO_2 and may contribute to the pore pressure increase.

In the last two decades research has been focused on studying fibre-reinforced concrete exposed to high temperatures. Polypropylene fibres – which melt at approximately 170°C – have been proven adequate to control the spalling of concrete. Their melting creates an additional capillary pore network inside the concrete, which allows the release of the water steam generated by dehydration reactions [10], [11]. Thus, the main objective of polypropylene fibres is not the structural reinforcement but rather the improvement of the fire resistance of concrete. On the other hand, steel fibres may be used in order to increase the residual strength of concrete after exposure to elevated temperatures [12], [13]. However, some authors [11] argue that the benefit of adding steel fibres may not be so obvious, since they contribute to speeding up the heating of the concrete and may be prone to producing internal micro-cracking because of thermal incompatibility with the concrete matrix. Hybrid fibre addition (steel and polypropylene) has been addressed in some recent research [10], [11], [14]–[17] and the experimental evidences suggest a synergy between both types of fibre, enhancing the mechanical properties of concrete.

Concrete fire design is covered by international codes and standards, such as the Model Code 2010 [18] and Eurocode 2 [19]. Simplified approaches are proposed as well as more advanced methods based on tables or curves that describe the evolution of the compressive strength of concrete and other properties at high temperatures. A basis for advanced numerical modelling methods is also provided.

Nonetheless, the loss of bond between the steel reinforcement and the concrete during or after exposure to elevated temperatures is not addressed in those standards. The structural strength of a RC element highly depends on an appropriate bond between reinforcement and concrete. This mechanism can be compromised during a fire, because important temperature gradients within the structural element may appear due to the difference in thermal conductivity of concrete and steel. Steel to concrete bond at high temperatures may be one of the least investigated phenomena in concrete research. The evolution of the bond strength between steel reinforcement and fibre reinforced concrete exposed to high temperature has been studied and reported in [20]–[23]. The objective of this paper is to analyse the experimental evidence reported in those four references and to work out a predictive model for the bond strength in fibre-reinforced concrete after exposure to a target temperature as a function of several variables. The predictive model is built using a methodology based on the finite element method combined with the formulation of the Galerkin method.

2 METHODOLOGY

This section is comprised of two parts. First, we present the four experimental studies on the bond strength in fibre-reinforced concrete exposed to elevated temperatures. The independent variables of the numerical models will be selected from those studies. The second part introduces the numerical model which constitutes the basis for the predictive model.

2.1 Experimental studies

The data used to study the evolution of the bond strength at high temperature are collected from the following experimental campaigns:

- Haddad et al. (2008) [20]. High strength concrete specimens were tested. The shape of the specimens was prismatic (10×10×40 cm) and they included longitudinal and transverse reinforcement. The following types of fibres were added to the concrete admixtures: polypropylene fibres (PP), steel fibres of the hook end type (S) and brass coated steel fibres (BCS, with a very small diameter of 0.15 mm and a length of 6 mm). Four fibre-reinforced admixtures were prepared: Ref-1-A using steel fibres, Ref-1-B using hybrid fibres (S and PP), Ref-1-C using hybrid fibres (S and BCS) and Ref-1-D using BCS fibres. Exposure at high temperature was tested at 350, 500, 600 and 700°C. The target temperature in each test was maintained for 2 hours, to reach a homogeneous thermal state in the specimens.
- Varona et al. (2015) [21]. Normal strength and high strength concrete specimens were tested. In order to facilitate a homogeneous heating, specimens were cylindrical. The geometry of the specimens was based on the RILEM pull-out test as described in Windisch [24]. Samples with 10 mm diameter bars had a diameter of 100 mm whilst samples with 12 mm diameter bars had a diameter of 120 mm. In all cases the height of the specimens was equal to the diameter. The following types of fibres were tested: hook end type steel fibres (S) and polypropylene fibres (PP, with a diameter of 0.035 mm and a length of 12 mm). Three fibre-reinforced batches were prepared: Ref-2-A was a normal strength concrete with steel fibres; Ref-2-B was a high strength concrete with PP fibres; and Ref-2-C was a high strength concrete using a mix of steel and PP fibres. Exposure at high temperature was tested at 450, 650 and 825°C. The target temperatures were maintained for 75 minutes.
- Lublóy and Hlavička (2017) [22]. High strength concrete cylindrical specimens were tested. The diameter and height of the specimens was 120 mm and 100 mm, respectively. Two fibre-reinforced concretes were prepared: Ref-3-A using steel fibres and Ref-3-B using polypropylene fibres (diameter of 0.032 mm and a length of 18 mm). Exposure at high temperature was tested at 50, 150, 300, 500, 600 and 800°C. The target temperatures were maintained for 2 hours.
- Varona et al. (2018) [23]. This study is similar to [21]. Only 12 mm diameter bars were considered (cylindrical specimens with a diameter of 120 mm and a height of 120 mm). Apart from polypropylene fibres, two types of hook end steel fibres were considered: one with a diameter of 0.75 mm and an aspect ratio of 47 (type S1) and

another with a diameter of 0.35 mm and an aspect ratio of 86 (type S2). Three normal strength fibre-reinforced concrete batches were tested: Ref-4-A using PP fibres, Ref-4-B using a mix of S1 and PP fibres and Ref-4-C using S2 and PP fibres. Additionally, three high strength fibre-reinforced concrete batches were tested: Ref-4-D using PP fibres, Ref-4-E using S1 and PP fibres and Ref-4-F using S2 and PP fibres. Exposure at high temperature was tested at 450, 650 and 825°C. The target temperatures were maintained for 75 minutes.

Further information is given in Tables 1 and 2. Apart from the temperature, eight possible independent variables are identified in Tables 1 and 2.

Table 1: Experimental studies. Variables 1 to 4.

Reference	Variable 1 Strength $f_{c,20°C}$ (MPa)	Variable 2 Type of fibre	Variable 3 Volume fraction	Variable 4 Bar diameter (mm)
Ref-1-A	89.8	S	2%	20
Ref-1-B	85.5	S+PP	2%	20
Ref-1-C	93.7	S+BCS	2%	20
Ref-1-D	103.6	BCS	2%	20
Ref-2-A	31.0	S	0.25%	10
Ref-2-B	84.9	PP	0.33%	12
Ref-2-C	89.5	S+PP	0.58%	12
Ref-3-A	63.6	PP	0.11%	12
Ref-3-B	75.1	S	0.45%	12
Ref-4-A	29.0	PP	0.16%	12
Ref-4-B	19.1	S+PP	0.41%	12
Ref-4-C	24.6	S+PP	0.41%	12
Ref-4-D	89.0	PP	0.27%	12
Ref-4-E	97.6	S+PP	0.52%	12
Ref-4-F	83.4	S+PP	0.52%	12

Table 2: Experimental studies. Variables 5 to 8.

Reference	Variable 5 Bond length to diameter (ratio)	Variable 6 Concrete cover to diameter (ratio)	Variable 7 Age at test (days)	Variable 8 Time at max. temperature
Ref-1-A to D	7.5	2	35	2 hours
Ref-2-A	5	4.5	60	75 min
Ref-2-B & C	4.17	4.71	60	75 min
Ref-3-A & B	3.33	4.5	28	2 hours
Ref-4-A to F	4.17	4.71	60	75 min

In all cases bond strength was measured after cooling to room temperature (around 20°C). Cooling was natural, not forced with water immersion or spraying. The bond strength τ_b was obtained through eqn (1):

$$\tau_b = \frac{P}{\pi D l_b}, \tag{1}$$

where P is the maximum pull-out load, D is the diameter of the embedded bar and l_b is the bond length.

2.2 Numerical modelling

The numerical methodology is based on Galerkin's formulation of the finite element method [25], [26] and allows us to obtain predictive models for the relationship between variables that have been defined a priori. As is usual in the finite element method, a geometric model of finite elements is generated so that the model is defined by its value at the nodes of the elements [27]. In this way, for each finite element mesh defined in a hyper-cube, a numerical model is generated. Galerkin's formulation of the finite element method improves the computational efficiency of algorithms from other methodologies based on geometric models of finite elements, as reported in [28], [29]. The finite element mesh is characterized by its complexity index, which is defined as the number of elements defined along each edge of the hyper-cube. The total number of nodes is thus a function of the complexity index. In this way, a numerical model is generated for each set of experimental data and for each complexity. The use of Galerkin's formulation of the finite element method and the corresponding improvement in computational efficiency allows us to analyse relationships with a greater number of variables and relationships in which complexity needs to be increased without involving much time in the execution of the software.

Bond strength at high temperature is the dependent variable. Amongst the independent variables, the fundamental one is the target temperature or maximum exposure temperature. The experimental methodologies applied in the reports introduced in previous section are consistent and it is possible to identify eight additional independent variables, presented in Tables 1 and 2. The number of independent variables to be considered in the numerical models strongly depend on the data available. The number of data points is calculated below.

- Each of the four references Ref-1-A to D provides bond strength values at the following temperatures: 20, 350, 500, 600 and 700°C (5 targets). Therefore, the number of data points is $4 \times 5 = 20$.
- Each of the three references Ref-2-A to C provides bond strength values at the following temperatures: 20, 450, 650 and 825°C (4 targets). Therefore, the number of data points is $3 \times 4 = 12$.
- Each of the two references Ref-3-A & B provides bond strength values at the following temperatures: 20, 50, 150, 300, 500, 600 and 800°C (7 targets). Therefore, the number of data points is $2 \times 7 = 14$.
- And each of the six references Ref-4-A to F provides bond strength values at the following temperatures: 20, 450, 650 and 825°C (4 targets). Therefore, the number of data points is $6 \times 4 = 24$.

This gives a total of $20+12+14+24 = 70$ data points, so the number of variables should not be greater than three – at least 15 to 25 data points per variable are usually recommended. Since the target temperature is the fundamental independent variable, two more variables may be chosen. The data points are shown in Fig. 1 as a function of the target temperature. Given that the study is focused on fibre-reinforced concretes, the volume fraction seemed to be a natural option. The third independent variable is the type of fibres, because they have different effects on the endurance of concrete exposed to high temperatures – the influence of polypropylene fibres is arguably radically different from the influence of steel fibres. Therefore, the following independent variables were selected:

- Variable 1: type of fibre. It is a categorical variable with three discrete values: 1 (steel fibre), 2 (polypropylene fibre) and 3 (hybrid fibre).
- Variable 2: volume fraction. It is a numerical variable, expressed in percentage.
- Variable 3: exposure temperature. It is a numerical variable, expressed in °C.

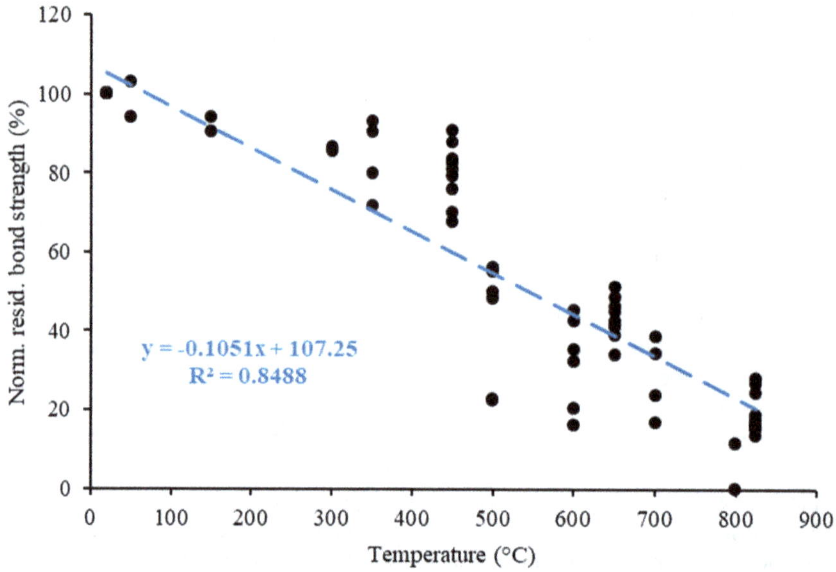

Figure 1: Experimental data points available and linear regression analysis.

3 RESULTS AND DISCUSSION

This section is divided into two parts. In the first part we apply regression analyses to study the evolution of the bond strength at high temperature based on experimental evidence. We also show the error parameters obtained with each technique in order to select the appropriate one. In the second part we present design curves of the evolution of the bond strength at high temperature, which are based on the regression method selected.

3.1 Regression analyses and error parameters

Linear regression was the first type of analysis carried out, with the exposure temperature as the independent variable. The 70 available data points and the linear regression analysis are presented in Fig. 1. The vertical axis in Fig. 1 is the normalised residual bond strength after exposure to high temperature, i.e. the percentage of the bond strength after heating and cooling with respect to the bond strength at ambient temperature.

Next, linear multiple regression was performed. The independent variables were the target or exposure temperature, the type of fibre and the fibre volume fraction. The result is shown in eqn (2):

$$\frac{\tau_{b,T}}{\tau_{b,20°C}} = \frac{1}{100}\left(99.59 - 0.1061\,T + 3.877\,FT - 0.5588\,VF\right), \tag{2}$$

where: $\tau_{b,T}$ is the residual bond strength after exposure to temperature T; $\tau_{b,20°C}$ is the bond strength at ambient temperature; T is the exposure temperature in °C; FT is the fibre type (1 for steel fibres, 2 for polypropylene fibres and 3 for hybrid fibres); and VF is the fibre volume fraction expressed as percentage. The R^2 coefficient was 0.857, slightly increasing with respect to the linear regression analysis (see Fig. 1).

Finally, we applied the numerical modelling based on Galerkin's formulation of the finite element method. In this particular case, the space of solutions is a 4-dimensional hypercube because there are four variables – the dependent variable plus three independent variables. Complexity index C was set at five levels: 40, 50, 60, 70 and 80. The meshing of the hypercube is controlled by index C and the mesh resolution was between $40^3 = 64000$ nodes and $80^3 = 512000$ nodes.

Apart from the R^2 coefficient, several error measurements were studied. These are calculated through eqns (3) to (7):

$$MAE = \frac{\sum_{i=1}^{N} \left| R_i - P_i \right|}{N}, \tag{3}$$

$$MAPE = \frac{\sum_{i=1}^{N} \frac{\left| R_i - P_i \right|}{R_i}}{N}, \tag{4}$$

$$RMSE = \sqrt{\frac{\sum_{i=1}^{N} (R_i - P_i)^2}{N}}, \tag{5}$$

$$\varepsilon = \sqrt{\frac{\sum_{i=1}^{N} (R_i - P_i)^2}{N - V}}, \tag{6}$$

$$\delta = \frac{\varepsilon}{\sqrt{\frac{1}{N}\sum_{i=1}^{N} R_i^2}}, \tag{7}$$

where: R_i are the real or experimental values of the dependent variable; P_i are the predicted values of the dependent variable; N is the number of data points (in this case, $N = 70$); V is the number of independent variables (in this case, $V = 3$); MAE is the mean absolute error; $MAPE$ is the mean absolute percentage error; $RMSE$ is the root mean square error; ε is an error measurement similar to $RMSE$ that penalises the number of independent variables; and δ is a normalised version of the ε error.

Table 3 illustrates how each method performed by showing the R^2 values and the aforementioned error measurements. *MAPE* was not finally considered because there were data points where the dependent variable's value was zero or close to zero. The performance of the different types of analysis is graphically compared in Fig. 2.

Table 3: R^2 and error measurements of the types of analysis considered.

Type of analysis		R^2	*MAE*	*RMSE*	ε	δ
Linear regression		0.8488	9.35	12.6	12.9	0.183
Linear multiple regression		0.8575	9.00	12.2	12.5	0.178
Finite element method	$C = 40$	0.9863	2.91	3.79	3.87	0.055
	$C = 50$	0.9885	2.61	3.47	3.54	0.050
	$C = 60$	0.9898	2.40	3.26	3.34	0.047
	$C = 70$	0.9902	2.31	3.21	3.28	0.047
	$C = 80$	0.9904	2.27	3.17	3.24	0.046

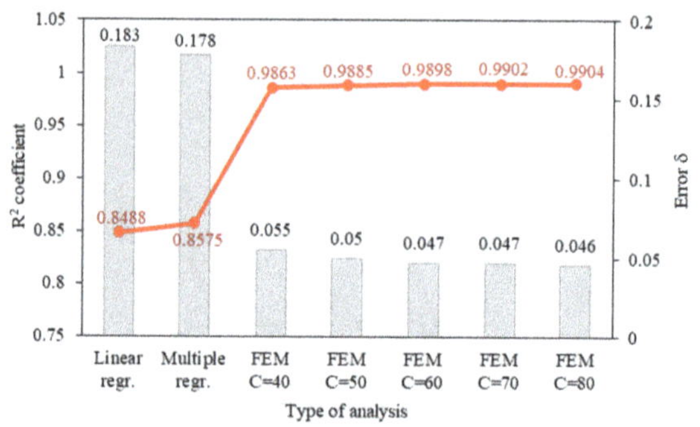

Figure 2: R^2 coefficient and error δ.

As expected, the numerical modelling techniques outperformed the linear regression methods. The five attempts with the finite element method were capable of explaining around 99% of the data points. With a complexity of C = 40, the numerical model with three independent variables exhibited a R^2 coefficient in excess of 98% and was more successful than the multiple linear regression with its R^2 around 86%. Regarding the complexity index, error δ seemed to get stabilised at $C = 60$. Therefore, the numerical model with $C = 60$ was chosen for application.

3.2 Application of the numerical model

The application of the numerical model was carried out for the following values of the independent variables:

- Variable 1 encodes the type of fibre and its three discrete values were considered: $v_1 = 1$ (steel fibre), $v_1 = 2$ (polypropylene fibre) and $v_1 = 3$ (hybrid fibre).
- Variable 2 corresponds to the fibre volume fraction percentage and the following values were applied: $v_2 = 0.25\%$, $v_2 = 0.5\%$, $v_2 = 1\%$, $v_2 = 1.5\%$ and $v_2 = 2\%$.
- Finally, variable 3 corresponds to the exposure temperature. In addition to the ambient temperature (20°C), the range between 100 and 800°C at intervals of 50°C was studied.

The application is illustrated in Fig. 3: steel fibres, Fig. 4: polypropylene fibres, and Fig. 5: hybrid fibres, e.g. steel and polypropylene.

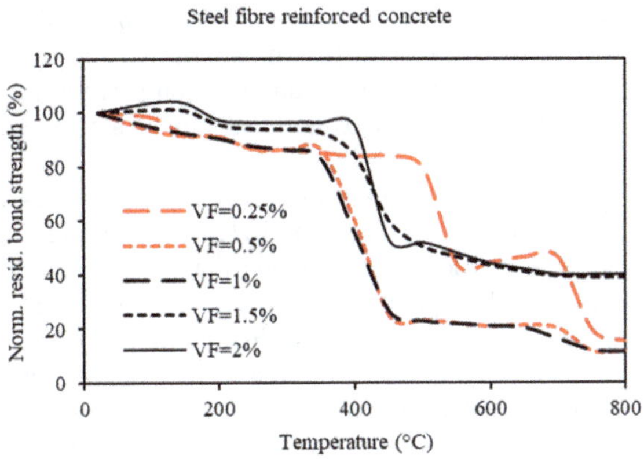

Figure 3: Application of the numerical model for steel fibre-reinforced concretes.

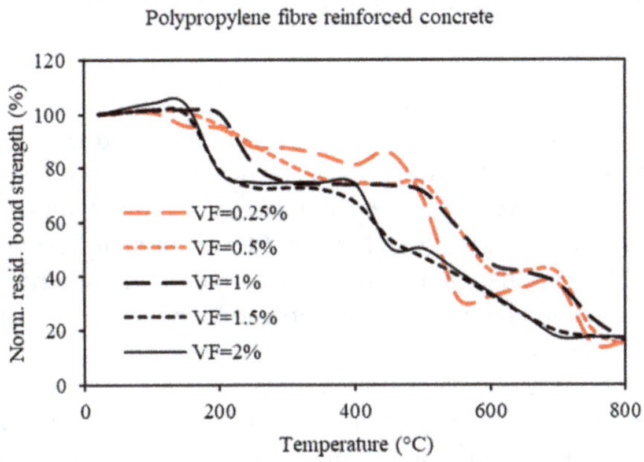

Figure 4: Application of the numerical model for polypropylene fibre-reinforced concretes.

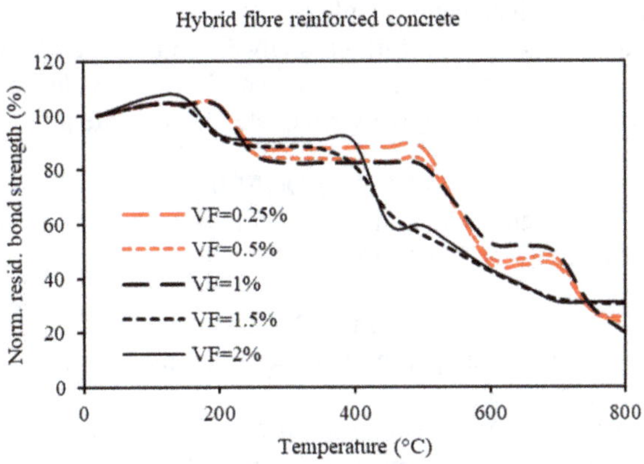

Figure 5: Application of the numerical model for hybrid fibre-reinforced concretes.

Regardless of the fibre volume fraction, Fig. 3 shows that steel fibre-reinforced concretes seemed to maintain their original bond strength up to temperatures around 400°C. Over 500°C, fibre content may have impact on the bond strength. Volume fractions of 1.5% and 2% may control the residual bond strength at around 40% up to 800°C, whilst concretes with smaller volume fractions dropped below 20%. An unexpected and strange result was found between 500 and 700°C, with the smallest volume fraction performing as good as contents of 1.5% and 2%. This result is difficult to explain and may be a consequence of a bias in the original source of experimental data.

The evolution of the bond strength with the other types of fibres considered was less abrupt (see Fig. 4 and Fig. 5). Both with polypropylene and hybrid fibres and in the 400–700°C range, moderate volume fractions not greater than 1% seemed to yield better values of the residual bond strength than greater volume fractions up to 2%. In the case of the polypropylene fibres this result is logical, because their melting at 170°C produces a capillary network in the concrete matrix. Although this effect is beneficial to the spalling control, an excess in fibre content may weaken the microstructure of the concrete after melting. This phenomenon is also likely to compromise the synergy between steel and polypropylene fibres for total volume fractions over 1%. After exposure to 800°C, the numerical model estimated that the steel fibre-reinforced concretes with at least 1.5% of fibre volume fraction exhibited the best residual bond strength (around 40%), followed by the hybrid reinforced concretes (around 30%).

4 CONCLUSIONS

The evidence sourced from four experimental studies was used to predict the evolution of the bond strength in fibre-reinforced concretes after exposure to high temperature. The number of data points available allowed us to apply a numerical simulation based on Galerkin's formulation of the finite element method. Different levels of complexity were tested and one of them was selected for application, based on the evolution of the error measurements.

The numerical application helped to bring forth some conclusions regarding the effect of fibre addition on the evolution of bond strength. In the case of steel fibres, volume fraction may have a positive effect at temperatures up to 800°C. However, in the case of polypropylene fibres, bond strength seemed to get compromised for volume fractions in excess of 1%, which could be attributed to their premature melting at around 170°C. In the range between 400 and 600°C, hybrid fibre-reinforced concretes may be expected to exhibit the best performance – in terms of the residual bond strength. Moreover, hybrid fibre content over 1% did not seem to produce better results.

Some unexpected results require further analysis and, most likely, more exhaustive experimental background. Nonetheless, as new experimental evidence becomes available, the methodology presented here is arguably adequate to refine the prediction models.

ACKNOWLEDGEMENT

The authors want to acknowledge Generalitat Valenciana, which has made possible this piece of research through the grant GV/2018/015.

REFERENCES

[1] Lea, F.C., The effect of temperature on some of the properties of materials. *Eng.*, **110**(3), pp. 293–298, 1920.

[2] Bazant, Z.P. & Kaplan, M.F., *Concrete at High Temperatures: Material Properties and Mathematical Models*, Longman Group Ltd.: Essex, 1996.

[3] Castillo, C. & Durrani, A.J., Effect of transient high temperature on high strength concrete. *ACI Mater. J.*, **87**(1), pp. 47–53, 1987.

[4] Kodur, V., Fire performance of high-strength concrete structural members. *Constr. Technol. Updat.*, p. 4, 1999.

[5] Sideris, K.K., Mechanical characteristics of self-consolidating concretes exposed to elevated temperatures. *J. Mater. Civ. Eng.*, **19**(8), pp. 648–654, 2007.

[6] Anagnostopoulos, N., Sideris, K.K. & Georgiadis, A., Mechanical characteristics of self-compacting concretes with different filler materials, exposed to elevated temperatures. *Mater. Struct.*, **42**(10), pp. 1393–1405, 2009.

[7] Foti, D., Prestressed slab beams subjected to high temperatures. *Compos. Part B Eng.*, **58**, pp. 242–250, 2014.

[8] Bailey, C.G. & Ellobody, E., Whole-building behaviour of bonded post-tensioned concrete floor plates exposed to fire. *Eng. Struct.*, **31**(8), pp. 1800–1810, 2009.

[9] Ahmed, A.E., Al-Shaikh, A.H. & Arafat, T.I., Residual compressive and bond strengths of limestone aggregate concrete subjected to elevated temperatures. *Mag. Concr. Res.*, **44(**159), pp. 117–125, 1992.

[10] Peng, G.F., Bian, S.H., Guo, Z.Q., Zhao, J., Peng, X.L. & Jiang, Y.C., Effect of thermal shock due to rapid cooling on residual mechanical properties of fiber concrete exposed to high temperatures. *Constr. Build. Mater.*, **22(**5), pp. 948–955, 2008.

[11] Ding, Y., Azevedo, C., Aguiar, J.B. & Jalali, S., Study on residual behaviour and flexural toughness of fibre cocktail reinforced self compacting high performance concrete after exposure to high temperature. *Constr. Build. Mater.*, **26**(1), pp. 21–31, 2012.

[12] Cheng, F.P., Kodur, V.K.R. & Wang, T.C., Stress-strain curves for high strength concrete at elevated temperatures. *J. Mater. Civ. Eng.*, **16**(1), pp. 84–90, 2004.

[13] Lau, A. & Anson, M., Effect of high temperatures on high performance steel fibre reinforced concrete. *Cem. Concr. Res.*, **36(**9), pp. 1698–1707, 2006.

[14] Poon, C.S., Shui, Z.H. & Lam, L., Compressive behavior of fiber reinforced high-performance concrete subjected to elevated temperatures. *Cem. Concr. Res.*, **34(**12), pp. 2215–2222, 2004.

[15] Varona, F.B., Baeza, F.J., Bru, D. & Ivorra, S., Influence of high temperature on the mechanical properties of hybrid fibre reinforced normal and high strength concrete. *Constr. Build. Mater.*, **159**, pp. 73–82, 2018.

[16] Alberti, M.G., Enfedaque, A.J., Gálvez, C., Cánovas, M.F. & Osorio, I.R., Polyolefin fiber-reinforced concrete enhanced with steel-hooked fibers in low proportions. *Mater. Des.*, **60**, pp. 57–65, 2014.

[17] Ganesan, N., Indira, P.V., & Sabeena, M.V., Behaviour of hybrid fibre reinforced concrete beam–column joints under reverse cyclic loads. *Mater. Des.*, **54**, pp. 686–693, 2014.

[18] International Federation for Structural Concrete (fib), *Fib Model Code for Concrete Structures 2010*, Ernst & Sohn: Berlin, 2013.

[19] European Committee for Standarisation, *Eurocode 2: Design of Concrete Structures – Part 1–2: General Rules – Structural Fire Design*. AENOR: Madrid, 2011.

[20] Haddad, R.H., Al-Saleh, R.J. & Al-Akhras, N.M., Effect of elevated temperature on bond between steel reinforcement and fiber reinforced concrete. *Fire Saf. J.*, **43**(5), pp. 334–343, 2008.

[21] Varona, F.B., Baeza, F.J., Ivorra, S. & Bru, D., Experimental analysis of the loss of bond between rebars and concrete exposed to high temperatures. *Dyna*, **90**(1), pp. 78–86, 2015.

[22] Lublóy, E. & Hlavička, V., Bond after fire. *Constr. Build. Mater.*, **132**, pp. 210–218, 2017.
[23] Varona, F.B., Baeza, F.J., Bru, D. & Ivorra, S., Evolution of the bond strength between reinforcing steel and fibre reinforced concrete after high temperature exposure. *Constr. Build. Mater.*, **176**, pp. 359–370, 2018.
[24] Windisch, A., A modified pull-out test and new evaluation methods for a more real local bond-slip relationship. *Mater. Struct.*, **18(**105), pp. 181–184, 1985.
[25] Hatami, M., *Weighted Residual Methods*, 1st ed., Academic Press, 2017.
[26] Zienkiewicz, O., Taylor, R., & Zhu, J.Z., *The Finite Element Method: Its Basis and Fundamentals*, 7th ed., Butterworth-Heinemann, 2013.
[27] Villacampa, Y.F., Navarro-González, J. & Llorens, J., A geometric model for the generation of models defined in complex systems. *Ecosystems and Sustainable Development VII*, 2009, pp. 71–82.
[28] Navarro-González, F.J. & Villacampa, Y., A new methodology for complex systems using n-dimensional finite elements. *Adv. Eng. Softw.*, **48**, pp. 52–57, 2012.
[29] Navarro-González, F.J. & Villacampa, Y., A finite element numerical algorithm for modelling and data fitting in complex systems. *Int. J. Comput. Methods Exp. Meas.*, **4**(2), pp. 100–113, 2016.

OPTIMISATION OF MODELS FOR THE DETERMINATION OF THE CREST OF BARS ON SANDY BEACHES

ISABEL LÓPEZ[1], YOLANDA VILLACAMPA[2], JOSE IGNACIO PAGÁN[1],
FRANCISCO JOSÉ NAVARRO-GONZÁLEZ[2] & LUIS ARAGONÉS[1]
[1]Department of Civil Engineering, University of Alicante, Spain
[2]Department of Applied Mathematics, University of Alicante, Spain

ABSTRACT

Sand bars are important morphological characteristics of beaches, and changes in their position and height are the main causes of profile variability. The cross-shore movement of the bars can be important for the artificial beach nourishments, because the success of the latter depends on its interaction with the bar position. Therefore, it is important to determine the location of the bars in the profile. The aim of this work is to evaluate the different existing models, and to obtain an optimized model that uses the least possible number of variables and obtains the best results. First of all, a total of 25 variables related to the characteristics of the waves, sediment and physical characteristics of the beach have been studied by means of a correlation analysis. Secondly, we have tried to generate linear models using the backward method, which generates successive models eliminating variables in each of them. These models, however, did not offer good results, with R2 values lower than 0.4. For this reason, different numerical models have been generated using among others the same variables used by different authors in their formulations or models. The numerical models of finite elements use Galerkin's methodology and show that the most influential variables on the location of the bars crest are: wave height, period and median sediment size. These variables are very similar to those proposed by other authors; however, the formulations proposed by these authors do not offer good results in the area of study, while with the models generated, the errors committed in absolute value are less than 8%. This leads us to the final idea that the influential variables in the bars are the same in any study area, but the degree of influence or relationship with the study parameter depends on the study area.

Keywords: bar crest, sand beaches, numerical models, Galerkin's methodology.

1 INTRODUCTION

Sand bars are an important morphological feature of natural beaches. In many places, changes in the position and height of the bar are the main causes of profile variation [1]. The bars also significantly affect the transformation of the wave and therefore the subsequent evolution of the profile [2], [3]. However, the formation and evolution of the sandbanks are little known. Holman and Sallenger [1] reviewed the hypotheses for sand bar formation and concluded that both the breaking point [4] and the infragravity wave [5] may be important.

Several researchers have suggested mechanisms that could cause bars to migrate to the coast. Trowbridge and Young [6] demonstrated that, under moderately energetic waves, the flow generated by the amount of movement in the boundary layer has a local maximum on the bar. They suggested that this would lead to a maximum in the transport of sediments towards the coast over the bar, and to the migration of the bars towards the coast. The Trowbridge and Young [6] model successfully predicted a bar-to-shore migration event observed on a natural beach, but this model excludes sediment transport by hangover and fails to predict inshore migration.

Among the different models developed for bar prediction, the most commonly used variables are wave height in deep water (H_o), period (T) or wavelength in deep water (L_o), and median sediment size expressed explicitly (D_{50}) or implicitly (fall velocity; $w = 14 \cdot D_{50}^{1.1}$)

WIT Transactions on Engineering Sciences, Vol 125, © 2019 WIT Press
www.witpress.com, ISSN 1743-3533 (on-line)
doi:10.2495/CMEM190201

[7]–[11]. Some of these authors also incorporate in their models the wave height at break (H_b) [8], [9], [11] and/or the slope (m) [8], [10]. It is worth noting the maximum wave height (H_{max}) used by Plant et al. [8], the wave incidence direction with respect to the beach normal (θ) used by Pape et al. [9], and the tidal range integrated by Masselink and Short [11]. In addition, most of these models also take into account the wave duration time (t).

Therefore, the aim of this work is to optimize the current models. For this purpose, a numerical model will be determined, which provides the position of the bar crest as precisely as possible and uses as few variables as possible. In addition, the most influential variables in the position of the bar crest will be determined, for which different models will be generated using some of the variables proposed by other authors and other new variables.

2 STUDY AREA

The area under study covers a length of 17.7 km of the Valencia coast (Spain) (Fig. 1). The area is divided into two distinct morphodynamic units to the North and South of the Port of Valencia. The beaches in these areas are of fine sand, with median sediment sizes between 0.172 mm and 0.452 mm. It is a micromareal zone, with astronomical tides ranging between 20 and 30 cm, and together with meteorological tides can reach up to 75 cm [12].

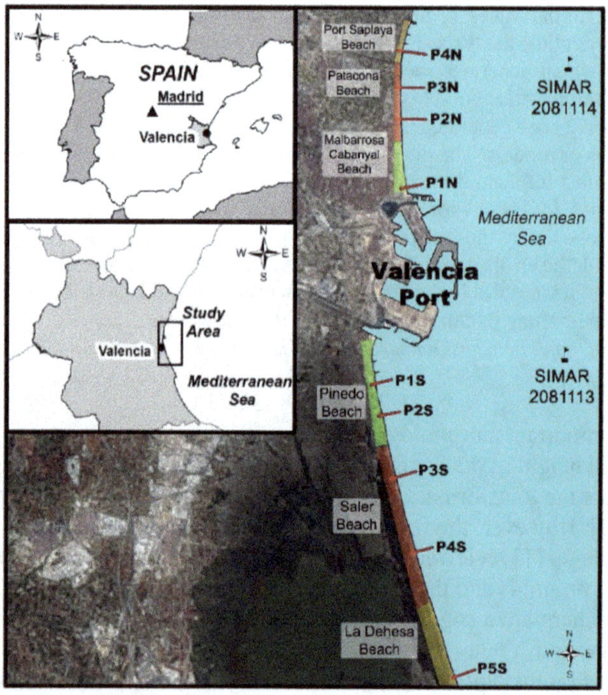

Figure 1: Location of the study area and the analysed profiles.

In Fig. 1, the nine representative points where the profiles of this study have been taken are observed (four profiles to the North of the Port and five to the South). In the northern area are the profiles P1N (Malbarrosa-Cabanyal Beach), P2N and P3N (Patacona Beach) and P4N (Port Saplaya Beach). In the southern zone that occupies about 13 km are the profiles P1S and P2S (Pinedo Beach), P3S and P4S (Saler Beach) and P5S (La Dehesa Beach).

3 METHODOLOGY

Below is a description of the procedure followed to obtain the data necessary for the study, as well as the procedure followed in the generation of the different numerical models.

3.1 Data collection

The methodology followed for the determination of the position of the bars crest and obtaining the different characteristics of the waves and sedimentology is presented.

3.1.1 Position of the bar crest

In order to determine the position of the bars, 209 precision profiles (error less than 2 cm) were obtained using the Profiling Bar method (BP) [13]. This method has a system of joinable bars with an articulated foot and a crown with two reflective prisms that make it possible to determine the height of the beach profile, independently of the mean sea level.

The frequency of profiling was not regular. In the period 1992–1994, the profiles were taken every 2 months, from 1994–1997 every 4 months and during the period 2005–2014 they were taken every 6 months, in April and October of each year.

Each of the profiles was carefully analysed and the distance to the coastline and the depth of the bar crest was obtained for each of them.

3.1.2 Maritime climate

Wave data were obtained from SIMAR (time series of wind and wave parameters from numerical modelling) node data (recorded every 3 hours since 1958) provided by the Puertos del Estado Public Agency. Specifically, the data of the nodes SIMAR 2081114 (North zone with 62489 wave height data since 1992) and SIMAR 2081113 (South zone with 62489 wave height data since 1992) were used.

For the study of the waves in each of the profiles studied, a sectorization of the incoming energy was carried out first, discarding those waves that do not affect the profile. Next, for each of the profiles and the period elapsed between the taking of each profile, Carol v1.0 [14] was used to obtain the wave height $H_{s,12}$ (wave height exceeded only 12 hours per year, with a probability of 0.137% regardless of the period studied) and the maximum wave height (H_{max}) produced in each period studied, as well as their corresponding periods (T) and directions (θ).

3.1.3 Sedimentology

Finally, the median sediment size (D_{50}) (UNE-EN 933-1:2012) corresponding to each of the studied profiles in the dry beach was obtained. For this purpose, the samples corresponding to these profiles collected during the sampling campaign carried out by the University of Alicante in 2013 were used.

3.2 Model generation

Initially, linear models were studied considering all the variables, for which the backward method was applied, starting with a linear model of 24 explanatory variables to eliminate one by one the variables. All models had a bad fit with a maximum value of R^2 of 0.4. This indicates that there are no linear relationships between the considered explanatory variables and the bars crest position. Therefore, it is necessary to analyse non-linear relationships.

Before generating the numerical models, a correlation study was carried out to see the influence of each of the variables. Given that the values of r obtained in the study of correlations between the different variables and the position (x and y) of the bar crest were very weak (0.0–0.2), it was decided to generate models using the variables most used by other authors and gradually adding variables. Six models were generated with the following input variables:

1. M1: Hs,12 and T
2. M2: Hs,12 and m
3. M3: Hs,12 and D50
4. M4: Hs,12, T and m
5. M5: Hs,12, T and D50
6. M6: Hmax, Hs,12, T and m

The numerical models were generated according to the methodology developed in Navarro-González and Villacampa [15]. This methodology is based on Galerkin's formulation [16], [17] of the finite element method and improves the computational efficiency of the algorithms of the numerical methodologies used by the authors in other research [15], [18]–[21]. A model is generated for each finite element or mesh model defined in a hyper cube. As usual, each finite element mesh is determined by its complexity (mesh density) which is defined as the number of elements defined along each edge of the hypercube. Therefore, for each data set (M1, M2, M3, M4, M5 and M6) families of models were generated from the selected complexity.

To evaluate the performance of each model, the results obtained were compared with the formulations proposed by Hsu [22], Silvester and Hsu [23], Günaydın and Kabdaşlı [24] and Kömürcü et al. [25], as well as the artificial neural network (ANN) proposed byLópez et al. [26]. The results were compared in terms of R^2, absolute error (eqn 1) and mean absolute percentage error (MAPE) (eqn (2))

$$\varepsilon = \sqrt{\frac{\sum_{i=1}^{N}(a^i - d^i)^2}{N-p}}, \quad (1)$$

$$\text{MAPE} = \frac{1}{N}\sum_{i=1}^{N}\left|\frac{a^i - d^i}{a^i}\right|. \quad (2)$$

4 RESULTS

First, the results for R^2 obtained for each of the models generated are shown (Table 1). As can be seen, the models with two variables (M1, M2 and M3) yield R^2 values lower than 0.8 for both x and y (distance to shore and depth of bar crest) in almost all complexities. While the other three models reach values higher than 0.8 with relatively low complexities, the M5 model reaches values higher than 0.9 from complexity 60. In addition, as complexity increases, the value of R^2 increases until it reaches values of 0.96 for complexity of 100 in x and 0.96 for complexity of 130.

On the other hand, if the absolute error (Table 2) and the MAPE (Table 3) are analysed, it can be seen that the errors are relatively small. If the 3-variable models (M4 and M5) are analysed, it can be seen how, for distance from the coast, errors are less than 20 m for complexity 40, and even less than 10 m for complexity 70 in the M5 model, which means an MAPE of 10.7%. In the case of depth (y), the M4 model is not capable of lowering the 19cm error while the M5 model goes down to 10 cm, which means an MAPE of 8.4% and 5.4%, respectively. This indicates that the influence of the sediment is greater than that of the slope, given that the results of the M5 model are better than those achieved with the M4, using much lower complexities. As for the results obtained by the M6 model with 4 variables, it can be

Table 1: R^2 values for each of the generated models.

Comp.	Results for X						Results for Y					
	M1	M2	M3	M4	M5	M6	M1	M2	M3	M4	M5	M6
30	0.188	0.503	0.400	0.636	0.691	0.666	0.192	0.353	0.405	0.560	0.678	0.603
40	0.234	0.559	0.475	0.708	0.796	0.741	0.237	0.420	0.493	0.645	0.796	0.683
50	0.260	0.608	0.541	0.756	0.866	0.789	0.259	0.481	0.569	0.698	0.867	0.730
60	0.278	0.650	0.587	0.796	0.905	0.826	0.274	0.532	0.624	0.739	0.906	0.764
70	0.290	0.686	0.625	0.821	0.926	0.857	0.286	0.579	0.665	0.767	0.925	0.793
80	0.297	0.717	0.652	0.843	0.940	0.875	0.293	0.619	0.693	0.790	0.937	0.810
90	0.304	0.744	0.674	0.860	0.951	0.890	0.299	0.654	0.714	0.808	0.945	0.826
100	0.309	0.765	0.688	0.872	0.957	0.902	0.303	0.682	0.727	0.821	0.950	0.839
110	0.313	0.786	0.702	0.886		0.914	0.307	0.712	0.740	0.836	0.954	0.852
120	0.316	0.802	0.713	0.894			0.309	0.734	0.751	0.846	0.957	0.861
130	0.319	0.815	0.719	0.903			0.311	0.754	0.756	0.857	0.960	

Table 2: Absolute error values for each of the generated models.

Comp.	Results for X						Results for Y					
	M1	M2	M3	M4	M5	M6	M1	M2	M3	M4	M5	M6
30	39.79	29.77	32.54	25.19	22.10	23.83	0.79	0.78	0.54	0.45	0.38	0.42
40	38.57	27.54	30.05	21.25	16.84	19.57	0.80	0.77	0.49	0.38	0.29	0.35
50	37.79	25.42	27.65	18.38	12.98	16.57	0.81	0.76	0.45	0.33	0.23	0.30
60	37.30	23.62	25.73	16.04	10.26	14.58	0.82	0.76	0.42	0.30	0.18	0.27
70	36.92	22.10	23.95	14.47	8.69	12.94	0.82	0.76	0.39	0.27	0.16	0.25
80	36.66	20.73	22.59	13.14	7.64	11.88	0.82	0.76	0.37	0.25	0.14	0.23
90	36.52	19.30	21.49	12.11	6.81	10.94	0.82	0.76	0.35	0.23	0.12	0.22
100	36.36	18.24	20.64	11.33	6.31	10.11	0.82	0.76	0.33	0.22	0.12	0.21
110	36.29	17.05	19.76	10.49		9.41	0.83	0.76	0.32	0.21	0.11	0.20
120	36.19	16.08	18.99	9.97			0.83	0.76	0.31	0.20	0.11	0.19
130	36.11	15.23	18.48	9.35			0.83	0.76	0.30	0.19	0.10	

Table 3: MAPE values for each of the generated models.

Comp.	Results for X						Results for Y					
	M1	M2	M3	M4	M5	M6	M1	M2	M3	M4	M5	M6
30	51.1%	28.1%	43.2%	23%	29.2%	21.5%	42.9%	41.5%	28.5%	20.5%	19.4%	19.0%
40	48.6%	25.3%	39.7%	18.7%	22.1%	17.1%	43.7%	41.0%	26.2%	168%	15%	15.3%
50	47%	23.1%	36.2%	15.8%	16.6%	14.3%	44.0%	40.6%	24.0%	14.5%	11.6%	13.3%
60	45.9%	21.2%	33.5%	13.6%	12.8%	12.5%	44.2%	40.3%	22.3%	12.9%	9.3%	12.1%
70	45.3%	19.8%	31.3%	12.2%	10.7%	11.1%	44.4%	40.2%	20.9%	11.9%	8.0%	11.1%
80	44.8%	18.4%	29.4%	11%	9.3%	10.2%	44.5%	40.2%	19.7%	11%	7.2%	10.4%
90	44.5%	17.1%	27.8%	10.1%	8.0%	9.4%	44.5%	40.1%	18.7%	10.2%	6.5%	9.8%
100	44.3%	16.1%	26.6%	9.4%	7.3%	8.7%	44.5%	40.1%	17.9%	9.7%	6.1%	9.3%
110	44.2%	15%	25.5%	8.7%		8.1%	44.5%	40.1%	17.2%	9.2%	5.8%	8.9%
120	44%	14.1%	24.4%	8.2%			44.5%	40.1%	16.4%	8.8%	5.6%	8.5%
130	43.9%	13.3%	23.8%	7.7%			44.5%	40.1%	16.0%	8.4%	5.4%	

observed that they are very similar to those obtained by the M4 model, which indicates that the inclusion of the maximum wave height has hardly any influence on the model.

Therefore, the M5 model with a complexity of 80 is selected as the optimal model. With this complexity, admissible errors of 7.5 m in the distance to the coast and 14 cm in the depth are reached, which corresponds to a MAPE of 9.3% and 7.2%, respectively. If this model is compared with the results obtained using the formulations or models proposed by other authors (see Section 3.2), it can be seen (Table 4) that the model generated (M5) considerably improves these results. If one analyses the formulations that use the same variables (S&H and G&K) for the coast distance, the model reduces the absolute error by 56.6 m and 53.8 m respectively, while for depth the error decreases in the order of 90cm. The smallest reduction in error occurs when comparing the M5 model with the ANN, and yet the reduction is 16 m in the distance to the coast and 29 cm in depth.

Table 4: Comparison of errors between the model generated (M5) and the formulations proposed by other authors.

Models	M5	H	S&H	G&K	K	ANN
Number of variables	3	4	3	3	4	7
Variables	$H_{s,12}$, T and D_{50}	$H_{s,12}$, T, Θ and m	$H_{s,12}$, T and m	$H_{s,12}$, T and m	$H_{s,12}$, T, m and D_{50}	Month, H_{max}/L_{omax}, Θ_{hmax}, Days, H_m, D_{50}, Beach width difference between profiles
R^2 for X	0.940	0.229	0.179	0.179	0.000	0.706
R^2 for Y	0.937	0.105	0.004	0.004	0.059	0.621
Absolute error X	7.6	315.3	64.2	53.8	203.7	23.6
Absolute error Y	0.138	1.20	1.05	1.04	1.01	0.42
MAPE in X	9.3%	289.2%	61.1%	47.0%	270.5%	28.5%
MAPE in Y	7.2%	48.2%	47.1%	49.3%	42.8%	23.3%

Where: H refers to Hsu [22], S&K to Silvester and Hsu [23], G&K to Günaydın and Kabdaşlı [24], K to Kömürcü et al. [25], and ANN to the artificial neural network proposed by López et al. [26].

5 DISCUSSION

Correct determination of the location of sand bars has important implications for coastal engineering. Therefore, if we are able to determine the exact position or range of motion of the bars, we will be able to determine with greater precision the influential elements in the calculation of the coastal protection elements or the volumes of sand needed for a nourishment. For this reason, this article has attempted to optimize existing formulations and models for the determination of the bar crest. Thus, different numerical models based on Galerkin's formulation [15] have been generated using different input variables. The use of Galerkin's formulation of the finite element method and the corresponding improvement in computational efficiency allows us to analyse relationships in which a greater number of variables are involved without this implying a high time in the execution of the software.

Comparing the results obtained using numerical models with existing formulations shows that there is a big difference. As for R^2, the proposed formulations do not have a good fit in the study area with values lower than 0.25, while the neural network proposed for the study area reaches values of 0.7. With the generated numerical model (M5), which has three variables and a complexity of 80, values of R^2 greater than 0.93 are reached. This difference in the adjustment is also observed in the errors. Thus, the numerical model selected results in an absolute error about three times lower than that of the ANN proposed by López et al. [26], which is the one that results in the least error of the different models proposed by other authors (Table 6).

Given that the selected model has been generated using variables proposed and used by other authors [22]–[26], and the errors obtained are much smaller, it can be deduced that the influential variables in the bars are the same in any area of study, but the degree of influence or relationship with the study parameter depend on the study area. Therefore, it is necessary to generate different models depending on the characteristics of the study area or, in order to be able to use a model globally, it would be necessary to generate a single model that includes (in its generation) data from all over the world.

Finally, the proposed new model associated with other similar tools such as the determination of erosion [27] or bar volume [28], becomes not only a tool for bar knowledge, but also an additional method for coastal control and management, which may imply a reduction in both construction and maintenance costs.

6 CONCLUSION

In this paper a numerical model based on Galerkin's formulation of the finite element method has been developed, which allows to reduce the computation time and to generate families of numerical models with high complexity. Among the different models generated it can be stated that:

- The use of three variables ($H_{s,12}$, T and D_{50}) is enough to generate models with high precision in the determination of the sand bar. It achieves errors of less than 10% for the distance to the coast and 7.5% for the depth of the bar crest.
- The use of the median sediment size (D_{50}) influences the improvement of the model results much more than the inclusion of the profile slope.
- The influential variables in the bars are the same in any study area, but the degree of influence or relationship with the study parameter depend on the study area.

ACKNOWLEDGEMENT

This work was partially supported by the Universidad de Alicante through the project "Estudio sobre el desgaste y composición de los sedimentos y su influencia en la erosión de las playas españolas" (GRE16-09).

REFERENCES

[1] Holman, R.A. & Sallenger Jr, A.H., Sand Bar Generation: A Discussion of the Duck Experiment Series. *Journal of Coastal Research*, pp. 76–92, 1993.

[2] Takeda, I. & Sunamura, T., Beach changes by storm waves. *Coastal Engineering Proceedings*, **1**(20), 1986.

[3] Seymour, R.J., An assessment of NSTS, Coastal Sediments (1987), ASCE, pp. 642–651, 1987.

[4] Dyhr-Nielsen, M. & Sorensen, T., Some sand transport phenomena on coasts with bars. *Proceedings of the 12th Coastal Engineering Conference*, American Society of Civil Engineers, pp. 855–865, 1971.

[5] Holman, R.A. & Bowen, A.J., Bars, bumps, and holes: Models for the generation of complex beach topography. *Journal of Geophysical Research: Oceans*, **87**(C1), pp. 45–68, 1982.

[6] Trowbridge, J. & Young, D., Sand transport by unbroken water waves under sheet flow conditions. *Journal of Geophysical Research: Oceans (1978–2012)*, **94**(C8), pp. 10971–10991, 1989.

[7] Gallagher, E.L., Elgar, S. & Guza, R., Observations of sand bar evolution on a natural beach. *Journal of Geophysical Research: Oceans*, **103**, pp. 3203–3215, 1998.
[8] Plant, N., Holman, R. Freilich, M. & Birkemeier, W., A simple model for interannual sandbar behavior. *Journal of Geophysical Research: Oceans (1978–2012)*, **104**(C7), pp. 15755–15776, 1999.
[9] Pape, L., Ruessink, B.G., Wiering, M.A. & Turner, I.L., Recurrent neural network modeling of nearshore sandbar behavior. *Neural Networks*, **20**(4), pp. 509–518, 2007.
[10] Różyński, G., Data-driven modeling of multiple longshore bars and their interactions. *Coastal Engineering*, **48**(3), pp. 151–170, 2003.
[11] Masselink, G. & Short, A.D., The effect of tide range on beach morphodynamics and morphology: a conceptual beach model. *Journal of Coastal Research*, pp. 785–800, 1993.
[12] ECOLEVANTE, Estudio ecocartográfico de las provincias de Valencia y Alicante, in: D.G.d. Costas (ed.) Ministerio de Medio Ambiente, 2006.
[13] Serra, J. & Medina, J.R., Beach monitoring program of Valencia (Spain). *Proceedings of the 25th Coastal Engineering*, pp. 590–591, 1996.
[14] IHCantabria, Análisis Matemático y Estadístico de Variables Medioambientales (AMEVA), in: U.d. Cantabria (ed.) Cantabria, Spain. Available online: http://ihameva.ihcantabria.com/, 2013.
[15] Navarro-González, F.J. & Villacampa, Y., A finite element numerical algorithm for modelling and data fitting in complex systems. *International Journal of Computational Methods and Experimental Measurements*, **4**(1), pp. 100–113, 2016.
[16] Zienkiewicz, O.C., Taylor, R.L. & Zhu, J.Z., *The finite element method: its basis and fundamentals*, Elsevier Butterworth-Heinemann, 2005.
[17] Hatami, M., *Weighted Residual Methods*, Academic Press, 2018.
[18] Navarro-González, F.J. & Villacampa, Y., A new methodology for complex systems using n-dimensional finite elements. *Advances in Engineering Software*, **48**(1), pp. 52–57, 2012.
[19] Aragonés, L., Villacampa, Y., Navarro-González, F.J. & López, I., Numerical modelling of the equilibrium profile in Valencia (Spain). *Ocean Engineering*, **123**, pp. 164–173, 2016.
[20] Aragonés, L., López, I., Villacampa, Y. & Navarro-González, F.J., Using the presence of seagrass Posidonia oceanica to model the equilibrium profile parameter A of sandy beaches in Spain. *Journal of Coastal Research*, **33**(3), pp. 642–652, 2017.
[21] Navarro-González, F.J. & Villacampa, Y., Generation of representation models for complex systems using Lagrangian functions. *Advances in Engineering Software*, **64**, pp. 33–37, 2013.
[22] Hsu, T.-W., Geometric characteristics of storm-beach profiles caused by inclined waves. *Ocean Engineering*, **25**(1), pp. 69–84, 1998.
[23] Silvester, R. & Hsu, J.R., Coastal stabilization. *Advanced Series on Ocean Engineering*, p. 14, 1997.
[24] Günaydın, K. & Kabdaşlı, M.S., Investigation of offshore bar geometry under regular and irregular waves. *Journal of Coastal Research*, pp. 374–382, 2005.
[25] Kömürcü, M., Özölçer, İ., Yüksek, Ö. & Karasu, S., Determination of bar parameters caused by cross-shore sediment movement. *Ocean Engineering*, **34**(5), pp. 685–695, 2007.
[26] López, I., Aragonés, L., Villacampa, Y. & Serra, J.C., Neural network for determining the characteristic points of the bars. *Ocean Engineering*, **136**, pp. 141–151, 2017.

[27] Tsekouras, G.E., Rigos, A., Chatzipavlis, A. & Velegrakis, A., A Neural-Fuzzy Network Based on Hermite Polynomials to Predict the Coastal Erosion, in: L. Iliadis, C. Jayne (eds), *Engineering Applications of Neural Networks: 16th International Conference*, EANN 2015, Rhodes, Greece, pp. 195–205, 2015.
[28] Demirci, M., Üneş, F. & Aköz, M.S., Prediction of cross-shore sandbar volumes using neural network approach. *Journal of Marine Science and Technology*, **20**(1), pp. 171–179, 2015.

Author index

Part II

Earthquake Resistant Engineering Structures XII

WIT*eLibrary*
Home of the Transactions of the Wessex Institute.
Papers published in this volume are archived in the WIT eLibrary in volume 185 of WIT Transactions on the Built Environment (ISSN 1743-3509).
The WIT electronic-library provides the international scientific community with immediate and permanent access to individual papers presented at WIT conferences.
Visit the WIT eLibrary at www.witpress.com.

Preface

This volume contains some of the papers presented at the 12th International Conference on Earthquake Resistant Engineering Structures (ERES), organised by the Wessex Institute, in Seville, Spain. The meeting was sponsored by the International Journal of Safety and Security Engineering, and the WIT Transactions on the Built Environment. The Conference was organised by the Wessex Institute, UK.

The conference began in Thessaloniki, Greece in 1997, followed by Catania, Italy (1999); Malaga, Spain (2001); Ancona, Italy (2003); Skiathos, Greece (2005); Bologna, Italy (2007), Cyprus (2009), Tuscany, Italy (2011), A Coruña, Spain (2013), Opatija, Croatia (2015) and Alicante, Spain (2017).

The meeting provided a unique forum for the discussion of basic and applied research in the various fields of earthquake engineering relevant to structures.

The problem of protecting the built environment in earthquake-prone regions involves not only the optimal design and construction of new facilities, but also the upgrading and rehabilitation of existing structures including heritage buildings. The type of highly specialized retrofitting employed to protect the built heritage is an important area of research and appropriate to the conference objective of designing better earthquake resistant buildings. Due attention has also been given to earthquake related phenomena where fluid-structure interaction is an important damaging factor: see recent major accidents caused by tsunamis.

This ERES conference strives to address these problems continuing to expand on the development of previous meetings in the series. Papers presented at ERES are an invaluable record of the state of the art in this field. All of them are now freely available on the Wessex Institute eLibrary (https://www.witpress.com/elibrary), where they are a permanent record demonstrating the quality of the research presented at the ERES conference series.

The Editors are grateful to the members of the International Scientific Advisory Committee and other colleagues who have helped to select the papers in this book, as well as to all authors for their collaboration.

The Editors
Seville, 2019

NUMERICAL MODEL OF TRM-REINFORCED MASONRY WALLS UNDER LATERAL IN-PLANE LOADS

SALVADOR IVORRA[1], DAVID BRU[1], FRANCISCO JAVIER BAEZA[1], BENJAMÍN TORRES[1] & DORA FOTI[2]
[1]University of Alicante, Spain
[2]Politecnico di Bari, Italy

ABSTRACT
The development of Textile Reinforced Mortar (TRM) has offered new potential for a number of applications in the field of structural strengthening. Its advantages over other strengthening techniques – such as compatibility with substrates, lower costs, better performance at high temperatures and high ductility – make the TRM vital for unreinforced masonry structures which are most vulnerable to seismic loadings. This paper presents a non-linear numerical simulation with a 2D finite element analysis (FEA) of TRM reinforced masonry walls under lateral in-plane loading. Simulations include the modelling of walls with and without openings; walls were modelled through a non-linear shell-layered element; brick masonry wall was modelled as a continuous material using two overlapping layers with different mechanical characteristics. The TRM was defined as a single shell of material for the mortar matrix and the mesh glass fibre applied through two layers that make up the reinforced plaster applied to the two faces of the wall. The increment of the maximum horizontal load and the high increment of ductility of the TRM retrofitted have been observed. Results provide a prediction of the plane's structural behaviour in order to propose a reinforcement method for masonry structures using a TRM.
Keywords: cement composite, masonry, push-over, FEM, retrofitting.

1 INTRODUCTION

Masonry buildings constitute most of the South European Heritage. Most of them are also artistic and valuable goods and are considered among the most popular structural systems in the European continent. The seismic retrofitting of these structures has become a topic of great relevance given their usually high vulnerability and, above all, the result of the latest and devastating seismic events. Therefore, the reinforcement of existing structures today is becoming one of the major activities in the field of civil engineering. The development of research and technology in recent decades has enabled us to use composite materials that best meet performance requirements. They are a valid alternative to traditional reinforcement and refurbishment techniques. To date, the use of composite materials is increasing in professional practice. However, being recently introduced, these materials require studies and experimental tests to carefully evaluate the benefits of their application [1], [2]. Innovative and sustainable materials have been recently proposed for the reinforcement of concrete, such as the use of PET from recycled waste bottles [3]–[5]. For masonry structures, Textile Reinforced Mortars (TRM) appeared as an alternative for the seismic retrofitting of URM walls [6], [7]. These materials comprise a fiber mesh and a cement mortar with different additives for higher ductility. The textile material is usually a bidirectional grid of glass, carbon or basalt fibers. The main advantage of TRM is a better compatibility with masonry, which could avoid the aforementioned bonding or permeability issues. In addition, TRM can also improve masonry's strength and ductility [6]. TRM seemed to provide a significant improvement of masonry's strength and deformability [8].

Numerical modelling of masonry structures with FEM is a very computationally time demanding task due to: (i) Complex typological characteristic of masonry structures, (ii) non-

WIT Transactions on The Built Environment, Vol 185, © 2019 WIT Press
www.witpress.com, ISSN 1743-3509 (on-line)
doi:10.2495/ERES190011

linearity in material behaviour and (iii) the lack of reliable experimental data to characterize the material can be counted as thereof the several reasons. Masonry consists of brick units and mortar as a bonding material. Because of its complex geometric nature, it is necessary to assume convenient material behaviour (stress–strain) and conduct the analysis with numerical techniques, to obtain the global response of the structure. The addition of some retrofitting elements can increment the computational time and the complexity of the analysis.

The present paper focuses on the use of textile-reinforced mortars (TRM). Some studies have also been carried out by the authors on the use of reinforced resins for the retrofitting of concrete columns [8]–[10]. A new simplified formulation and implementation of a FEM model for the analysis of walls reinforced by mean of textile-reinforced mortar (TRM) is here proposed. The objective of such numerical analysis is, of course, the prediction of the behavior of real walls in order to propose a reinforcement method for masonry structures using a TRM composite.

The research is part of a wider experimental project at the University of Alicante (Spain) with the aim of analyzing the behavior of masonry walls against seismic actions in order to propose a reinforcement method using TRM composite material [11].

2 FE MODEL

Two specimens with 0.25 m thickness have been considered. Both have the same dimensions, but one has an opening for a door. Fig. 1 shows the geometric characteristics of the specimens.

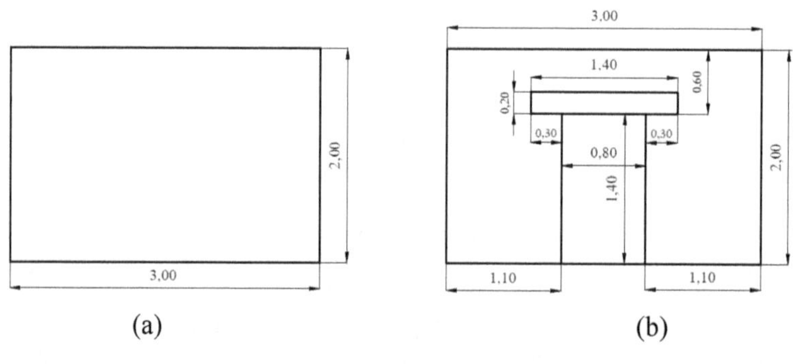

Figure 1: Dimensions of the specimens. (a) Without openings; and (b) With an opening for a door.

In this study, in-plane behaviour of the specimens has been simulated. Two different loading conditions of compressive vertical loads have been analyzed. Both can be similar to the gravity load caused by the weight of several floors. For this reason, 150 kN and 300 kN have been considered. Under these load conditions, maximum horizontal load and its displacement have been analyzed. For boundary conditions, the base of the specimens is constrained (each node has restrained its vertical and horizontal displacements) and the top where load is applied is free.

There are several modeling strategies that can be classified into micro-modeling and macro-modeling.

Micro-modeling considers the masonry as a material composed of bricks and mortar joints molded continuously, in which the interface represents a point of discontinuity. Therefore, there is a large number of parameters that come into play in the construction of the model: (i) properties of blocks and mortar, (ii) block geometry, (iii) characteristics of joints, (iv) block-mortar interface, etc. Therefore, the application of micro-models for the entire structure is prohibitive due to the too high number of variables [12]. Results with these numerical simulations can present a very approximation to the crack pattern of a real masonry wall. The macro-modeling, instead, considers the masonry as a homogeneous continuous that replaces the two materials (blocks–mortar).

In this paper, a macro-modelling strategy has been considered. The aim of this study is to evaluate the effect of TRM on the in-plane behavior of masonry walls by a macro-elements FEM developed with a commercial software [13]. To this aim, masonry wall has been simulated with a non-linear behavior and the structural response of the specimen loaded in its own plane has been obtained. The following section provides a description of the modelling procedure and some theoretical aspects related to the constitutive laws of the materials adopted.

2.1 Modelling strategy

The wall was modeled through a non-linear shell-layered element. In particular, two different layers have been considered, each of which is characterized by its own constitutive equation (stress–strain curve). The shell-layered allows defining any number of layers in the thickness direction; each layer is identified by its position, thickness. Fig. 2 shows the arrangement of the material that is characterized by its own stress–strain relationship in the case of masonry wall with reinforcement. This modeling strategy implies a perfect adhesion between the reinforcing material and the masonry.

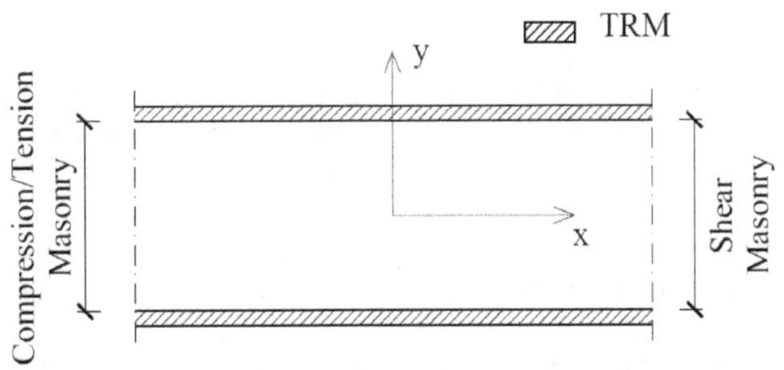

Figure 2: Numerical model layered shell for the TRM-reinforced wall.

The anisotropy of masonry will be modeled by two different stress strain curves. Each of them will represent respectively vertical and horizontal stresses, and shear stress: S11 and S22 stress–strain curves will show the same behavior.

In both specimens, the constitutive equations of the masonry have been considered as a nonlinear layered shell. For specimen with an opening door, the beam has been modelled as an elastic material to simulate timber material.

Fig. 3 shows a general view of the FEM developed with a 0.10 m squared mesh. The wall has been constrained at the base, each node has restrained its vertical and horizontal displacements. Moreover, in order to reduce local effects during the pushover analysis a singular element has been modelled with the same thickness as the wall but of a material having the following characteristics:

- Zero self-weight;
- Same modulus of elasticity as the masonry;
- Elastic behaviour.

With this configuration, an incremental horizontal displacement has been applied in point B according the recommendations of Foti [14] and Bilgin and Korini [15].

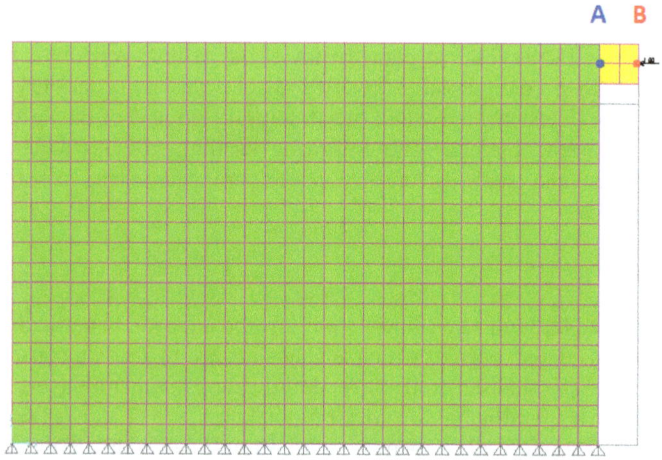

Figure 3: Numerical model of the complete wall.

2.1.1 Brick masonry material

The brick masonry wall was modelled as a continuous material. For it, two overlapping layers with different mechanical characteristics in compression/tension and shear were used. According the Italian seismic code [16], Appendix C8.A of this Standard identifies that it is possible to define the values of the mechanical parameters of the masonry by referring to Table C8A.2.1: "Reference values of the mechanical parameters (minimum and maximum) and average specific weight for different typologies of masonry". The use of this table presupposes the framing of the specific case in one of the pre-constituted types of masonry. For this case study, the values related to the type of masonry "solid brick and lime mortar masonry" was decided to use. For it, the mean of the suggested range of values were adopted for compressive and shear strength, respectively.

The density of masonry was considered of 1800 kg/m^3 according these Standards. Fig. 4 shows the compression and tension stress–strain diagram of the masonry used in this paper. Fig. 5 show the shear stress–strain diagram.

2.1.2 Timber material

Timber beam is considered as an elastic and linear material with a maximum tension/compression strength of 14.0 MPa and an ultimate strain for tension and compression of 0.0019 mm/mm.

2.1.3 TRM material

TRM reinforcement (mortar matrix and mesh glass fiber) was defined as a single shell and was applied at two faces of the wall. Table 1 shows the mechanical properties for the TRM reinforcement and Fig. 6 shows its constitutive behavior.

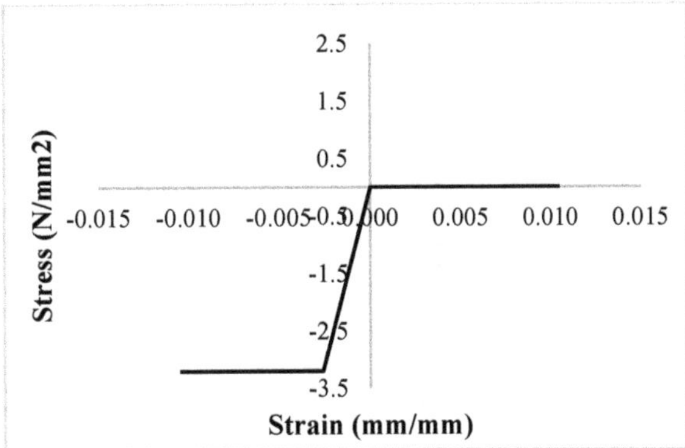

Figure 4: Compression-tension constitutive behaviour of masonry.

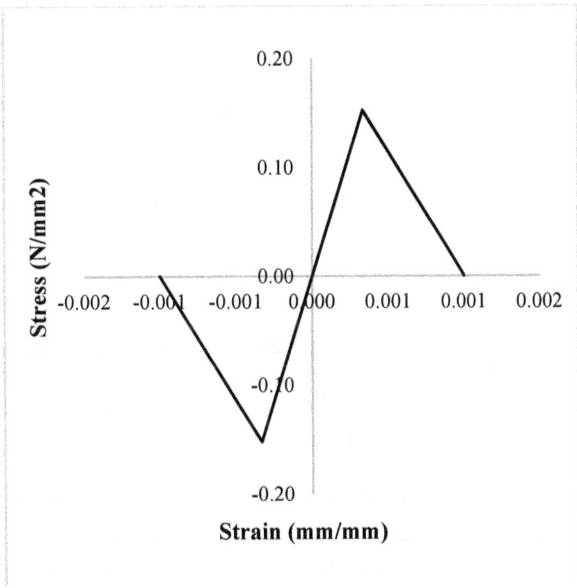

Figure 5: Shear constitutive behaviour of masonry.

Table 1: Mechanical characteristics of the TRM reinforcement.

	Fibre				Mortar		TRM			
	σ_f	E_f	V_f^*	ε_f	σ_{mt}	E_m	σ_{TRM}	ε_1	$\varepsilon_2 \equiv \varepsilon_f$	E_{TRM}
	MPa	MPa			MPa	MPa	MPa	-	-	MPa
Tension	1200	7.2E+3	3.75E-3	16.7E-3	1.1	8E+3	5.60	0.679E-3	16.7E-3	8240
Compression	1200	7.2E+3	3.75E-3	16.7E-3	16	8E+3	20.44	2.481E-3	4.96E-3	8240

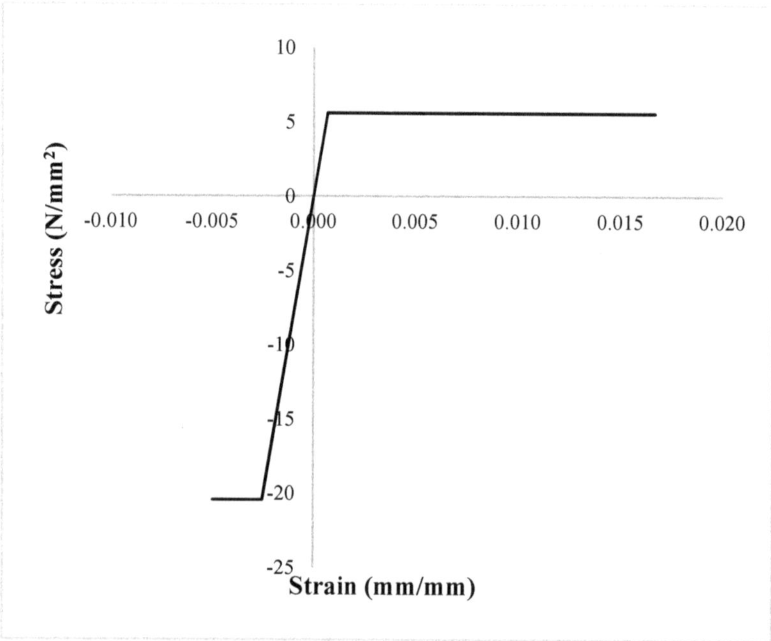

Figure 6: Constitutive behaviour of TMR reinforcement.

2.2 Results of structural analysis

A push-over analysis has been performed. Results show the structural behavior for two specimens (complete wall and wall with an opening) under two different gravity loads (Vertical 1 are 150 KN and Vertical 2 are 300 KN). Table 2 show the values for forces (F_u) and displacements (δ_u) for both specimens and for both gravity loads.

Table 2: Load/displacement results for the pushover analyses.

	Complete wall				Wall with an opening			
	Vertical 1 150 kN		Vertical 2 300 kN		Vertical 1 150 kN		Vertical 2 300 kN	
	δ_u mm	F_u kN	δ_u mm	F_u kN	δ_u mm	F_u kN	δ_u mm	F_u kN
Without TRM	1.04	66	1.04	73	1.09	39	1.3	47
With TRM	70.2	430	180	510	29.7	320	108	420

From Table 2, it can be observed that a TRM reinforcement has a clear improvement in structural response. The improvement is higher for "vertical 2" load condition. In the specimen without TRM, however, there is no significant difference in the behaviour of the walls depending on the two vertical load levels.

Figs 7 and 8 show the relation between shear and displacement obtained in numerical models. The simulations on the unreinforced wall exhibited a brittle failure mode (Fig. 7), whereas reinforced specimens have a high ductility (Fig. 8).

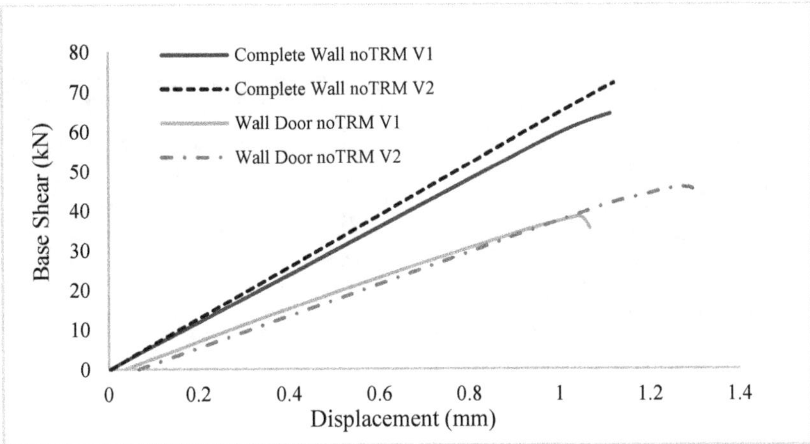

Figure 7: Base shear-displacement curves of un-reinforced walls.

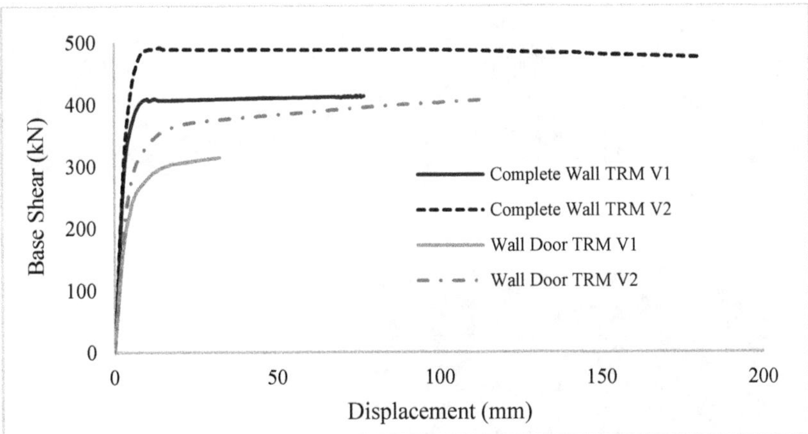

Figure 8: Base shear-displacement curves of TRM-walls.

It is observed that the wall with a higher stiffness is the complete one, the level of compression increases the stiffness and the maximum load value reached. Reinforced specimens have a higher load and a higher displacement. Fig. 9 shows the distribution of the maximum stresses in the different layers of the masonry wall. It is observed how the reinforcement mesh redistributes the levels of stresses; this effect can be clearly observed in Fig. 9(a) and (e). The redistribution of the maximum stresses in the layers related to the modeling of wall shear behavior is also clearly shown in Fig. 9(c) and (d).

Fig. 10 shows the distribution of the maximum stresses in the different layers of the wall with an opening for the door, both without reinforcement and those reinforced with TRM. In this case, the effects of the concentration of stresses close to the opening are clearly observed. As in the complete wall, the effect of TRM improves the redistribution of the concentrated stresses throughout the wall. Fig. 10(a) and (e) are clear examples of this.

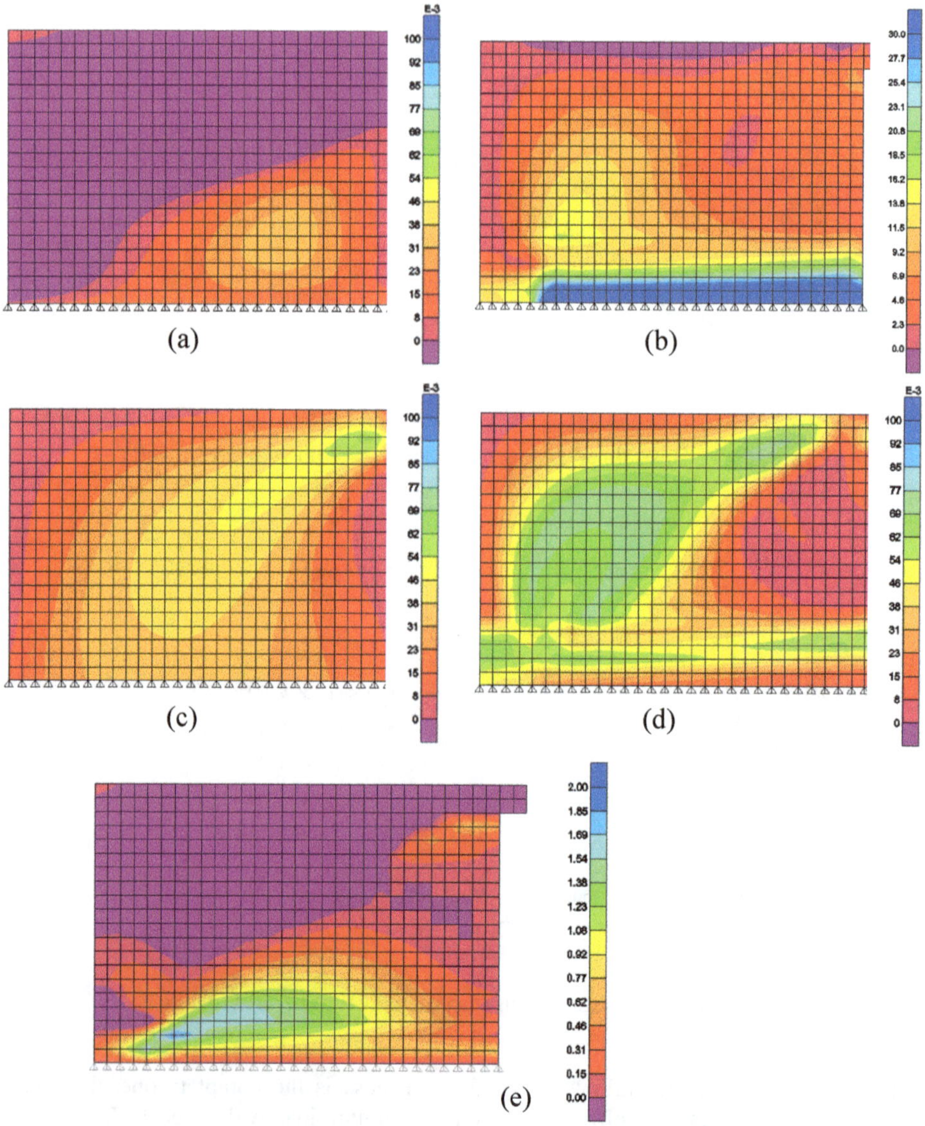

Figure 9: Maximum stresses on a complete wall with and without retrofitting with TRM. Vertical load of 300 kN. (a) Unreinforced wall; (b) Reinforced TRM stresses; (c) Unreinforced shear-layer; (d) Reinforced shear-layer; and (e) Reinforced compression-layer.

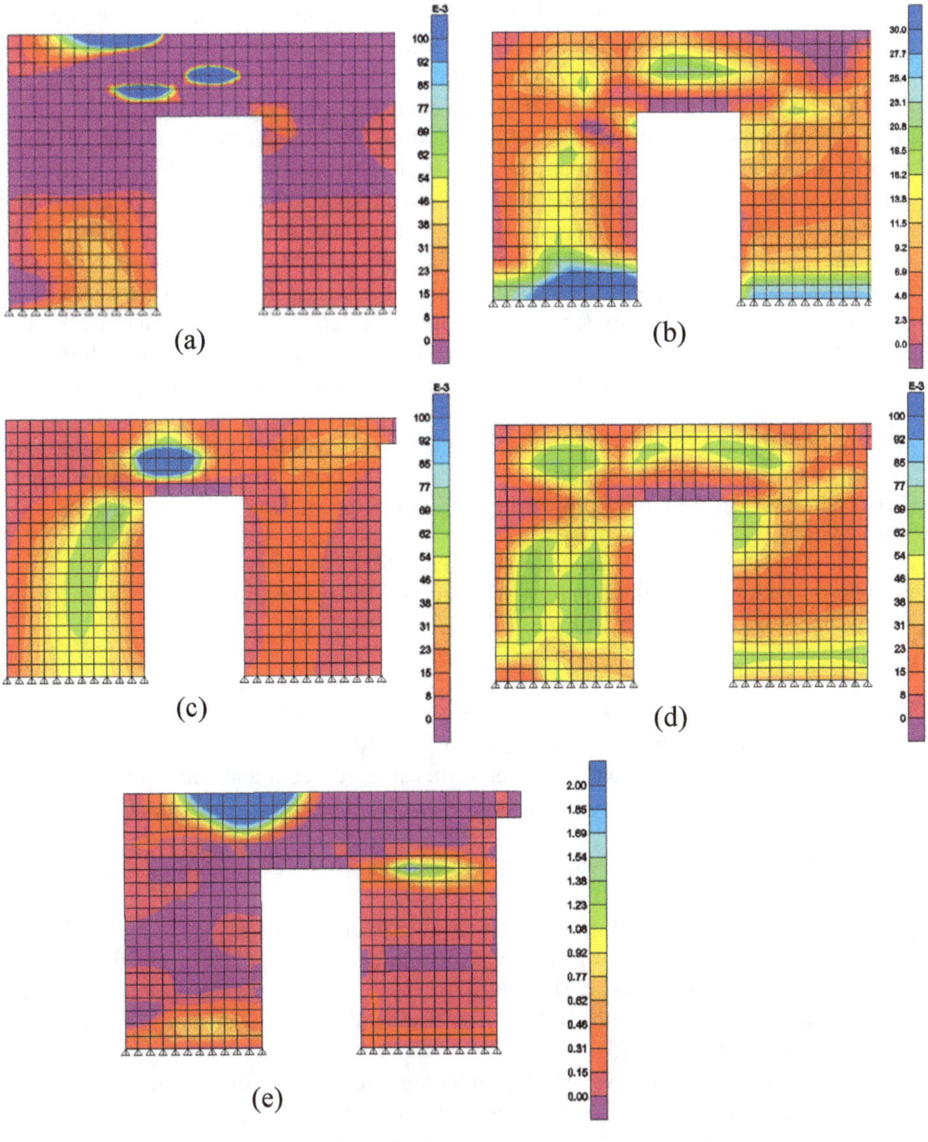

Figure 10: Maximum stresses on a door wall with and without retrofitting with TRM. Vertical load of 300 kN. (a) Unreinforced wall; (b) Reinforced; (c) Unreinforced-shear layer; (d) Reinforced shear layer; and (e) Reinforced compression-layer.

3 CONCLUSIONS

The scientific community has shown that the use of TRM in masonry structural elements brings great benefits. From a practical point of view, it is well suited to be applied above all in those cases where the monolithic nature of the walls is missing. In fact, this absence undermines the basis of a good seismic behavior of a structure, since it can make even null the presence of suitable characteristics, such as a good cohesion between elements.

The more traditional reinforcement techniques aim to increase the seismic capacity by improving the connection between the different elements, reducing the action of any pushing roofs or the excessive deformability of the floors. TRM therefore, for what previously obtained, is configured as a valid alternative or addition among the available possibilities.

In this work layered numerical shell elements models have been presented as an alternative to other complex numerical models to simulate masonry walls reinforced with TRM. The purpose of these numerical models is to estimate the structural behavior in masonry walls, with and without door opening. The layered non-linear shell element has been presented as a good option to simulate this type of retrofitting. The increment of the maximum horizontal load and the high increment of ductility of the TRM retrofitted walls have been simulated with a high level of coherence for different walls and different vertical compression loads.

The full reliability of the utilized modelling technique will be verified by comparing the above results with the results obtained by an ongoing experimental campaign on samples tested in laboratory at the Civil Engineering Laboratory of the University of Alicante.

ACKNOWLEDGEMENT

The authors acknowledge the Spanish Ministry of Economy and Competitiveness who partially supported this study by the grant BIA2015-69952-R

REFERENCES

[1] Foti, D., Innovative techniques for concrete reinforcement with polymers. *Construction Building Materials,* **112**, pp. 202–209, 2016.

[2] Foti, D. & Vacca, S., Comportamiento mecánico de columnas de hormigón armado reforzadas con mortero reoplástico. *Materiales de Construcción,* **63**(310), pp. 267–282, 2013.

[3] Foti, D., Use of recycled waste pet bottles fibers for the reinforcement of concrete. *Composite Structures,* **96**, pp. 396–404, 2013.

[4] Foti, D. & Paparella, F., Impact behavior of structural elements in concrete reinforced with PET grids. *Mechanics Research Communications,* **57**, pp. 57–66, 2014.

[5] Koutas, L., Bousias, S.N. & Triantafillou, T.C., Seismic strengthening of masonry-infilled RC frames with TRM: Experimental study. *Journal of Composites Construction,* **19**(2), 2015.

[6] Babatunde, S.A., Review of strengthening techniques for masonry using fiber reinforced polymers. *Composite Structures*, **161**, pp. 246–255, 2017.

[7] Bernat, E., Gil, L., Roca, P. & Escrig, C., Experimental and analytical study of TRM strengthened brickwork walls under eccentric compressive loading. *Construction and Building Materials*, **44**, pp. 35–47, 2013.

[8] Papanicolaou, C.G., Triantafillou, T.C., Karlos, K. & Papathanasiou, M., Textile reinforced mortar (TRM) versus FRP as strengthening material of URM walls: in-plane cyclic loading. *Materials and Structures*, **40**(10), pp. 1081–1097, 2007.

[9] Bournas, D.A., Triantafillou, T.C., Zygouris, K. & Stavropoulos, F., Textile-reinforced mortar versus FRP jacketing in seismic retrofitting of RC columns with continuous or lap-spliced deformed bars. *Journal of Composites for Construction,* **13**(5), pp. 360–371, 2009.

[10] Bernat, E., Roca, P. & Gil, L., Experimental study of brick masonry walls strengthened with textile reinforced mortar. *Key Engineering Materials,* **624**, pp. 397–404, 2016.

[11] Ivorra, S., Bru, D., Galvañ, A., Silvestri, S., Apera, C. & Foti. D., TRM reinforcement of masonry specimens for seismic areas. *International Journal of Safety Security Engineering,* **7**(4), pp. 463–474, 2017.
[12] Gisbert, J.I., Bru, D., Gonzalez, A. & Ivorra, S., Masonry micromodels using high order 3D elements. *Procedia Structural Integrity,* **11**, pp. 428–435, 2018.
[13] CSI, SAP2000, *Computers and Structures,* Inc. Structural Software for analysis and Design, 2014.
[14] Foti, D., A new experimental approach to the pushover analysis of masonry buildings. *Computers & Structures,* **147**, pp. 165–171, 2015.
[15] Bilgin, H. & Korini, O., A new modeling approach in the pushover analysis of masonry structures. *International Students' Conference Civil Engineeing,* ISCCE, EpokaUniversity, Tirana, Albania, 10–11 May 2012.
[16] NTC2008, Norme tecniche per le costruzioni, D.M. 14/01/2008, Gazz. Uff. 29, 2008.

LATERAL BEHAVIOUR OF STEEL FRAMES WITH PRECAST REINFORCED CONCRETE EXTERIOR INFILL PANELS

ASHRAF OSMAN & MOHAMED G. KHERD
Structural Engineering Department, Cairo University, Egypt

ABSTRACT

This study investigates the lateral behaviour of steel moment resisting frames cladded externally with precast reinforced concrete panels, under the action of both wind and earthquake loads. The aim is to explore the possibility of achieving extra saving in the steel structure weight by utilizing the external cladding to engage in resisting the applied lateral loads. To achieve these objectives, several hypothetical steel moment-resisting frames were initially developed. Then, these frames were analysed using a push-over technique. In performing these analyses and for sake of comparison, the frames were considered as bare steel frames once and in other cases the contribution of the external reinforced concrete cladding panels was included. The structural elements of the analysed moment resisting frames (beams and columns) were modelled analytically using a beam-column element that allow the formation of plastic hinges at elements ends, while the external reinforced concrete precast cladding panels were modelled using equivalent diagonal struts that can yield and possess a criterion for collapse that simulates the possibility of failure of connection between panels and steel frames. For all performed analyses, lateral stiffness, strength, and sequence of plastic hinging were evaluated and compared. It was concluded that utilizing precast concrete external panels' contribution in designing moment resisting frames against wind loads can improve significantly frames' initial stiffness and strength, leading to significant reduction in their steel tonnage. For designing against earthquake loads, the panels limited frames' lateral deformation and reduced their weights which was considered beneficial in resisting minor and moderate earthquakes and limiting the damage to non-structural elements. However, at ultimate state under severe earthquakes, to maximize the contribution of these panels, the connections between the steel frames and the panels needs to be sufficiently strong to ensure engagement of panels in resisting the lateral loads, and frames should be detailed for a strong column–weak beam concept.

Keywords: seismic behaviour, steel moment resisting frames, precast reinforced concrete panels, exterior cladding.

1 INTRODUCTION

In today's building practice, steel moment resisting frames with exterior precast reinforced concrete infill panels are commonly used in construction. However, engineers often ignore the existence of these panels in the design process because they are regarded as non-structural elements. This is attributed to lack of knowledge concerning the contribution of these panels to frames lateral stiffness and strength. Historically, many experimental and theoretical research studies tried to reach an adequate expression for the behaviour of structural frames integrating exterior cladding panels [1]–[3]. They utilized finite element models through analyses to evaluate the effects of considering the exterior panels in the analysis on frames lateral stiffness, roof displacement and sequence of plastic hinges formation. In general, it was noted that exterior panels tend to decrease frames lateral sway, increase initial lateral stiffness and in some cases increase frames strength, but how this can be beneficial to these frames?

In practice, separation between these panels and the frames is not usually provided which leads to the panels engaging in resisting the lateral loads acting on these frames. On the other

WIT Transactions on The Built Environment, Vol 185, © 2019 WIT Press
www.witpress.com, ISSN 1743-3509 (on-line)
doi:10.2495/ERES190021

hand, several design codes identify serviceability sway limits for frames under the action of wind and seismic loads. Therefore, in this study the lateral behaviour of hybrid structural system consisting of a steel moment resisting frame combined with exterior reinforced concrete precast infill panels is investigated. The main objectives of this study are to assess the contribution of these panels in resisting the applied lateral loads on frames, exploring the benefits that can be gained from utilizing them in the analysis, and providing design recommendations to maximize these benefits.

2 HYPOTHETICAL STRUCTURE

To achieve these objectives, a hypothetical residential building having (12.0 x 16.0 m) in plane dimensions and story height of 4.0 m was configured as shown in Fig. 1. The structure is formed by intersection of three parallel two bay plan frames disposed along each of the two orthogonal directions of the building. Each building floor was assumed to support gravity loads consist of dead load equal to 665 kg/m^2 accounting for the weight of concrete slab, partition, and flooring and live load equal to 250 kg/m^2 in addition to its self-weight.

For resisting lateral loads (wind or earthquake loads) in east-west direction, only exterior moment resisting frames along axes A and C were considered, while frame B was assumed to be leaning frame. In north-south direction, the frames along axes 1 and 3 were braced and the frame along axis 2 was assumed to be leaning frame. The structure was cladded externally with precast panels that were connected at each level with steel anchorage connections at the steel building edge beams. These connections are vertically slotted at the top level of panels and hinged at the bottom level of panels. The study is concerned with investigating the behaviour of moment resisting frames along axes A and C.

The intensity of wind lateral pressure acting on the frames was calculated in accordance with the Egyptian code of practice for loads [4] assuming that the wind speed is 36 m/s in building location, while the seismic load was calculated in accordance with the Egyptian code

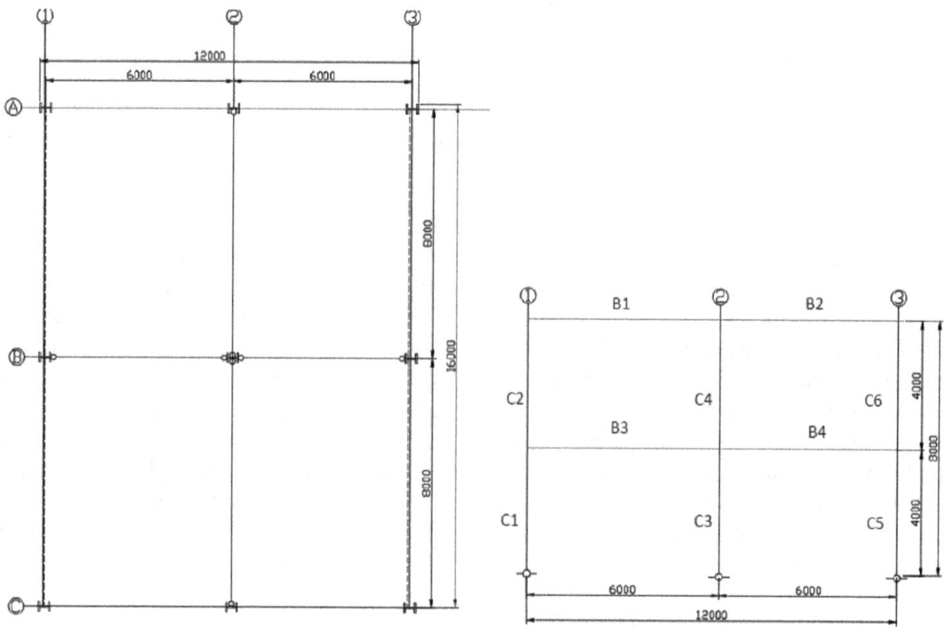

Figure 1: Plan and section elevation to the hypothetical building.

of practice for loads assuming that the peak ground acceleration is equal to 0.2 g, soil class is A and the moment resisting frames are of limited ductility (intermediate moment resisting frames) with reduction coefficient equal to 5.

Four steel moment resisting frames were designed representing different scenarios. In the first scenario, the frame, designated in this study as model A, was designed to support gravity and wind loads only with no consideration to code drift limit which is equal to H/500, where H is the floor height. In other words, it is designed to satisfy only strength requirements neglecting drift requirements. In the second scenario, the frame, designated in this study as model B, was designed to support gravity and wind loads and satisfying the code drift limits. Consequently, it was designed to satisfy both strength and drift limitations. Table 1 shows beams and columns cross-sections for the two frames. As can be noted, the weight of frame A is 35% less than that of frame B.

Table 1: Cross-sections of structural elements for frames A and B.

Beam/Column sections	Frame A	Frame B
Beams B1, B2, B3 and B4	HEB240	HEA300
Columns C1, C2, C5 and C6	HEA220	HEA400
Columns C3 and C4	HEA260	HEA400
Total tonnage per frame	3.34	5.2

In the third scenario, the frame, designated in this study as model C, was designed to support gravity and earthquake loads only with no consideration to code drift limit which is equal to H/50 during inelastic action, where H is the floor height. In other words, it is designed to satisfy only strength requirements with no consideration to drift limits. In the fourth scenario, the frame, designated in this study as model D, was designed to support gravity and earthquake loads and satisfying the drift limits. Table 2 shows the cross-sections for the beams and columns for the two frames. As can be noted, the weight of frame C is 31% less than that of frame D.

For all frames, the beams and columns were high-grade steel S 355 with yield stress equal to 3.45 t/cm^2 and ultimate strength of 4.1 t/cm^2.

Table 2: Cross-sections of structural elements for frames C and D.

Beam/Column sections	Frame C	Frame D
Beams B1, B2, B3 and B4	HEA300	HEA300
Columns C1, C3 and C5	HEB260	HEA500
Columns C2, C4 and C6	HEA260	HEA500
Total tonnage per frame	4.053	5.86

3 STRUCTURAL ANALYSIS MODELLING

To evaluate the structural behaviour of the case study frames, the mathematical model for each frame was analysed using the non-linear version of finite element package SAP2000 [5]. Frames' columns and beams were modelled using the beam-column element available in the programme that allows the development of concentrated plastic hinge at members ends. The precast infill panels were simulated by defining a pair of diagonal crossed tension and compression struts. The compression struts were acquired equivalent dimensions representing the effective stiffness of the panels. For such purpose, the thickness of each compression strut was taken equal to the panel thickness and its width was calculated based on the results extracted from a simple finite element model for the panel wall as will be shown later in details. On the other hands, the tension strut was simulated to represent an equivalent area for the resultant of the horizontal and vertical component of the reinforcement mesh used to reinforce the panel. The crossed tension and compression struts area connected through multi-plastic links to the beams. These links are representing the connection of the panels to steel supporting beams. Fig. 2 shows the finite element idealization to analysed frames with blow-up to connections between beam and tension/compression struts through multi-elastic links.

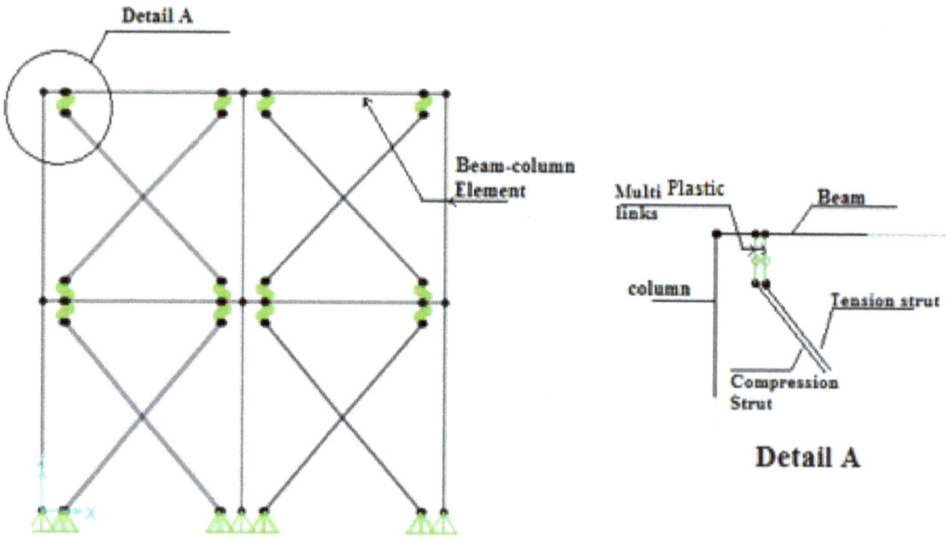

Figure 2: Finite element idealization to studied frames.

3.1 Effective area and stiffness of compression strut

The thickness of the compression strut can be taken equal to wall thickness, while the equivalent width of compression strut can be computed by modelling the panel using finite element [6]–[8]. For such purpose, the panel was modelled using shell element with support articulation assigned for compression corners only as hinged support in bottom compression corner and roller support in the upper corner, so the resultant of the reaction forces can be replaced with inclined strut subjected to compression forces. Joint load of 1 ton was applied in lateral direction at the location of roller support nodes and the results was obtained as shown in Fig. 3(a) which shows the panel model. For the studied frames with 6 x 4 m dimension panels, the effective panel width (L) was calculated 6m representing frame width reduced by column width and 400 mm gap which gives L equal to 5.1 m. The effective panel height (H) was calculated 4m representing frame height reduced by half the beam height and half the supporting angle leg which gives H equal to 3.8 m. The panel thickness was assumed to be 100mm reinforced with 5Φ10 per meter for both longitudinal and transversal direction. The cylinder compressive strength of concrete utilized for the panel was assumed to be 30 N/mm^2. the models consist of two columns an outer 500 and beam with max depth 300 mm. Fig. 3(b) shows the results of the panel analysis. As can be noted, the stresses within the wall were concentrated along the wall diagonal forming a member of the rectangular cross-section. The area outside this diagonal had minor stresses which can be ignored. From this the effective width of the compression member that replace the precast wall can be obtained using the following equations:

$$\delta_{\text{diagonal}} = \delta_h \cos(\theta), \tag{1}$$

$$K_P = (P_h / \cos(\theta)) / (\delta_{\text{diagonal}}), \tag{2}$$

where K_P is equivalent strut stiffness and the equivalent strut width shall be calculated through the below equation:

$$K_P = \frac{EA}{L} \tag{3}$$

where, E is the elastic modulus of concrete, A is the cross-section area of the equivalent compression strut and L is the diagonal inclined strut length.

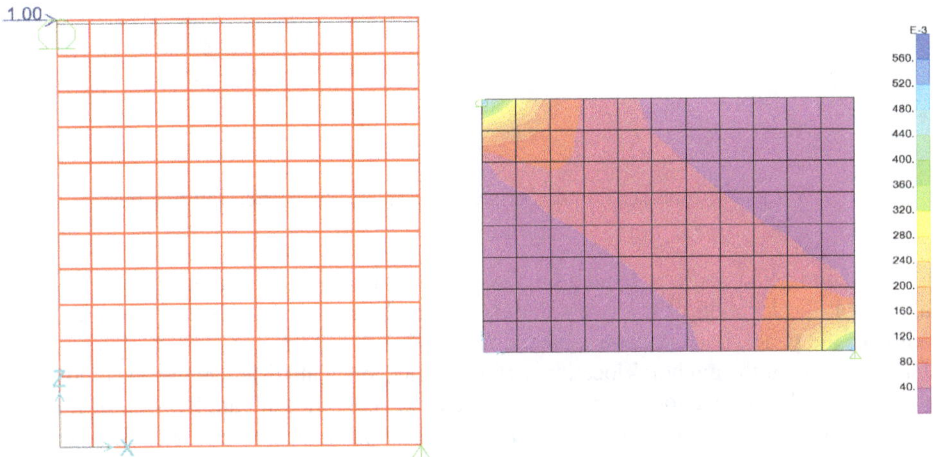

(a) (b)

Figure 3: (a) Idealization of precast panel; and (b) Finite element results.

Determining that δ_h was estimated as 0.00004077 m, the stiffness (K_p) of the compression strut was calculated as 35434.343 t/m and its effective width as 1.01 m.

3.2 Tension strut

Tension strut can be calculated simply by considering area equivalent to the effective area of steel reinforcement which provide tension stiffness of the precast cladding panel wall.

3.3 Connection between concrete panel and steel frame

As previously indicated, each panel wall was modelled using two pairs of compression and tension struts extending in both diagonal directions of the panel and connected to the frames' steel beams. To simulate the connection between steel beam and the diagonal struts, Multi-plastic links were assigned. The properties of these links such as nonlinear stiffness, failure load, and displacement criteria were computed utilizing Finite element model. For such purpose, the Steel beam was modelled using shell elements considering the nonlinear properties for the steel material. The concrete panel was assumed to be connected to the steel beam through steel angle (100 x 100 x 10) attached from one side to the concrete panel by two anchors which are represented as two points on the steel angles through which the panel loads are transferred to the steel beam and from the other side, the steel angle was welded to steel beam as shown in Fig. 4.

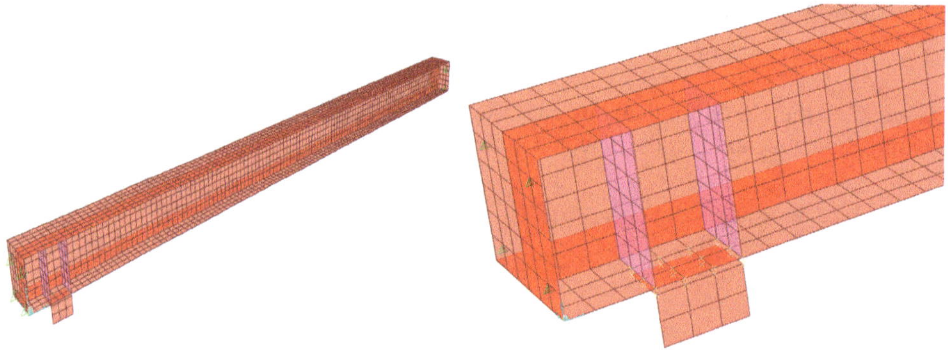

Figure 4: Idealized mode for steel beam and its connection to panel.

The connecting angle is loaded laterally and vertically with incremental increased load applied at the location of the two steel anchors and having an angle equal to the diagonal angle of the panel. Each anchor is loaded with 50% of the total applied load. To ensure that failure occurs at the anchors' locations, 10 mm thick vertical stiffeners were provided at the locations of the connections, each on one side of connecting steel angle. Nonlinear analysis was performed to obtain the force-displacement curve for the connection. Stiffness, failure loads and corresponding deformation in both horizontal and vertical directions for the connection based on utilizing bolt diameter 12 mm assuming the beam profile is HEA 300 are given in Tables 3 and 4 which show the failure loads and associated deformations in case of using bolts diameters 16 mm and 20 mm, too.

4 ANALYSIS RESULTS

4.1 Analysis for wind loads

Both frames A and B were loaded with incremental lateral load having a uniform pattern representing wind. Frame A was studied twice, once with the contribution of wall panels and another without panel contribution. Fig. 5 shows the top floor lateral deformation against base shear for analysed frames. As can be noted, frame A with the contribution of the panel was having lateral stiffness exceeding that of frame B which is designed to satisfy the code lateral drift. Consequently, great saving in steel weight can be achieved by considering the contribution of wall panels.

Table 3: The horizontal stiffness of the connections and its failure criteria.

Stiffness (t/m)	Failure load for insert diameter M12 (ton)	Deformation corresponding to Failure load for insert diameter M12 (m)	Failure load for insert diameter M16 (ton)	Deformation corresponding to Failure load for insert diameter M16 (m)	Failure load for insert diameter M20 (ton)	Deformation corresponding to Failure load for insert diameter M20 (m)
21,562	7.3	0.00035	12.6	0.00059	19.9	0.000925

Table 4: The vertical stiffness of the connections and its failure criteria.

Stiffness (t/m)	Failure load for insert diameter M12 (ton)	Deformation corresponding to Failure load for insert diameter M12 (m)	Failure load for insert diameter M16 (ton)	Deformation corresponding to Failure load for insert diameter M16 (m)	Failure load for insert diameter M20 (ton)	Deformation corresponding to Failure load for insert diameter M20 (m)
16,421	5.5	0.00033	9.7	0.0006	15.22	0.0009

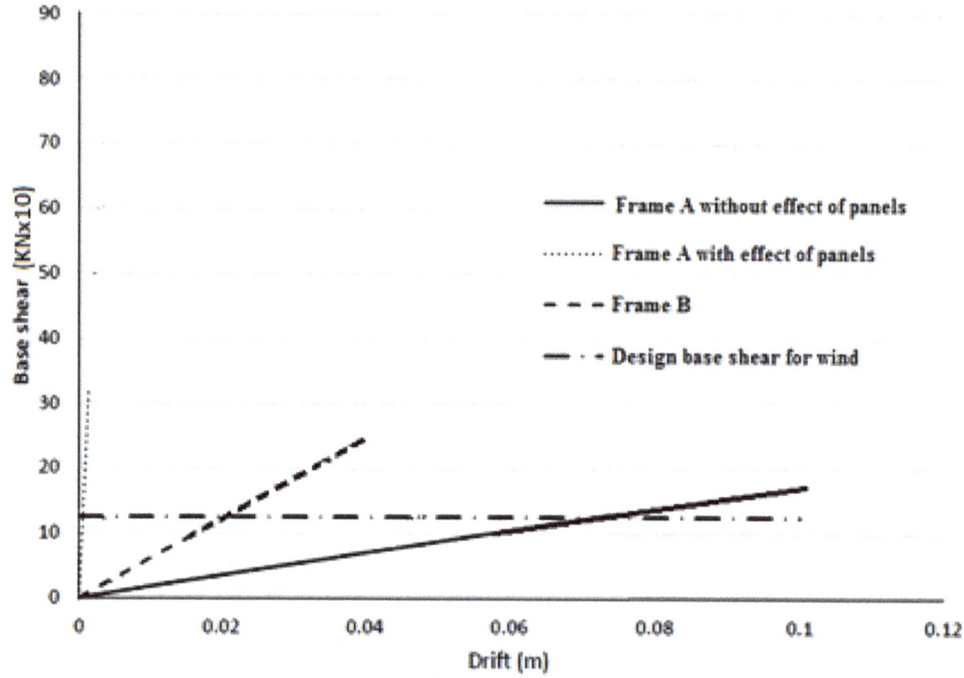

Figure 5: Base shear – top floor deformation under incremental uniform lateral loads.

4.2 Analysis for seismic loads

Both frames C and D were loaded with incremental lateral load having an inverted triangle loading pattern representing the lateral earthquake load. Frame C was studied twice, once with the contribution of wall panels and another without wall panel contribution. Fig. 6 shows the top floor lateral deformation against base shear for analysed frames. As can be noted, frame C, which is mainly designed for strength requirements, ignoring drift limits, with the contribution of the panel was having initial lateral stiffness exceeding that of frame D which was designed to satisfy both code strength and drift requirements. However, once the panels connections to steel frame start to fail, the frame stiffness start degrading and at ultimate stage started to behave as bare frame. In other words, started to behave as if frame C is bare frame without contribution. This behaviour indicated that:

1. For minor and moderate earthquakes when controlling drift is the key issue, the contribution of the exterior precast concrete panels can improve significantly the behaviour of the frames by controlling their drift. This was clear from comparing the behaviour of Frames C with and without wall panel contribution.
2. The need for controlling drift in frames result in increasing the strength of the frames significantly due to the increase in steel cross-sections. This can be detected from examining the behaviour of frame d and frame C.
3. For major earthquake when strength is the key issue, behaviour of frames with exterior precast panels can be significantly improved by utilizing strong connections between the wall panels and the steel frames.

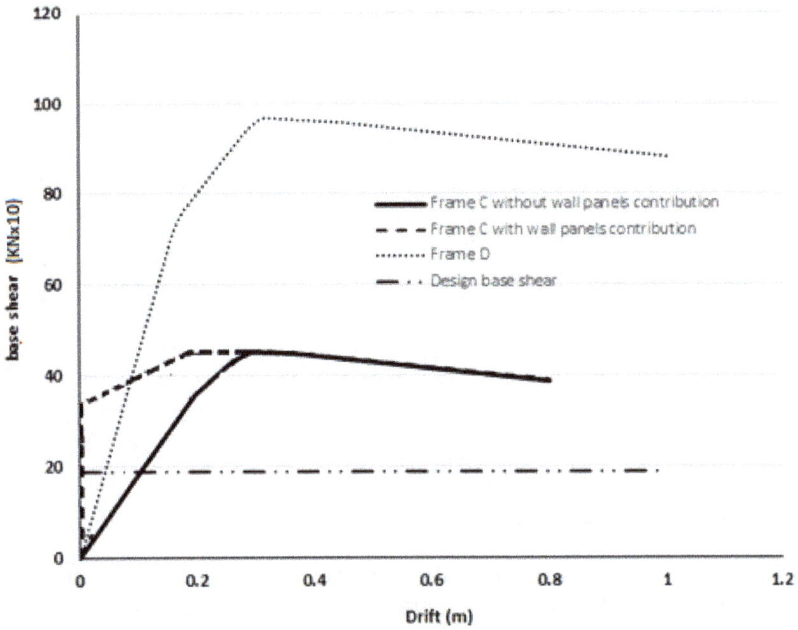

Figure 6: Base shear – top floor deformation under inverted triangular lateral loads.

With regards to plastic hinging pattern, Figs 7 and 8 show the plastic hinges formed in both frames prior to failure. As can be noted, frame C failed by forming story mechanism at 1st story, while frame D failed by forming beam-column sway mechanism. Although the failure of frame D is recommended compared to the failure of frame C, it should be noted that both frames were designed as intermediate moment resisting frames with same R value equal to 5 according to Egyptian code [4] and with no strong column–weak beam special ductility requirements. The only difference as stated before was that frame D satisfied drift limits of 2% as per code requirements.

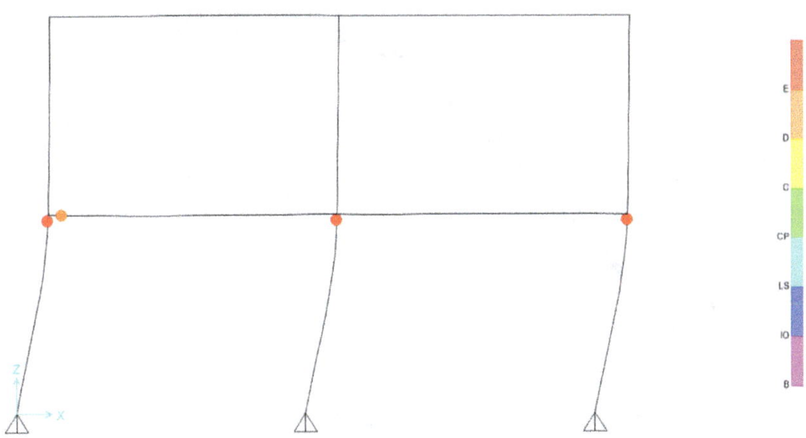

Figure 7: Plastic hinging pattern for Frame C before failure.

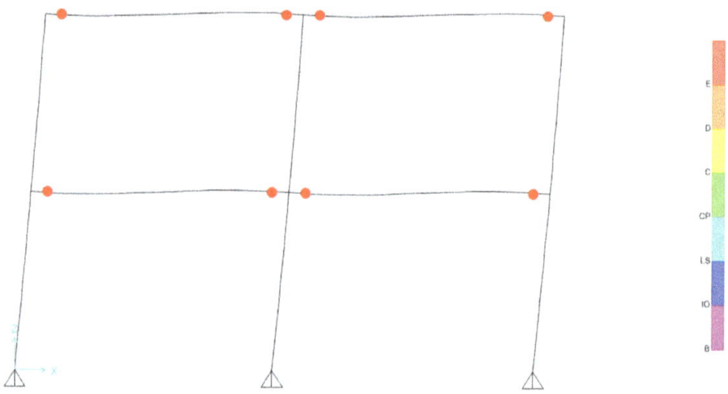

Figure 8: Plastic hinging pattern for Frame D before failure.

4.3 Parameters affecting the frames behaviour

To further explore this problem and identify the parameters that affects such frames types, it was decided to examine the behaviour of frames in case of increasing the strength of the wall panel to steel member connection through increasing the bolt diameter. For such purpose the same frame was subjected to monotonically increasing lateral loads having an inverted triangular loading pattern, but with using connections utilizing different bolts diameter (12 mm, 16 mm, and 20 mm) to increase connection strength. The values given in Tables 3 and 4 were used for such purpose. Fig. 9 shows the behaviour of these frames. Obviously, increasing the connection strength increased the frames strength for same lateral drift. In other words, improved the frames seismic behaviour significantly. Consequently, it can be concluded that wall panels improve significantly the frames lateral behaviour and such improvement can be further magnified by using stronger connection between wall panel and the frame.

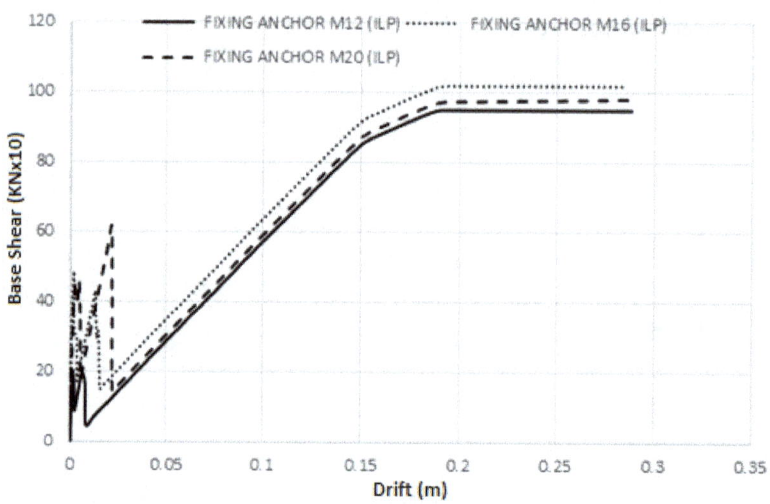

Figure 9: Base shear – top floor lateral displacement for different connection strength.

WIT Transactions on The Built Environment, Vol 185, © 2019 WIT Press
www.witpress.com, ISSN 1743-3509 (on-line)

5 CONCLUSION

Based on provided study, it can be concluded that:

1. The precast concrete external panels can engage efficiently with steel moment resisting frames in resisting lateral loads and controlling their deformation which can lead to reduction in frames weight.
2. For moment resisting frames precast cladding panel can be modelled simply as compression and tension struts and the analysis parameters can be obtained using a simple finite element model with commercial software.
3. For case of wind loads, the contribution of the panels increased frames' initial stiffness, controlling the lateral drift, and limiting the design requirement to satisfying strength requirements only, which will reduce the weight of steel structures significantly.
4. For case of earthquake, the contribution of cladding panels again controlled the lateral drift of the frames, reducing the expected non-structural damage during minor and moderate earthquakes, However, for major earthquake, the connection of panels failed, and the steel frames relayed on resisting lateral earthquake loads by its own ductility. Consequently, it is recommended that the connections between steel frames and panels designed for higher reduction factor than that used for designing the frames and requirements related to strong column–weak beam applied even if the frames area designed as intermediate or ordinary moment resisting frames.

REFERENCES

[1] Barua, H. & Mallick, S., Behaviour of mortar infilled steel frames under lateral load. *Building and Environment*, **12**(4), 263–272, 1977.
[2] Gaiotti, R. & Smith, B., Stiffening of moment-resisting frame by precast concrete cladding. *PCI Journal*, **37**(5), 80–92, 2014.
[3] Goodno, B. & Craig, J., Modelling of advanced, pre-cast cladding connections for seismic design. *Building Research & Information*, **20**(5), 313–319, 2007.
[4] Housing Research Building Center, Egyptian code for calculating loads and forces for structural and building works (ECP 201). Code of practice, Egypt, 2012.
[5] Computers and Structures, Inc., SAP2000 v20. Integrated software for structural analysis and design, USA, 2018.
[6] Goodno, B.J., Meyyappa, M., Nagarajaiah, S., Meyyappa, M. & Nagarajaiah, S.B., A refined model for precast cladding and connections. *Ninth World Conference on Earthquake Engineering*, pp. 2–9, 1988.
[7] Liauw, T.-C. & Kwan, K.-H., Nonlinear behaviour of non-integral infilled frames. *Computers and Structures*, **18**(3), 551–560,1984.
[8] Negro, P. & Lamperti Tornaghi, M., Seismic response of precast structures with vertical cladding panels: The SAFECLADDING experimental campaign. *Engineering Structures*, **132**, 205–228, 2017.

BEHAVIOR OF PRECAST REINFORCED CONCRETE BEAM-COLUMN EXTERNAL CONNECTIONS UNDER CYCLIC LOADING

AHMED TARABIA[1], SAID ALLAM[1], EMAD ETMAN[2] & MOHAMED ABOELHASSAN[1]
[1]Department of Structural Engineering, Faculty of Engineering, Alexandria University, Egypt
[2]Department of Structural Engineering, Faculty of Engineering, Tanta University, Egypt

ABSTRACT

One of the main problems associated with beam-column precast connections is the lack of ductility and low shear strength. The main objective of this paper is to study experimentally the behavior of precast reinforced concrete beam-column external connections under cyclic loading and to improve the performance of shear resistance capacity for this connection type. This experimental study focused on the cyclic behavior of the bolted connection type, which is suitable for low level capacity of shear forces. In the proposed connection system, two factors were studied; firstly, introducing a shear key between column and beam in the connection zone, and secondly, studying the effect of adding shear reinforcement to this zone. Five external beam-column specimens were tested in the experimental program, including one monolithic specimen and four precast bolted connection specimens. Three variables studied in this paper were the effect of shear reinforcement in concrete shear key zone, the length of connection measured from the column face, and the location of threaded bars. All specimens were designed according to the concept of strong column and weak beam, and the specimens scale factor was two-thirds of a prototype structure in geometry according to ACI code. All specimens were loaded on the top of the column by a constant axial compressive load at the beginning of each test, and the reversed cyclic loading in accordance to a pre-described displacement history at the beam tip. The performance of the precast and monolithic connections was evaluated and compared on the basis of flexural strength, ductility, energy dissipation, and lateral drift capacity. The precast connections, with shear reinforcement at the shear key zone, exhibited higher flexural strength and initial stiffness when compared to the similar precast connection without shear reinforcement and the monolithic specimen.
Keywords: precast concrete, reinforced concrete, beam-column connection, cyclic loading, exterior joint.

1 INTRODUCTION

The main concept of precast reinforced concrete structures (sometimes called prefabricated structures) includes a building in which the most structural components are prepared and produced in a factory away of the building, and after that, all components are moved to the project site for installation. In order to apply this concept to build a large number of structures at low cost and in a short time, these structural components are produced in the factory according to industrial methods with a mass production. The basic advantages of precast concrete structures are the high construction speed; which reduces total construction time of the project; a better quality control, and less requirement for labor. However, the basic weakness is the vulnerability of connections, particularly under seismic effects. A lot of precast concrete structures were heavily damaged by earthquakes. The main reason for this widespread damage was due to the poor performance of precast concrete connections. There are two types of beam-column joint connections, hinged connections and rigid connections (hard connections). In hinged connections, the beams are supported directly on columns by seating on corbels, for making the construction process easier and to transfer the shear forces from the beam to the column. However, rigid connections are used to make its performance similar to monolithic connections. Hard connections are those in which tensile, compressive,

WIT Transactions on The Built Environment, Vol 185, © 2019 WIT Press
www.witpress.com, ISSN 1743-3509 (on-line)
doi:10.2495/ERES190031

and shear stresses are transferred through special steel reinforcement and concrete details. There are three main requirements for precast concrete connections, which are the structural adequacy, the architectural function, and economy and speed [1].

Numerous experimental and analytical investigations have studied the performance and capacity of precast concrete connections, especially the detailing and location of precast connections. The behavior of welded precast beam-column connection was studied by Seckin and Fu [2]. In this study, simple connectors were introduced by using welded horizontal and vertical steel plates. It was concluded that precast connections performed as well as the monolithic connection and provided adequate ductility and maintained strength. Another experimental investigation for welded beam-column connections was presented by Ochs and Ehsani [3]. The results of this experimental program answered many questions about precast concrete connections and the relocation of plastic hinges, and the overall performance of the test specimens was found to be satisfactory. The post-tensioned precast concrete beam-column connections were presented in the experimental research conducted by Cheok and Lew [4] at NIST laboratory. It appeared that the performance of this connection type was strong and ductile as those of the monolithic connections, and the energy dissipation characteristics could be improved. A new precast technology was developed by Englekirk [5], where the major key to the concept was a ductile connector rod, which was embedded in a precast column, the overall performance of the ductile connector was better than that of the monolithic specimen. Stanon et al. [6] developed a new hybrid beam-column connection for precast frames, in which precast elements connected by unbonded post-tensioning steel and bonded reinforcing threaded bars. This connection proved to be at least similar to the monolithic connection in almost all respects, and superior in some aspects. Nakaki et al. [7] and Priestley et al. [8] presented an experimental study as a part of the PRESSS research program. A five-story precast concrete building with a sixty percent scale model was tested by using pseudo-dynamic test procedures. The building consisted of a wall system in one direction, and four different frame systems in the other direction. Hollow core slabs were used on two floors, while the other three floors were topped by double tees. It was reported that the behavior of the structure was extremely satisfactory, and the damage of the building for frame systems was much less than that of an equivalent monolithic building.

Another two-phase research program was developed for the same type of connections, in the first phase, the performance of four types of ductile precast beam-column connections and a reference monolithic specimen were studied by Ertas et al. [9]. It was concluded that the bolted connection response was more satisfying than the monolithic specimen in terms of energy dissipation. In the second phase, Ozden and Ertas [10] presented the results of tests performed on hybrid precast concrete connections, focused on the percentage of non-pre-stressed steel. The results showed that all precast specimens had an adequate flexural strength that could be sustained up to a 4% story drift without major strength degradation, and the design assumptions for using a 20% to 30% non-pre-stressed steel in a hybrid connection was better satisfied. Moreover, Ozden and Ertas [11] proposed a hysteretic moment-rotation modelling and an alternative section analysis. It was observed that the hysteretic behavior of the analytical model and experimental results were the same. Alcocer et al. [12] tested two full scale precast beam-column connections under unidirectional and bi-directional cyclic loading, this wet connection type had already been used in Mexico for commercial buildings with height up to 25 m. The test results showed that the structural response of the precast frame was satisfactory. Khaloo and Parastesh [13] carried out a new proposed precast beam-column connection type, and their experimental investigation focused on the connection length and the transverse bars presence. It was concluded that the overall performance of the proposed connection was at least equivalent to the monolithic specimen.

Similarly, Parastesh et al. [14] developed a new ductile precast beam-column connection, the proposed precast connections exhibited higher response compared to the similar monolithic. Also, Guan et al. [15] presented a similar experimental study, in which pre-stressing strands were placed at the bottom of the beam instead of steel bars. Based on the obtained results, the proposed precast connection could provide adequate seismic resistance similar to monolithic connections. Guan et al. [16] introduced a similar connection without using pre-stressing. In this study and to achieve continuity of the reinforcement of beam, longitudinal bar anchoring and lap splicing were used. The results proved that the performance of precast connection was better than monolithic connection. Khoo et al. [17] introduced another concept in their experimental study by using a modified assembled configuration for precast concrete frames in which the connections were constructed on the beam span and kept away from column faces, and the overlapping inside the connection was 90-degree (L-bent) and 180-degree (U-bent) hooks.

The results showed that the connections could perform satisfactorily. Similarly, Wahjudi et al. [18], [19] conducted an experimental and analytical study of constructible precast reinforced concrete connections with U-bent and L-bent hooks outside the beam-column joint. A numerical procedure had been carried out in the analytical study using the computer program software SeismoStruct, to predict the hysteretic response behavior of precast specimens. Generally, the analytical results showed a fair conformity with the experimental ones, and the precast connections demonstrated higher ductility than the equivalent monolithic specimen. Also, Yan et al. [20] presented a similar experimental study using precast pre-stressed concrete, where the beam end at column side had L-bent hooks and additional U-bent bars were placed at the beam-column joint. The results showed that the failure mode of precast and monolithic specimens was similar. Yuksel et al. [21] experimentally investigated two types of precast exterior beam-column connections, and also numerically modelled in OpenSees software platform. There were two types of connections industrial and residential connections, and the classification was according to the existence of corbels on the column. The results proved that the industrial type and residential type joint connections could be classified as low and medium ductility type connections, respectively. Similarly, Girgin et al. [22] presented a similar beam-column connection using steel plates, which were welded to the lower beam reinforcement. The results showed that while increasing the embedded upper reinforcement of beam, the relative energy dissipation ratio was raised.

The main objective of this paper was to study experimentally the possibility of improving the beam-column performance and the shear resistance capacity of bolted beam-column external connection under cyclic loading. The bolted connection was presented by Ertas et al. [9], with modifications were implementer to increase the shear resistance of this bolted connection. A shear key between column and beam was created in the connection zone, and shear reinforcement was added to this area.

2 EXPERIMENTAL PROGRAM

2.1 Test specimens and connection details

Five two-third scale exterior beam-column Specimens were tested in this experimental study, including one monolithic specimen (SP-M) and four precast bolted connections (SP-PC1, SP-PC2, SP-PC3, and SP-PC4). All specimens were designed according to the Egyptian Code ECP 203-2007 [23], and to satisfy the recommendation of ACI 318-14 [24]. In addition, all the precast concrete connections were followed the recommendations of the PCI *Design Handbook* [25]. The studied variables in this research are summarized in Table 1, which are:

1. The presence of shear reinforcement placed at the shear key zone.
2. The connection length (distance from the column face).
3. The location of the bolted threaded bars.

Table 1: Test variables and type of the test specimens.

Specimen	Type	Reinforcement at the shear key (2 row 2ϕ16@100 mm)	L, mm	Location of threaded bars with respect to beam cross-section
SP-M	Monolithic	-	-	-
SP-PC1	Precast	Available	400	Middle
SP-PC2	Precast	Not Available	400	Middle
SP-PC3	Precast	Available	600	Middle
SP-PC4	Precast	Available	400	Corner
where L is the connection length (distance from the column face).				

To make the assembly process easier and to minimize tolerance, a hidden steel system was imbedded in the beam and column. This steel system consisted of two steel pipe sleeves with 36 mm diameter and 1.5 mm thickness located at the top and bottom of beams cross-section and through the column, to allow installation of the threaded bars during the assembly process. In addition, these steel pipes were welded at their end with steel plates at the top and bottom of beam cross-section to prevent concrete crushing at the beam face, and used as washers for the threaded bars on each side of the connection zone, these steel plates were connected to each other by two 10 mm diameter steel bars by welding. Also, to prevent the steel pipes from any possible sliding, 8 mm diameter steel bars were welded around the steel pipes as ribs and also steel bolts 10 mm diameter with 80 mm length were welded with the pipes as anchors. The tested specimens are:

- Specimen (SP-M): This monolithic specimen was used as a reference unit, the top and bottom longitudinal reinforcement of beam cross-section were continually passed through the column and bent at 90 degrees inside the column region (Fig. 1(a)).
- Specimen (SP-PC1): In this precast specimen, the shear reinforcement in the shear key area was added to improve the shear stiffness of the proposed precast concrete connection, and the length of connection was 400 mm (Fig. 1(b)).
- Specimen (SP-PC2): This test specimen was similar to SP-PC1, except that it was without shear reinforcement at the shear key zone, to study the effect of shear reinforcement (Fig. 1(b)).
- Specimen (SP-PC3): To study the effect of buckling of the bolted threaded bars, the precast connection length was increased to be 600 mm. However, all the other details of this specimen were the same as Specimen SP-PC1 (Fig. 1(b)).
- Specimen (SP-PC4): The main purpose of this specimen was to study effect of the location of the bolted threaded bars in the beam cross-section, this specimen was similar to SP-PC1, except the location of threaded bars was changed to be at the outside corner of beam cross-section instead of the middle (Fig. 1(b)).

The proposed precast connection is detailed in Fig. 2.

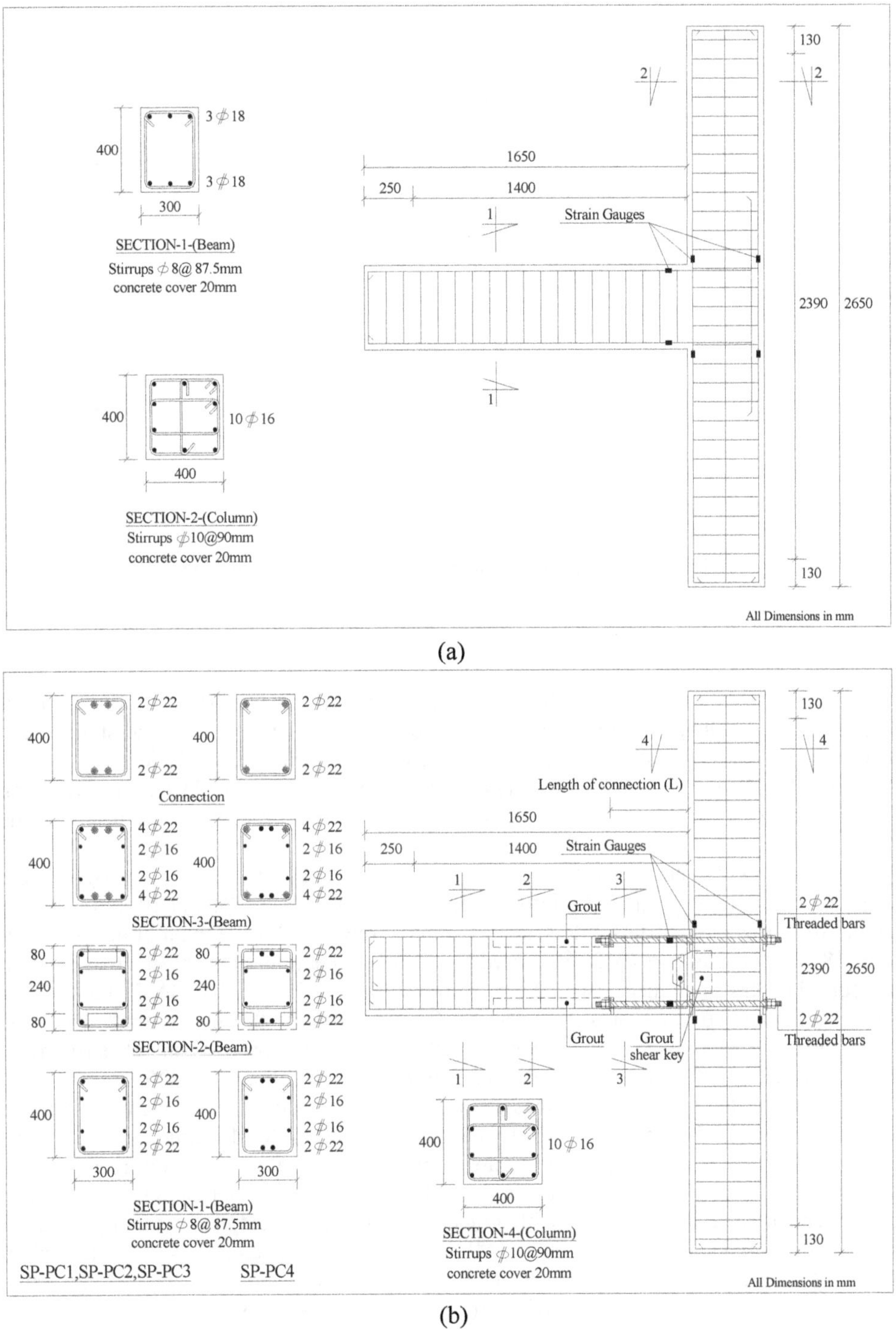

Figure 1: Reinforcement details of the specimens. (a) Monolithic specimen (SP-M); (b) Precast concrete specimens (SP-PC1, SP-PC2, SP-PC3, and SP-PC4).

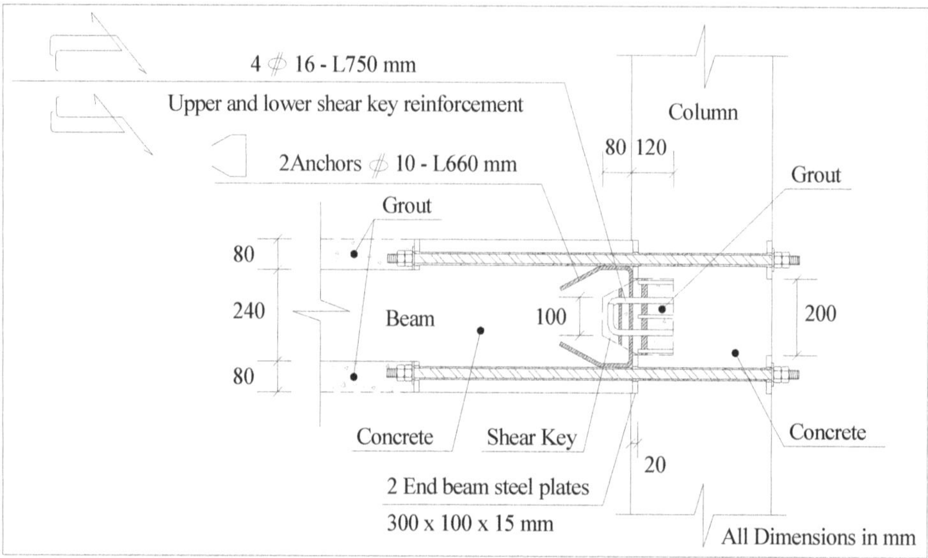

Figure 2: Detailing of the proposed precast connection.

2.2 Design basis and geometry

All specimens were designed according to the strong-column and weak-beam design concept. The prototype building had a plan of 20×20 m, column's spacing of 5 m, and story height 3.50 m with total height 35 m. The seismic design forces were calculated according to ECP 201-2012 [26], and the building was assumed to have enough ductile moment resisting frames. The scale factor was two-thirds of a prototype structure in geometry based on the ACI's T1.1R-01 [27].

For all monolithic and precast concrete specimens, ten 16 mm diameter bars (ϕ16) were used in the column longitudinal reinforcement with a ratio of 1.25%, and 10 mm diameter bars (ϕ10) horizontal stirrups with spacing 90 mm. However, for the monolithic specimen, three 18 mm diameter bars (ϕ18) were used for the top and bottom of the beam cross-section, and for all precast specimens, two 22 mm and two 16 mm diameter bars (ϕ22 and ϕ16) for the top and bottom of the beams cross-section. Closed stirrups with 8 mm diameter bars (ϕ8) with spacing 87.5 mm were used in beams for all monolithic and precast concrete specimens. The concrete cover thickness in all specimens for beams and columns was 20 mm.

Six 150 mm standard concrete cubes were cast with each specimen and were cured in the same environmental conditions. These cubes were tested to determine the 28 days compressive strength (f_{cu}), and the results are summarized in Table 2. The reinforcement steel bars used in all columns were with grade 400/600 for both the longitudinal and lateral reinforcement, and in all beams with grade 400/600 and 280/450 for the longitudinal reinforcement and stirrups, respectively. All precast concrete specimens were connected by a free-flowing micro-concrete grout (Sikacrete114), and the actual 28 days compressive strength (f_{cu}) are listed in Table 2. Also, all the spaces between steel pipe sleeves and the bolted threaded bars were injected by a cement mortar supplied by an expanding aid additive, (IntraplastZ) with the cement mortar, and the tested 28 days compressive strength (f_{cu}) are presented in Table 2.

Table 2: Compressive strength of concrete and grout materials.

Specimen	Average f_{cu} of concrete (N/mm^2)	C_v (%)	f_{cu} of micro-concrete grout (N/mm^2)	f_{cu} of cement mortar injection (N/mm^2)
SP-M	41	4.83	49	41
SP-PC1	45			
SP-PC2	46			
SP-PC3	46			
SP-PC4	46			
f_{cu} is the 28 days compressive strength, and C_v is the coefficient of variation				

All five specimens were cast in a horizontal position, and a level and smooth wooden mold was prepared for each specimen. In connection zone, the precast beam had a hollow shear key with the total width of the beam cross-section, and it was continuous through the column for 120 mm long. Also, hollow rectangular voids were made at the end of the connection zone at the top and bottom of the beam cross-section to allow the assembly process for bolting threaded bars. All voids were prepared using high density foam sheets for large spaces and spray foam in narrow places, which were removed before assembly process. All specimens were manufactured and poured in the Reinforced Concrete Laboratory, Faculty of Engineering, Alexandria University. The assembly process was made after 28 days from casting date. First, the beam was connected to column by placing the four threaded bars inside the steel sleeve pipes, and then, an initial pre-tensioning torque was applied to the threaded bars until the strain of threaded steel bars was about 10% of its yield value. After that, the shear key and the other voids of the connection were filled with the free-flowing micro-concrete grout, and all the steel pipe sleeves were injected with the cement mortar. After curing time, the specimens were transported and tested in the Reinforced Concrete and Heavy Structures Laboratory, Faculty of Engineering, Tanta University.

2.3 Test setup and instrumentation

All five specimens were tested while the column and beam in vertical and horizontal positions, respectively. The column was supported at its base on the strong beam of the lab structural frame, and the bottom and top ends of the column were prevented from horizontal movements but kept free to rotate as hinged supports. A vertical load equal to 498 KN was applied using a vertical hydraulic jack with a load cell at the top of the column to apply the constant axial load, and a hydraulic ram actuator with a displacement range of $\pm$ 250 mm was used at the beam end to apply the lateral vertical displacement (Fig. 3).

Six Linear variable displacement transducers (LVDTs) were used to measure the joint rotation, gab opening and shear deformation at the connection zone and to ensure that no lateral horizontal movement occurred at column ends. In addition, six and eight strain gauges were installed on the longitudinal reinforcement of beams and columns at connection zone for the monolithic and precast concrete specimens, respectively (Fig. 1). These strain gauges were placed to measure uniaxial strain of steel reinforcement at critical locations, and to define the yield lateral displacement of the connection.

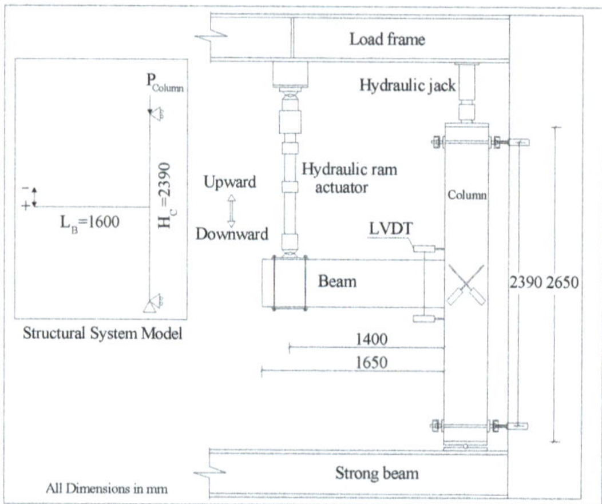

Figure 3: Test setup and boundary conditions for all specimens.

2.4 Test procedure and loading sequence

All experiments were conducted under the reserved displacement control lateral loadings at the beam tip according to ACI's T1.1R-01 [27]. These lateral loads were applied using a hydraulic ram actuator which moved in the upward and downward directions. The reverse cyclic loading was in accordance with a prescribed displacement history, consisted of displacement cycles of ±2, ±3, ±4, ±5, ±7, ±10, ±15, ±20, ±25, ±30, ±50, ±70 mm. Three fully reserved cycles were applied at each displacement level (Fig. 4). Also, to take into account the effect of the loads from upper floors, all specimens were loaded on the top of the column with a constant axial compressive load 498 KN by a hydraulic vertical jack, where this axial load was equal to 10% of the cylinder compressive strength of concrete times the column gross cross-sectional area.

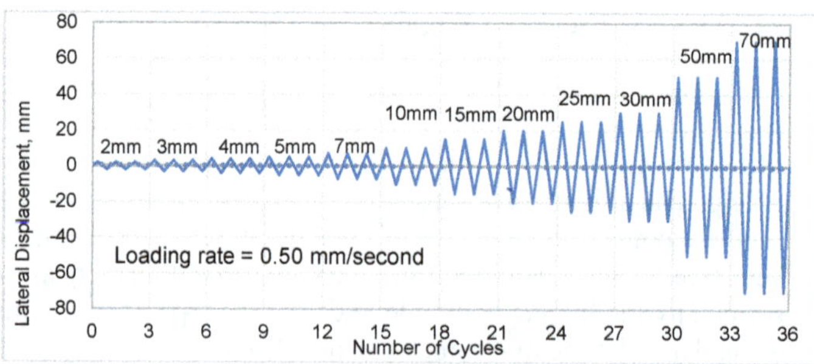

Figure 4: Cyclic loading history.

In this study, due to limitations of the test setup and to prevent damage of the hydraulic ram actuator, the test was terminated after completing three cycles at lateral displacement of ±70 mm (more than 4% lateral drift ratio: see eqn (1)), where, Δ is the beam vertical displacement at the loading point and L_B is the beam length from the equator centerline to the column centroid. All data were collected by a data acquisition system and loading rate was equal to 0.50 mm/second. Specimens were considered collapsed when the applied lateral load was less than 85% of the maximum actuator applied load, which is higher than the value (75% of the maximum load) for the acceptance criteria based on ACI's T1.1R-01 [27].

$$\text{Drift ratio} = \frac{\Delta}{L_B} \times 100. \quad (1)$$

3 EXPERIMENTAL TEST RESULT AND ANALYSIS

3.1 Hysteretic behavior and failure mode

The crack pattern and failure modes of all monolithic and precast specimens at the end of each test, as well as the hysteretic load displacement curves, are presented in Figs 5 and 6, respectively. The crack patterns show that no diagonal or horizontal cracks occurred in columns at the beam-column zone for the all precast specimens, except in the monolithic specimen, visible horizontal hair cracks were observed at the column edge close to the beam.

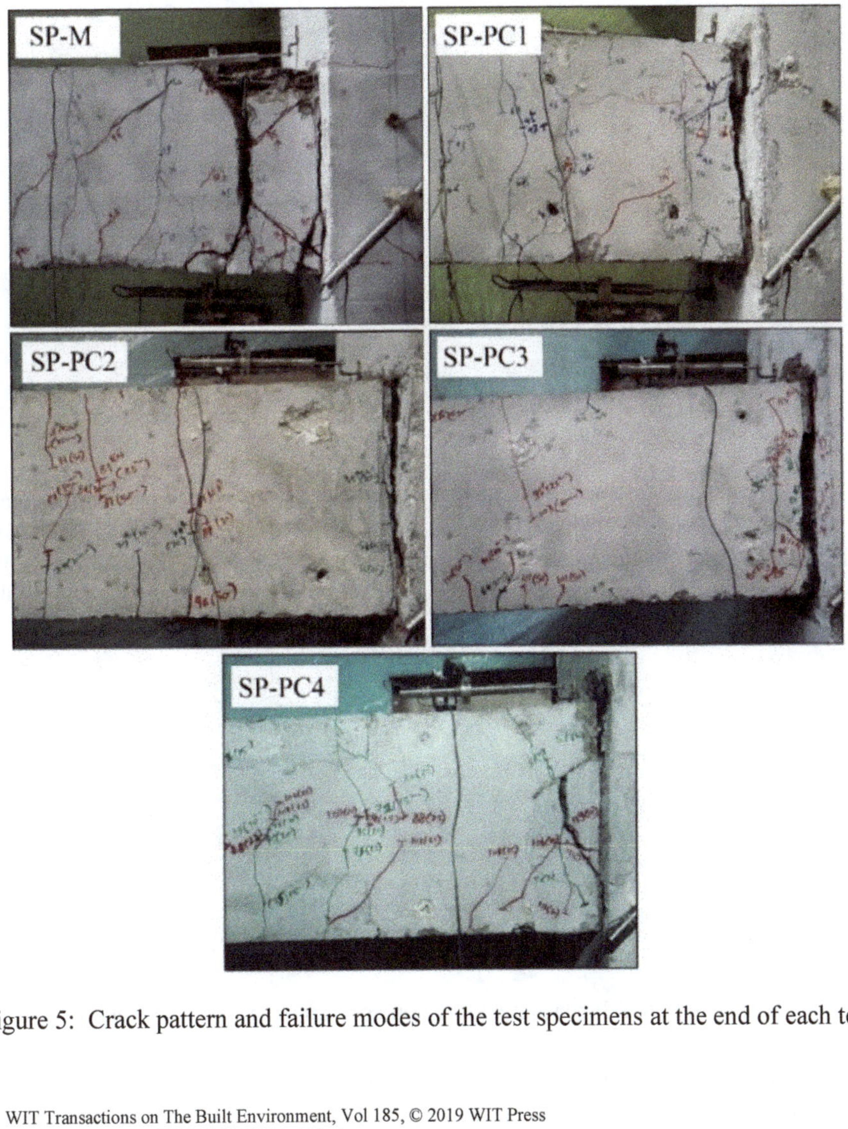

Figure 5: Crack pattern and failure modes of the test specimens at the end of each test.

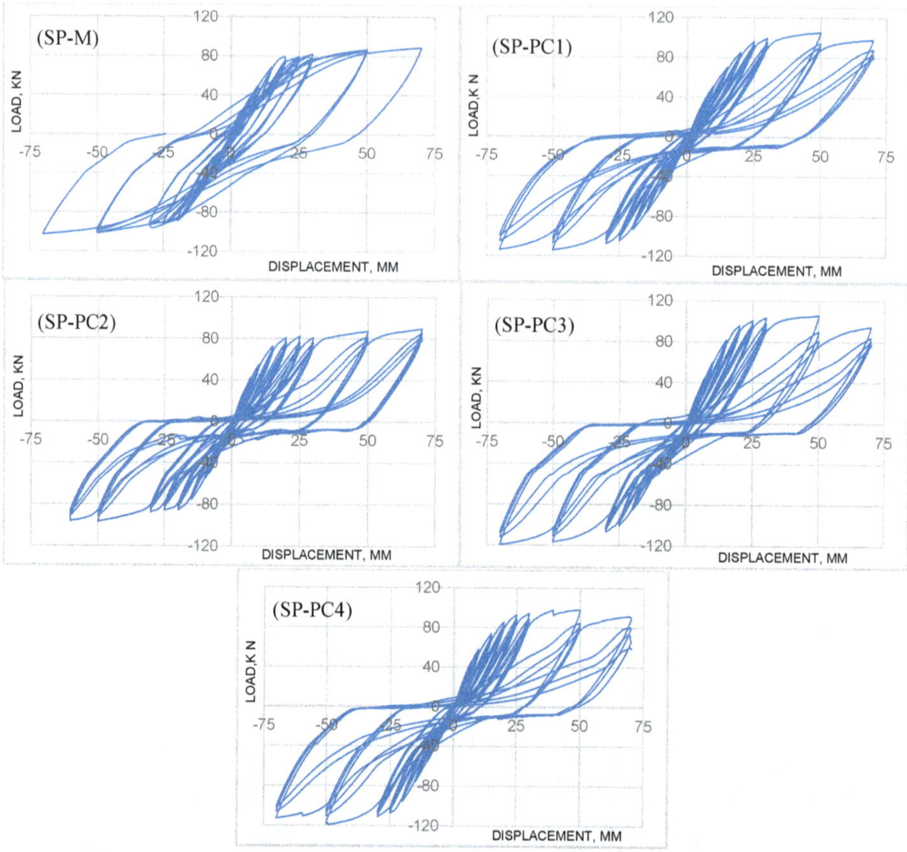

Figure 6: Hysteretic load displacement curves of the test specimens.

In addition, the hysteretic curves indicate that the pinching effect in monolithic specimen was lower than those of the precast specimens. The load displacement envelope curves show that there was a strength degradation in all precast specimens except SP-PC2 (Fig. 7). The yield and maximum vertical loads for upward and downward directions are listed in Table 3.

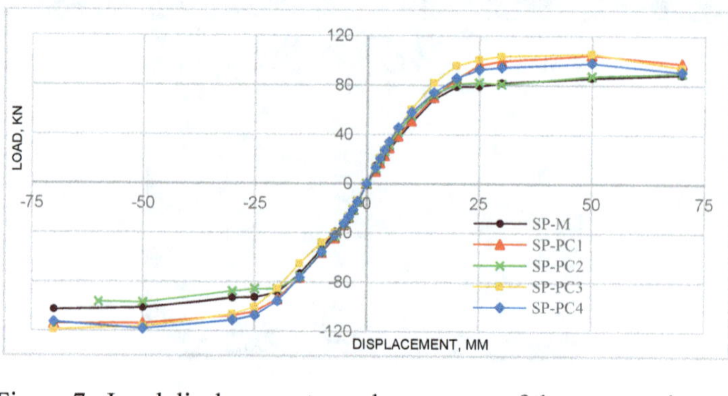

Figure 7: Load displacement envelope curves of the test specimens.

Table 3: Test results and the ratio between experimental and calculated capacities.

Specimen	(f_{cu}) N/mm^2	Loading direction	P_y KN	M_y KN.m	P_{max} KN	M_{max} KN.m	R_y	R_{max}
SP-M	41	Downward	78.25	109.55	88.66	124.13	1.00	1.08
		Upward	88.94	124.52	102.76	143.87	1.14	1.25
SP-PC1	45	Downward	83.53	116.94	104.59	146.42	1.06	1.25
		Upward	93.36	130.70	114.06	159.68	1.19	1.37
SP-PC2	46	Downward	80.69	112.97	89.75	125.65	1.03	1.08
		Upward	84.08	117.71	96.60	135.24	1.07	1.16
SP-PC3	46	Downward	86.05	120.47	105.56	147.78	1.10	1.27
		Upward	102.41	143.37	119.02	166.63	1.30	1.43
SP-PC4	46	Downward	75.62	105.87	97.97	137.16	0.96	1.18
		Upward	96.81	135.53	118.14	165.40	1.23	1.42

f_{cu} is the compressive strength of concrete
P_y and P_{max} are yield and maximum experimental vertical lateral load, respectively.
M_y and M_{max} are yield and maximum experimental bending moment, respectively.
$R_y = M_{y\ experimental} / M_{y\ calculated}$, and $R_{max} = M_{max\ experimental} / M_{max\ calculated}$

In the monolithic specimen SP-M, a minor flexural crack was observed at the top and bottom of the beam section at 0.625% drift ratio at a distance of 200 mm from the column face. The first diagonal crack in the beam was occurred at 1.25% drift ratio. Spalling of concrete at the beam near to the column face started at 3.1% drift level, and the beam upper longitudinal reinforcement buckled at 4.3% drift ratio.

In the precast specimen SP-PC1, when the drift ratio reached 0.45% and the actuator load was 38 KN, a hairline crack occurred at the beam-column interface, and this crack widened with the increase of the drift ratio. A hairline crack around shear key was observed at 0.65% drift ratio, and this crack widened from top side at 1.88% drift ratio. The first flexural cracks observed at a distance 450 mm from the column face at 0.5% and 0.94% drift ratio in upward and downward directions, respectively. Flexural cracks increased and were observed in other locations in the beam while the load and lateral drift ratio increased. A hairline diagonal crack beside the shear key was occurred at 3.5% drift ratio.

The experimental performance of specimen SP-PC2 was similar to the monolithic specimen. The first visible crack was observed at o.625% drift ratio and the load was 36 KN at the beam-column interface. This hair crack was widened at 0.94% drift ratio. No flexural or diagonal cracks were occurred at beam inside the connection region. The first flexural crack was observed at the lower part of beam cross-section at 1.25% drift ratio and was located 400 mm away from the column face. The flexural cracks were continually observed at the top and bottom of the beam cross-section with increasing loads and lateral drift ratio.

In specimen SP-PC3, the initial visible crack in the specimen was observed at 0.45% drift ratio at the beam-column interface. At 1.25% drift ratio and 92 KN actuator load, a hair crack around the shear key was observed. Minor flexural cracks occurred at distance 600 mm from the column face at 1.56% drift ratio, and they spread when the drift ratio was 3.1%. Also, at the same drift ratio level, the beam-column interface crack width increased to be more than 5 mm. Crushing of the concrete cover was observed at 4.30% drift ratio near failure. It should be noted that the flexural cracks in this precast specimen (SP-PC3) was less than all other monolithic and precast specimens.

In specimen SP-PC4, the first visible crack was observed at 0.625% drift ratio level, and the actuator load was 51 KN at the beam-column interface. At drift ratio 0.94%, a hairline crack around shear key and flexural cracks at both the upper and lower part of beam cross-section were observed at mid span of the beam, and this cracks widened with increasing drift ratio. A hairline diagonal crack near to the shear key occurred at 1.56% drift ratio. The width of crack around the shear key was more than 6 mm at 3.1% drift ratio.

3.2 Flexural strength

Based on the test setup, the bending moment of the specimen can be calculated as the applied load times the lever arms. The experimental values of yield and maximum bending moment for both directions are presented in Table 3. In addition, the yield and ultimate moment capacities of each specimen were calculated using a rectangular concrete stress block and actual yield and ultimate strength of steel reinforcement bars. The calculated values of the yield and ultimate bending moment capacities for monolithic specimen were 109.43 KN.m and 115.44 KN.m, respectively. However, the yield and ultimate bending moments for precast specimens were 109.90 KN.m and 116.69 KN.m, respectively. Based on the experimental results and from the ratio between the experimental bending moments to the calculated capacity, all connections achieved their calculated ultimate moment capacities. Also, the experimental results proved that the precast connections with shear reinforcement at the shear key zone (SP-PC1, SP-PC3 and SP-PC4) were about 18% higher in flexural strength when compared to specimens SP-PC2 and SP-M (Fig. 7). In addition, it was noted that the load displacement envelope curves of specimens SP-PC2 and SP-M were identical.

3.3 Displacement ductility

Displacement ductility is a measure of the ability of the element or the structure to undergo plastic deformations without significant loss of strength. This concept is a critical parameter in the seismic design to avoid brittle failure, and it is defined as the ratio of the maximum ultimate displacement (Δ_u) to yield displacement (Δ_y). The yield displacement is defined by utilizing the load displacement envelope relationship based on the criteria used by Park (Fig. 8) [15]. The yield displacement is determined by satisfying that the area enclosed by the idealized bilinear curve up to peak point is equal to that by envelope curve.

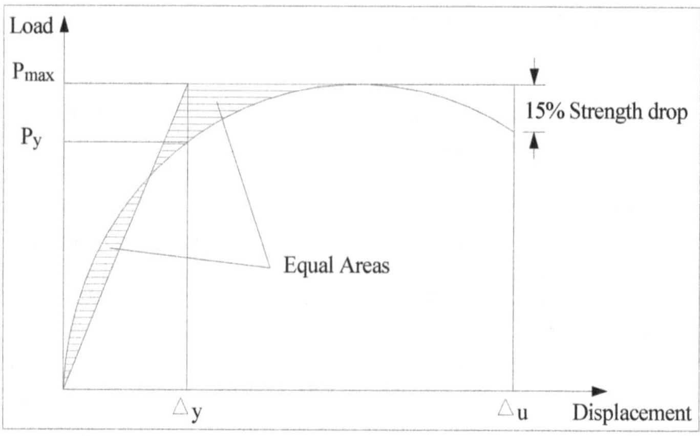

Figure 8: Equal energy principle [15].

The yield and ultimate displacements and displacement ductility are listed in Table 4, and these results showed that the all precast specimens behaved in a ductile manner as monolithic specimen. Also, Table 4 presents the ultimate drift ratio of each specimen (drift ratio at failure). According to ASCE 41-06 [28], reinforced concrete framed should resist inter-story drift ratios of 2% and 4% to reach life safety (LS) and collapse prevention (CP) performance level, respectively.

Table 4: Yield and ultimate displacements, displacement ductility, ultimate drift ratio, and the relative energy dissipation ratio.

Specimen	Loading direction	Δ_y (mm)	Δ_u (mm)	μ	μ average	Ultimate Drift ratio (%)	β (%) at 3.50% drift ratio
SP-M	Downward	19.52	70.00	3.59	3.38	4.38	36.40
	Upward	22.03	70.00	3.18			
SP-PC1	Downward	19.62	70.00	3.57	3.51	4.38	16.35
	Upward	20.27	70.00	3.45			
SP-PC2	Downward	19.59	70.00	3.57	3.55	4.38	14.86
	Upward	19.89	60.00	3.52			
SP-PC3	Downward	14.23	68.76	4.83	3.91	4.30	15.98
	Upward	23.42	70.00	2.99			
SP-PC4	Downward	15.84	64.06	4.04	3.74	4.00	17.82
	Upward	20.40	70.00	3.43			

Δ_y and Δ_u are the yield and maximum ultimate displacements, respectively.
μ is the displacement ductility, and β is the relative energy dissipation ratio.

Based on the results, the precast specimens SP-PC1 and SP-PC2 had the same drift ratio of the monolithic connection, and it was higher than that of SP-PC4, which indicates that the location of threaded bars in the middle of beam cross-section is better than placing them in the side corners. However, all specimens could achieve the CP performance level, and as a result, they could be used in high-seismic zones.

3.4 Stiffness degradation

The secant stiffness (K_{sec}) was used for the comparison of the stiffness degradation. The secant stiffness is defined as the slope of the straight line between the maximum drift levels of upward and downward direction at the last load cycle of each successive story drift level (Fig. 9), and it is called peak-to-peak stiffness. In order to easily compare of all test specimens, each secant stiffness value was normalized (K_{norm}) by dividing the initial secant stiffness of each specimen. It was observed that stiffness degradation of all specimens was similar, especially specimen SP-PC1, SP-PC3, and SP-PC4 were very similar due to the shear reinforcement at the shear key zone (see Fig. 10).

On the other hand, specimens SP-PC2 and SP-M were identically, especially at a drift ratio more than 1.5%. According to the acceptance criteria based on ACI's T1.1R-01 [27], the loss of initial stiffness at 3.5% drift ratio should have been not more than 95%. The loss of initial stiffness of all specimens was approximately 79% to 83% at the end of the test, which is satisfying the acceptance criteria of the loss of initial stiffness.

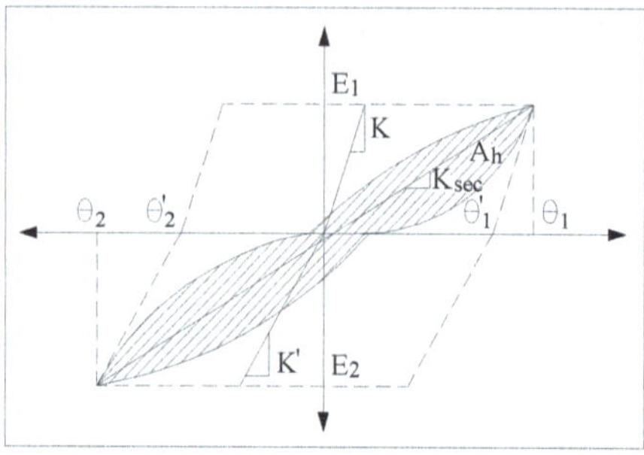

Figure 9: Secant stiffness and relative energy dissipation ratio [27].

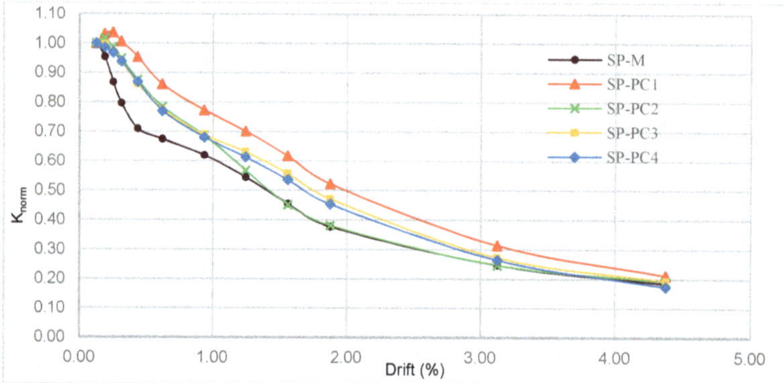

Figure 10: Normalized stiffness degradation of the specimens.

3.5 Energy dissipation

The relative energy dissipation ratio β (%) is defined by ACI's T1.1R-01 [27] as an acceptance criteria for design and evaluation of structures assigned to satisfy high seismic performance (Fig. 9). The dissipated energy can be measured as the hatched area A_h of the hysteresis loop for the third cycle of a specific story drift level. The initial stiffness (K and K′) values and peak loads (E_1 and E_2), at the drift ratio θ_1 and θ_2, may be different for the forward and reserve cycles. The relative energy dissipation ratio β is the area A_h divided by the area of the effective circumscribing parallelograms: see eqn (2). As an acceptance criteria based on ACI's T1.1R-01 [27], the relative energy dissipation ratio must not be less than 12.5% for the value of the third cycle at the 3.5% drift level.

$$\beta\ (\%) = \frac{A_h}{(E_1+E_2)(\theta_1'+\theta_2')} \times 100 \tag{2}$$

It was observed that the energy dissipation characteristic of all specimens was very similar until drift ratio 1.25% (average yield limit of the specimens), and then the monolithic specimen recorded a higher value (Fig. 11). The values of β were high at the beginning of the test due to the linear response and the cracks were closed. At 2% drift ratio, β values were in the range of 11.50% to 14% for the precast specimens, while β value reached to 28% for the monolithic specimen. The calculated β for specimens SP-M, SP-PC1, and SP-PC2 exceeded

12.5% at 2% drift ratio, while specimens SP-PC3 and SP-PC4 were slightly less than this required value. The β values for SP-PC3 and SP-PC4 were 11.68% and 11.91%, respectively. Based on the results, all monolithic and precast specimens satisfied the acceptance criteria for the relative energy dissipation ratio at 3.5% drift ratio. The β value for all specimens at 3.50% drift ratio is listed in Table 4.

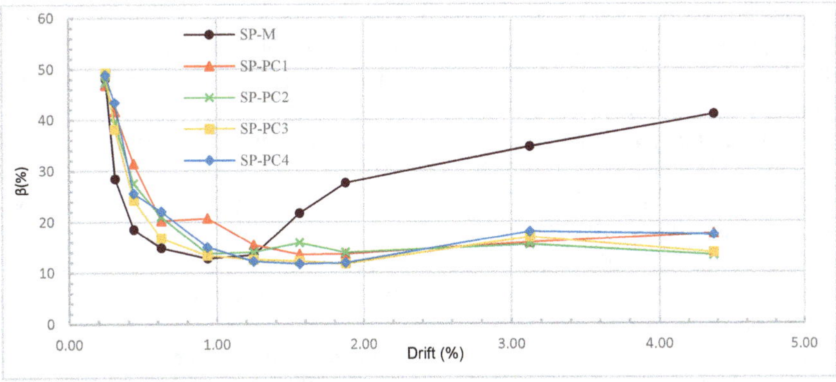

Figure 11: The relative energy dissipation ratio of the specimens.

4 CONCLUSIONS

The aim of this research was to study the cyclic experimental behavior of precast reinforced concrete bolted beam-column external connection, and to improve the performance and the shear resistance capacity. From the results obtained from the experimental cyclic loading testing on one monolithic and four precast specimens, the following conclusions may be drawn:

- All monolithic and precast specimens had adequate flexural strength that can be sustained up to 4% drift ratio. Moreover, the precast connections; with shear reinforcement at the shear key zone; exhibited higher flexural strength and initial stiffness when compared to the similar precast connection without shear reinforcement and the monolithic specimen.
- Considering the ductility factor, the results indicated that the all precast specimens behaved in a ductile manner similar to monolithic specimen, and all specimens could achieve the collapse prevention (CP) performance level.
- The stiffness degradation of all monolithic and precast specimens was similar, and the loss of initial stiffness of all specimens was approximately 79% to 83% at the end of the test.
- The energy dissipation characteristics of the monolithic and precast specimens satisfied the acceptance criteria for the relative energy dissipation ratio at 3.5% drift ratio according to ACI's T1.1R-01 [27], while the specimens SP-PC3 and SP-PC4 were slightly less than the required at 2% drift ratio.

According to all above, the proposed bolted connection with shear reinforcement placed at the created shear key zone can provide adequate flexural strength, ductility, energy dissipation, and lateral drift capacity. Thus, this connection type can be used in high-seismic zones.

REFERENCES

[1] FIB Bulletin 27, *Seismic Design of Precast Concrete Building Structures*, 2003.

[2] Seckin, M. & Fu, H.C., Beam-column connections in precast reinforced concrete construction. *ACI Structural Journal*, **87**(3), pp. 252–261, 1990.

[3] Ochs, J.E. & Ehsani, M.R., Moment resistant connections in precast concrete frames for seismic regions. *PCI Journal*, **38**(5), pp. 64–75, 1993.

[4] Cheok, G.S. & Lew, H.S., Performance of precast concrete beam-to-column connections subject to cyclic loading. *PCI Journal*, **36**(3), pp. 56–67, 1991.

[5] Englekirk, R.E., Development and testing of a ductile connector for assembling precast concrete beams and columns. *PCI Journal*, **40**(2), pp. 36–51, 1995.

[6] Stanon, J., Stone, W.C. & Cheok, G.S., A hybrid reinforced precast frame for seismic regions. *PCI Journal*, **42**(2), pp. 20–32, 1997.

[7] Nakaki, S.D., Stanton, J.F. & Sritharan, S., An overview of the PRESSS five-story precast test building. *PCI Journal*, **44**(2), pp. 26–39, 1999.

[8] Priestley, M.J.N., Sritharan, S., Conley, J.R. & Pampanin, S., Preliminary results and conclusions from the PRESSS five-story precast concrete test building. *PCI Journal*, **44**(6), pp. 42–67, 1999.

[9] Ertas, O., Ozden, S. & Ozturan. T., Ductile connections in precast concrete moment resisting frames, *PCI Journal*, **51**(3), pp. 66–76, 2006.

[10] Ozden, S. & Ertas, O., Behavior of unbonded, post-tensioned, pre-cast concrete connections with different percentages of mild steel reinforcement. *PCI Journal*, **61**(4), pp. 32–44, 2007.

[11] Ozden, S. & Ertas, O., Modeling of pre-cast concrete hybrid connections by considering the residual deformations. *International Journal of the Physical Sciences*, **5**(6), pp. 781–792, 2010.

[12] Alcocer, M., Carranza, R., Perez-Navarrete, D. & Martinez, R., Seismic tests of beam-to-column connections in a precast concrete frame. *PCI Journal*, **47**(3), pp. 70–89, 2002.

[13] Khaloo, A. & Parastesh, H., Cyclic loading of ductile precast concrete beam-column connection. *ACI Structural Journal*, **100**(3), pp. 291–296, 2003.

[14] Parastesh, H., Hajirasouliha, I. & Ramezani, R., A new ductile moment resisting connection for precast concrete frames in seismic regions – An experimental investigation. *Engineering Structures*, **70**, pp. 144–157, 2014.

[15] Guan, D., Guo, Z., Xiao, Q. & Zheng, Y., Experimental study of a new beam-to-column connection for precast concrete frames under reversal cyclic loading. *Advances in Structural Engineering*, **19**(3), pp. 529–545, 2016.

[16] Guan, D., Jiang, C., Guo, Z. & Ge, H., Development and seismic behavior of precast concrete beam-to-column connections. *Journal of Earthquake Engineering*, **22**, pp. 234–256, 2016.

[17] Khoo, J., Li, B. & Yip, W., Tests on precast concrete frames with connections constructed away from column faces. *ACI Structural Journal*, **103**(1), pp. 291–296, 2006.

[18] Wahjudi, D.I. et al., Behaviour of precast concrete beam-to-column connection with U and L-bent bar anchorages placed outside the column panel – Experimental study. *Procedia Engineering*, **95**, pp. 122–131, 2014.

[19] Wahjudi, D.I. et al., Behaviour of precast concrete beam-to-column connection with U and L-bent bar anchorages placed outside the column panel – Analytical study. *Procedia Engineering*, **95**, pp. 112–121, 2014.

[20] Yan, X., Wang, S., Huang, C., Qi, A. & Hong, C., Experimental study of a new precast prestressed concrete joint. *Applied Sciences*, **8**(10), p. 1871, 2018.

[21] Yuksel, E. et al., Seismic behavior of two exterior beam-column connections made of normal-strength concrete developed for precast construction. *Engineering Structures*, **99**, pp. 157–172, 2015.

[22] Girgin, S.C., Misir, I.B. & Kahraman, S., Experimental cyclic behavior of precast hybrid beam-column connections with welded components. *International Journal of Concrete Structures and Materials*, **11**(2), pp. 229–245, 2017.

[23] Ministry of Housing, *Egyptian Code for Design and Construction of Reinforced Concrete Structures*, ECCS203-2007, Egypt, 2007.

[24] American Concrete Institute, *Building Code Requirements for Structural Concrete (ACI 318-14) and Commentary (ACI 318R-14)*, American Concrete Institute: Detroit, Michigan, USA, 2014.

[25] Precast/Pre-stressed Concrete Institute, *PCI Design Handbook/7th Edition – (MNL-120-10)*, Precast/Pre-stressed Concrete Institute: Chicago, USA, 2010.

[26] Ministry of Housing, *ECP201-2012, Egyptian Code for Calculating Loads and Forces*, Ministry of Housing: Egypt, 2008.

[27] American Concrete Institute (ACI) Innovation Task Group 1 and Collaborators and ACI Committee 374, *T1.1R-01: Acceptance Criteria for Moment Frames Based on Structural Testing*, ACI: Farmington Hills, MI, 2001.

[28] ASCE 41-06, *Seismic Rehabilitation of Existing Buildings*, 1st ed., American Society of Civil Engineering, 2007.

VULNERABILITY ASSESSMENT OF MODULAR STEEL PLATFORM STRUCTURES UNDER SEISMIC LOADS

YASSER S. SALEM[1], ZHANG JIN[2], YANYAN HUANG[2] & TIFFANY YOO[3]
[1]California State Polytechnic University Pomona, USA
[2]Hubei University, China
[3]FWC Structural Engineers, USA

ABSTRACT
Modular steel structures are the standard method of construction in the arctic region. In this study, the feasibility of retrofitting these structures, using friction dampers, is evaluated. These structures possess some unique detailing that is the result of the method of construction used for them. This detailing is believed to contribute to the vulnerability of these structures during major seismic events. Analytical fragility curves are developed to evaluate the benefit of using friction dampers in reducing the risk of damages during significant earthquakes. The methodology that was used to develop the curves is based on calculating the maximum story drift using nonlinear time history analysis for different ground motions with different intensities and frequencies. The produced fragility curves highlight the advantages of using friction dampers in reducing the level of damage these structures can experience during earthquakes.
Keywords: modular steel structures, analytical fragility curves, friction dampers.

1 INTRODUCTION

Modular steel structures are getting more popular in the oil and gas industry in which cast-in-place methods are replaced by standard construction methods. Using modular structures, construction project teams not only can reduce the number of workers on sites, but also can speed up their construction process, control project quality and avoid the influence of weather and other unknown factors.

These modular structures are often pre-assembled offshore and are transported to their final location. They are mounted on top of support steel piles, with a connection that would allow for thermal expansion as well as clearance for lateral adjustment at their final location. The superstructure is constructed from structural steel where gravity loads are carried by steel columns and beams and the floor assembly is made from built-up plate girders. The lateral resisting system is concentric steel brace elements. The plate girders span over the width of the module and are supported on steel piles. The piles usually project from the finish grade a distance that ranges from ten to fifteen feet to the bottom of the plate girders. This is to allow for the free movement of the wildlife below the platform level. Fig. 1 shows an example of a typical modular steel structure that is considered in this study.

Piles are usually made from tubular steel. The piles are usually installed in a drilled shaft in the permafrost soils. The sources of strength for the pile are driven from the tangential adfreeze bond strength between the slurry and the steel pile after the slurry reaches subfreezing temperature.

The adfreeze strength provide a resistance to the pile's lateral movement, this resistance increases as the piles are embedded deeper into the permafrost. The piles are considered to behave as cantilever columns with a point of fixity at the location of the pile longitudinal axes that has zero lateral deformation. The extreme cold together with the cantilever action of the piles impact the ductility of these structures especially for old structures built prior to seismic provisions were implemented by modern building codes.

WIT Transactions on The Built Environment, Vol 185, © 2019 WIT Press
www.witpress.com, ISSN 1743-3509 (on-line)
doi:10.2495/ERES190041

Figure 1: Example of a modular steel structure during land transportation to final site location.

Friction dampers is an attractive retrofit solution to these types of structures when it is compared to other damper types such as viscous and visco-elastic dampers as they are less sensitive to the extreme temperature in the arctic region. Friction devices primarily use automotive brakes as an analogy [1] to their function on dissipating energy. Friction devices generate energy rectangular hysteric loops similar to the characteristic of coulomb friction. Fig. 2 illustrates a schematic diagram of pall friction device and Fig. 3 shows the typical force displacement relationship for friction dampers.

Fig. 4 shows the acceleration response spectra of the N-S components of 1985 Mexico City Earthquake under various damping ratios. The reduction of the spectral acceleration with the increase of damping is readily seen in this graph. One important observation from this graph is the acceleration reduction with the increase of damping is effective only in the

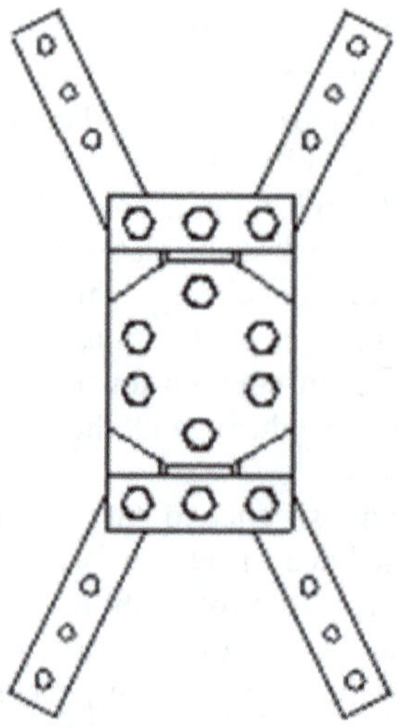

Figure 2: Pall friction device [1].

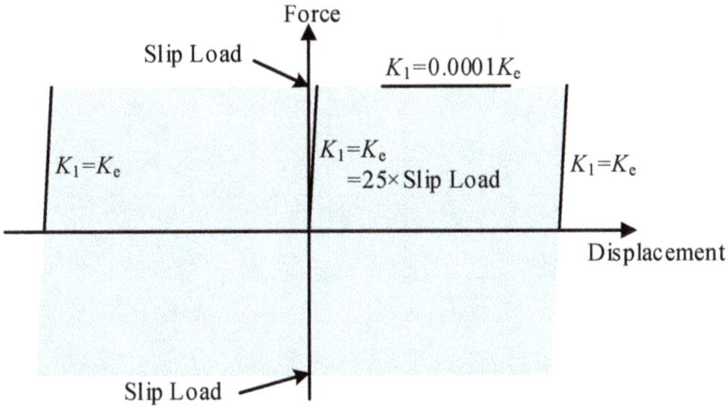

Figure 3: Theoretical hysteresis loops for a typical friction damper [2].

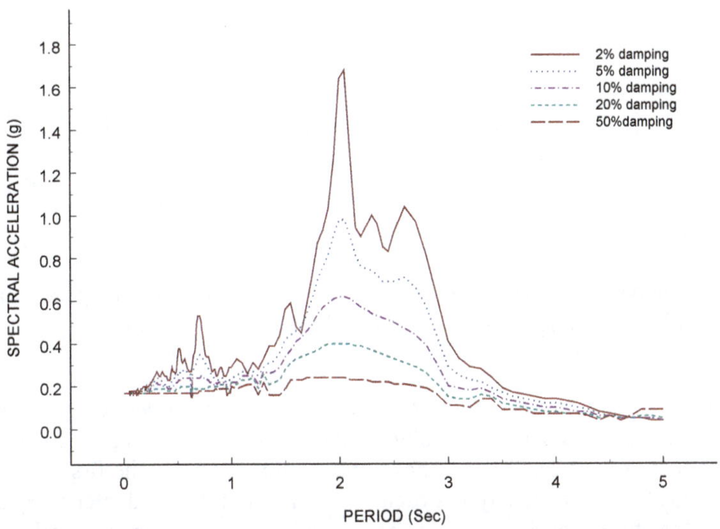

Figure 4: Response spectra acceleration (SA) Mexico earthquake, September 1985 [3].

range of 1.5 to 3.0 seconds of the natural period. This indicates that increasing the system damping has the potential to reduce the seismic demands to these structures. The supplemental damping devices can also reduce the story drift which is an important metric in determining the risk of damage to framed structures during earthquakes [3]. In this study, analytical fragility curves were used to assess the effectiveness of friction dampers to retrofit these structures.

2 METHODOLGY

A typical archetype modal structure is considered within the scope of this study. The same building is retrofitted by adding a brace–damper system within the structure in the space between the bottom of the cantilever steel pile and the bottom of the steel platform. 3D and elevation views of both buildings are shown in Figs 5 and 6.

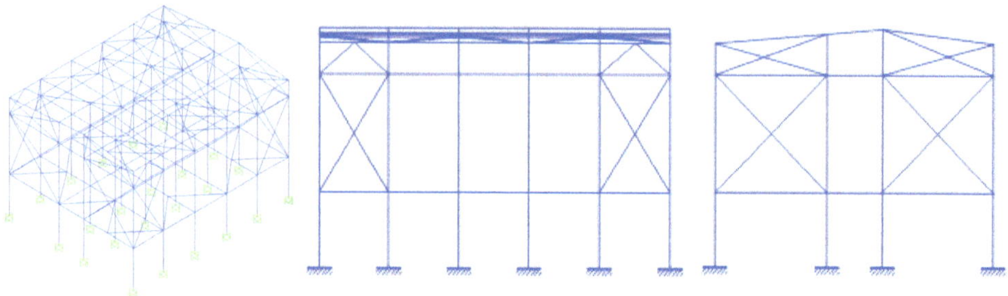

(a) (b) (c)

Figure 5: Archetype modular structure. (a) 3-D view; (b) View in long direction; and (c) View in trans. direction.

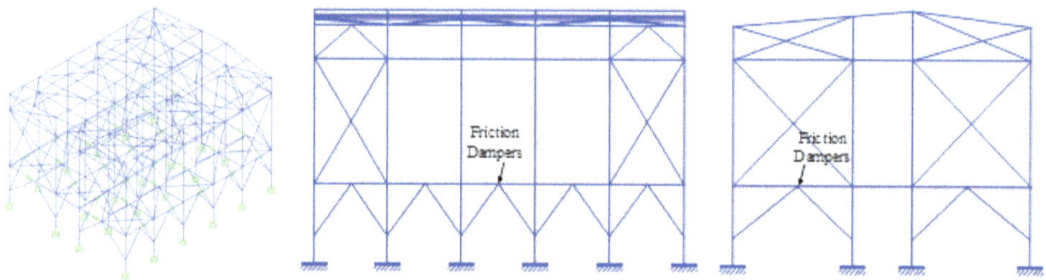

(a) (b) (c)

Figure 6: Archetype modular structure retrofitted with friction dampers. (a) 3-D view; (b) View in long direction; and (c) View in trans. direction.

The archetype structure was modelled based on the nonlinear behaviour of materials and its damping characteristics. An Eigen value modal analysis were conducted to determine the structure dynamic properties. Table 1 shows the natural period of the first 10 modes before and after the retrofit for the archetype structure. Pushover curves and inter-story drift ratio of the archetype structure at each step of pushover analysis were determined using SAP2000© software package [4]. A time history analysis was carried out to determine the response of structure based on 22 far field ground motions [5], as well as near field ground motions, selected from (PEER) ground motion database. Fig. 7 shows the response spectra for the 22 ground motions that were used in this study. From the graph, the selected ground motions represent a wide verities of earthquake records that have different peak ground accelerations (PGA) and frequency contents.

2.1 Analytical model of sample structure

To perform the pushover analysis, the nonlinear time history dynamic analysis, and to evaluate the vulnerability of the considered buildings, the building frames have been modelled using SAP 2000© software. The nonlinear or inelastic behaviour of various structural members, such as beams, columns, and bracing elements, plastic hinge has been introduced to the software based on the FEMA 356 guidelines [7]. For beams and columns, the general behaviour is shown in Fig. 8, and for bracing elements that carry the friction

Table 1: Structure dynamic properties.

Mode No.	Period (sec)				
	No Retrofit	5% Damping	10% Damping	15% Damping	20% Damping
1	0.9438	0.5883	0.5802	0.5778	0.5773
2	0.8846	0.4668	0.4186	0.4109	0.4079
3	0.8472	0.4326	0.3846	0.3811	0.3809
4	0.8471	0.4041	0.3809	0.3806	0.3805
5	0.7514	0.3861	0.3802	0.3802	0.3802
6	0.6789	0.3802	0.3802	0.3802	0.3802
7	0.6426	0.3802	0.3802	0.3802	0.3802
8	0.6163	0.3802	0.3802	0.3802	0.3802
9	0.6011	0.3802	0.3801	0.3801	0.3801
10	0.5070	0.3802	0.3801	0.3801	0.3801

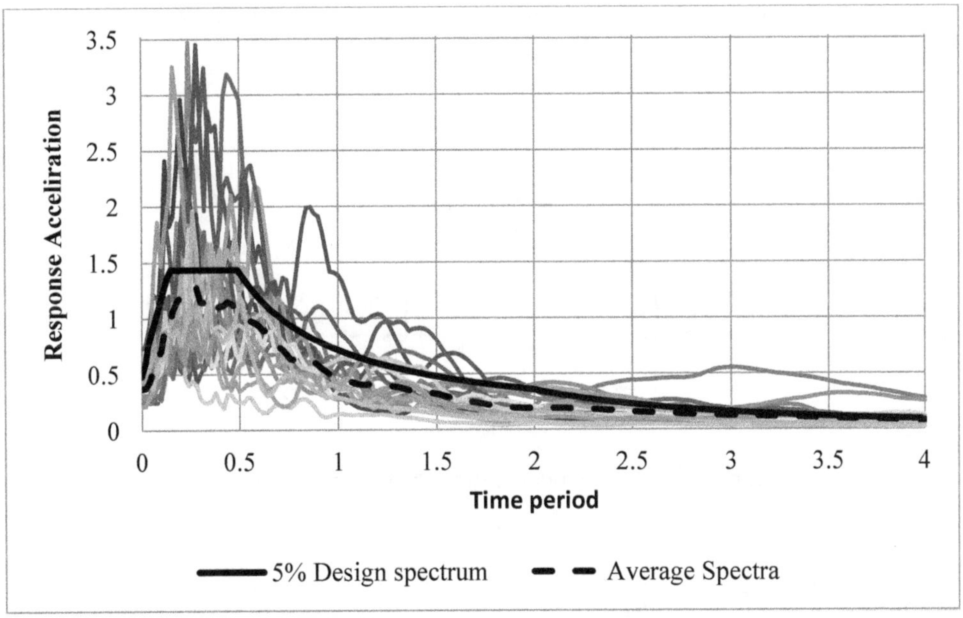

Figure 7: Response spectra acceleration (SA) for the 22 ground motion records used in the dynamic analysis.

dampers, the model that is shown in Fig. 9 was used. The hysteric force displacement relationship for a typical WT brace member which is the cross section commonly used in these structures is shown in Fig. 10. It can be seen from Fig. 11 that the effect of brace buckling on the overall ductility performance of the system is considerably low compared to that of other ductile systems. The parameters used in Fig. 9, which defines the inelastic behaviour of bracing elements, are calculated by the following formulas. The parameters for this plastic hinge model can be determined by using eqn (1):

$$P_c = 1.7\, F_a\, \mathrm{A}, \tag{1}$$

where F_a is the allowable axial stress obtained from AISC Table 4, A is the area of the brace cross section, and Pc represents the compressive axial force of the brace member. With P_c, deformation of the brace member can be calculated by using eqn (2):

$$\Delta_c = \frac{P_c\, L}{E\, A}, \tag{2}$$

where L is the length of the brace member.

For the tension side of the force-deformation relation model, the axial force and the deformation is defined as P_y and Δ_y are the brace axial force and displacement, respectively, where the F_y is the yield stress of the steel.

$$P_y = F_y\, \mathrm{A}, \tag{3}$$

$$\Delta_y = \frac{P_y\, L}{E\, A}. \tag{4}$$

Fig. 5 depicts the plastic hinge model of the brace frame used in the structure which is WT12X52 section.

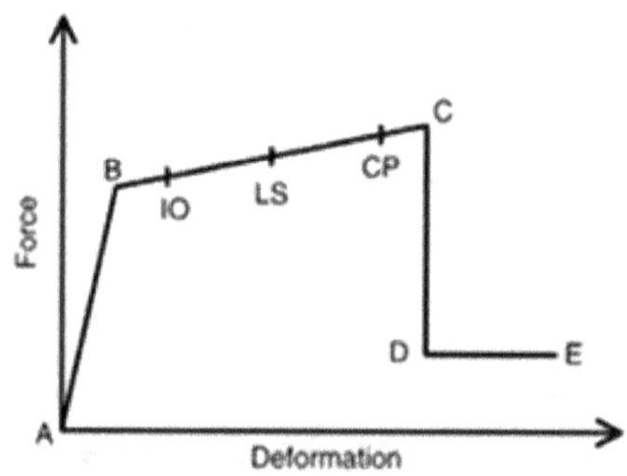

Figure 8: Inelastic model used for beams and columns [5].

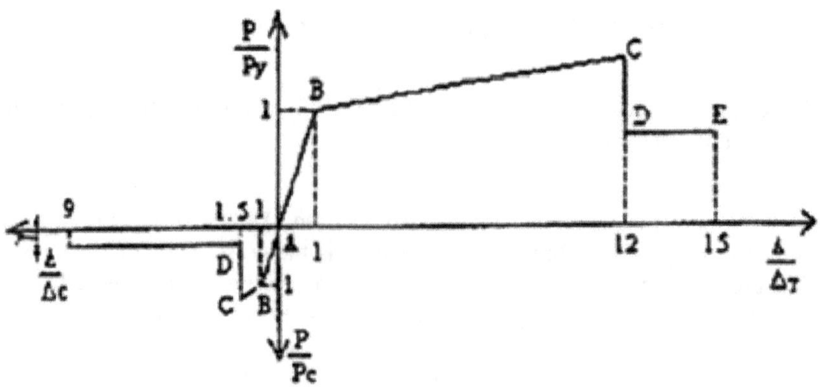

Figure 9: Inelastic model used for bracing elements [7].

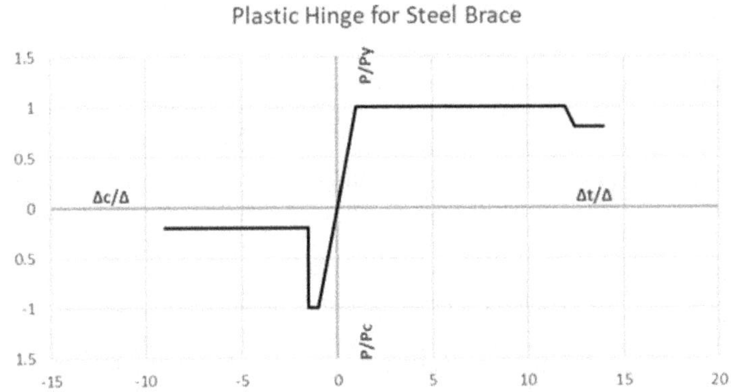

Figure 10: Plastic hinge properties for a typical brace member.

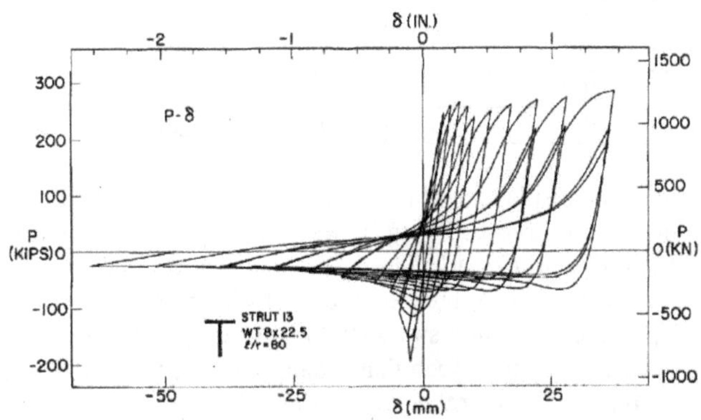

Figure 11: Force displacement relationship for a WT brace member under cyclic loading [6].

2.1.1 Friction dampers

To account for the additional system damping produced from supplemental damping devices, FEMA 356 specifies a damping modification factor to reduce the seismic effect (pseudo lateral load in a given horizontal direction) on the structure. The damping modification factor comes from the estimated effective damping ratio β_{eff}, which is expressed as follows for a structure with linear viscous dampers [7].

$$\beta_{eff} = \beta + \frac{\sum_j W_j}{4\pi W_k}, \tag{5}$$

$$W_k = \frac{1}{2}\sum_i F_i \delta_i, \tag{6}$$

where β is the damping ratio of the structure without any devices, W_j is the work done by device j in a complete cycle corresponding to floor displacement δ_i. W_k is the maximum strain energy in the frame and F_i is the inertia force in floor i [7].

The force-displacement relationship of the nonlinear damper is expressed as follows:

$$F = K_{eff} D, \tag{7}$$

where F is the damper force, D is the displacements across the damper, and K_{eff} is the stiffness of the brace and the damping device which can be calculated from the force displacement diagrams of the device.

Using eqns (5) and (6), the approximate damping ratios of the system can be estimated, and the supplemental damping devices can be sized to achieve the desired damping ratios. In this study damping ratios of 5%, 10%, 15% and 20% of the entire system were considered. Table 2 shows the structure properties of the dampers that are used in this study at different damping ratios.

Table 2: Friction damper properties.

Damping ratio	Dampers properties in X direction			Dampers properties in the Y direction		
	Slip load (kip)	Stroke length (in)	Weight	Slip load (kip)	Stroke length (in)	Weight
5%	30	7	0.5	40	7	0.5
10%	90	7	0.5	131	7	0.5
15%	208	7	0.5	230	7	0.5
20%	270	7	0.5	240	7	0.5

Fig. 12 shows the force displacement relationships for the dampers that were used in this research at different damping ratios. It can be seen that as the damping ratio increased the energy absorbed by the dampers increased. Fig. 13 shows the push over curves for the structure in both the transverse and longitudinal directions. The buckling load in the braces in relationship to the other limit states (failure criteria) that were used in this study are shown in this figure where it can be seen that the buckling of the braces is the most critical limit state compared to pile failure or the story drift.

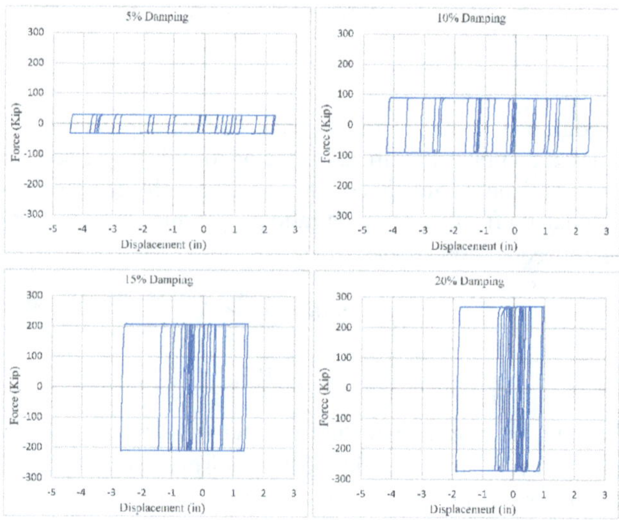

Figure 12: Damper force displacement relationship for different damping ratios.

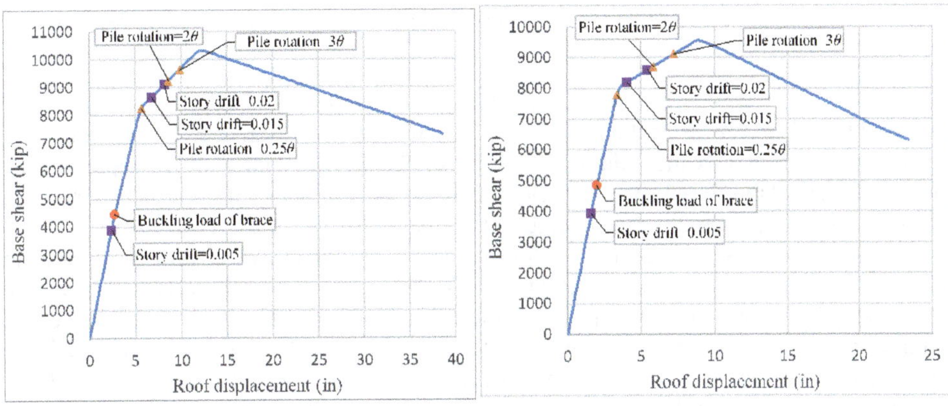

Ground Motion in UX direction　　Ground Motion in UY direction

Figure 13: Push over curves for the modular structure.

2.1.2 Analytical fragility curves

Fragility curves express the probability of reaching or exceeding the specific limit state at a given level of ground shaking intensity. Log-normal distribution for damage probability is a common assumption in the literature [8].

A log-normal distribution is used because all values are expected to be positive. The probability of reaching or exceeding a Limit State (LS) at a given earthquake intensity is given by eqn (8):

$$P[DS_i|S_a] = \varphi\left[\frac{1}{\beta_{S_{ds}}} ln\left(\frac{S_a}{S_{a,ds}}\right)\right], \tag{8}$$

where φ is the standard log-normal cumulative distribution function; S_a is the spectral acceleration amplitude (PGA in this case); $S_{a,ds}$ is the median value of spectral acceleration (PGA) at which the building reaches the threshold of damage state, *ds*; and β_{Sds} is the normalized composite log-normal standard deviation respectively. Demand spectra and capacity curves are described probabilistically by median properties and variability parameters, β_D and β_C, respectively [8].

$$S_{a,ds} = \delta_{R,S_{ds}}\alpha_2 h\delta_{R,S_{ds}}. \tag{9}$$

To account for variability, FEMA uses eqn (10):

$$\beta_{S_{ds}} = \sqrt{\left(CONV\left[\beta_{C,}\beta_{D}, S_{d,S_{ds}}\right]\right)^2 + \left(\beta_{M(S_{ds})}\right)^2}, \tag{10}$$

where β_{Sds} is the log-normal standard deviation that describes the total variability for structural damage state, *ds*; β_C is the log-normal standard deviation parameter that describes the variability of the capacity spectrum; β_D is the log-normal standard deviation parameter that describes the variability of the demand spectrum; $\beta_{M(Sds)}$ is the log-normal standard deviation parameter that describes the uncertainty in the estimation of the median value of the threshold of structural damage state, ds [8].

The variability of building response depends jointly on demand and capacity since capacity curves are nonlinear. The function "CONV" implies a complex process of convolving probability distributions of the demand spectrum and the capacity curve, respectively.

3 RESULTS AND ANALYSIS

To assess the effectiveness of the retrofit using friction dampers, three criteria were used in this study. Criterion 1 is the axial deformation of the braced frames which represent the buckling failure of the WT shapes. Criterion 2 is the story drift between the base and the plate girder platform. Criterion 3 is the bending of the steel piles.

The fragility curves were developed for the three main performance levels prescribed in FEMA 356 which are immediate occupancy (IO), life safety (LS) and collapse prevention (CP). The 0.5 PGA was used as a point of reference to compare structure performances for all three criteria for the different performance levels. The 0.5 PGA represents the design basic earthquake (DBE) which represent an earthquake that has a 10% probability of exceedance in 50 years. For the axial deformation of the vertical braces (criterion 1), Figs 14–18 show that increasing the system damping reduced probability of failure from 57% to 30% for the life safety performance level while the collapse prevention was reduced from 40% to 9%. On the other hand, increasing the damping ratio does not have any effect on reducing the probability of failure in the Immediate Occupancy performance level. Tables 3–5 summarize the structure performance as it relates to criterion 1. For the story drift criterion, increasing damping ratio from 2% to 20% reduced the probability of failure from 100% to 63% for the immediate occupancy level, 67% to 1% for the life safety performance level and 24% to 0% for the collapse prevention performance level. For the pile bending (criterion 3), increasing the damping ratio reduce the probability of failure from 68% to 3% for the 2% damping to 20% damping ratio, while it reduced from 12% to 1% for the life safety performance level and from 7% to 0% for the collapse prevention performance level. It should be noted that the structure performance at 20% damping is even worse than the result at 15% damping. This can be explained due to the fact that the damper slip load of 15% is 208 kip (925 kN) while for 20% is 270 kip (1,201 kN).

The results indicated that increasing the slip load in the damper is not an effective solution especially for criterion 2, because larger force is needed to initiate the damping mechanism of the friction damper, leading to larger deformation of the braces members. Therefore, the 15% damping ratio is the optimal value for the retrofit of this system.

4 CONCLUSIONS

Fragility analysis has been widely used in performance-based design (PBD) over the last decades to assess the response of different structure systems. In the context of PBD, fragility assessment modular steel structures subjected to earthquake hazards is conducted in this study. This study revealed that the friction dampers were effective in reducing the risk of damages for modular steel structures.

The fragility curve analysis shows clear improvements in increasing the damping ratios. It is noticed that there is no significant improvement when the damping ratio of the system was increased from 10% to 20%. It also found that the 15% damping ratio is the optimum value for improved performance for all criteria considered in this study.

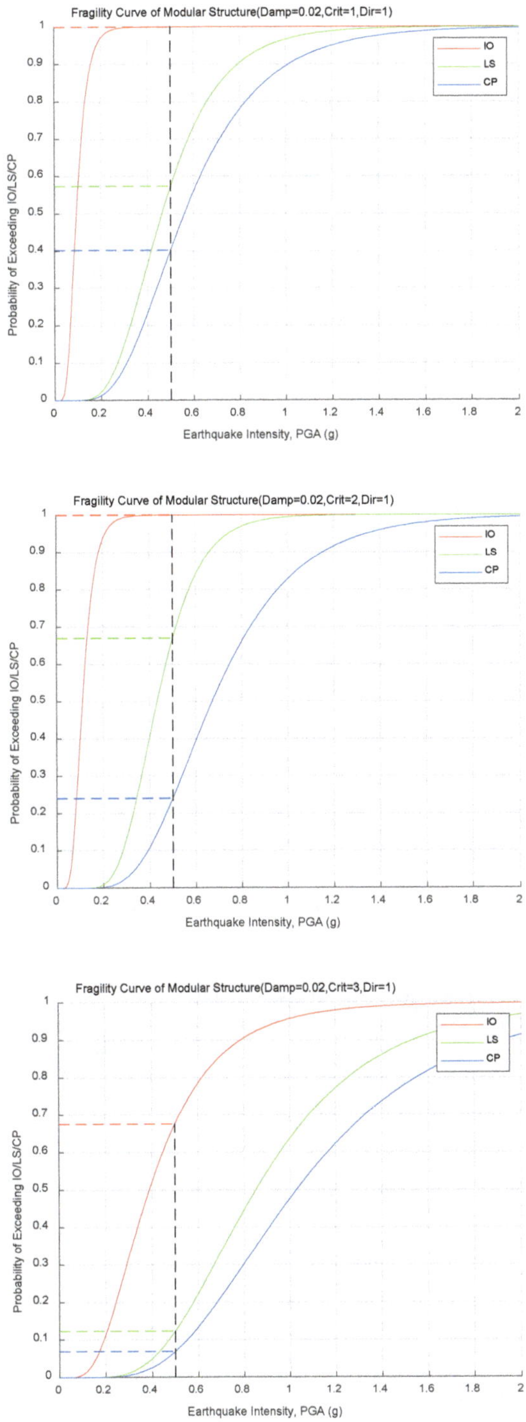

Figure 14: Fragility curves for existing structure (2% modal damping).

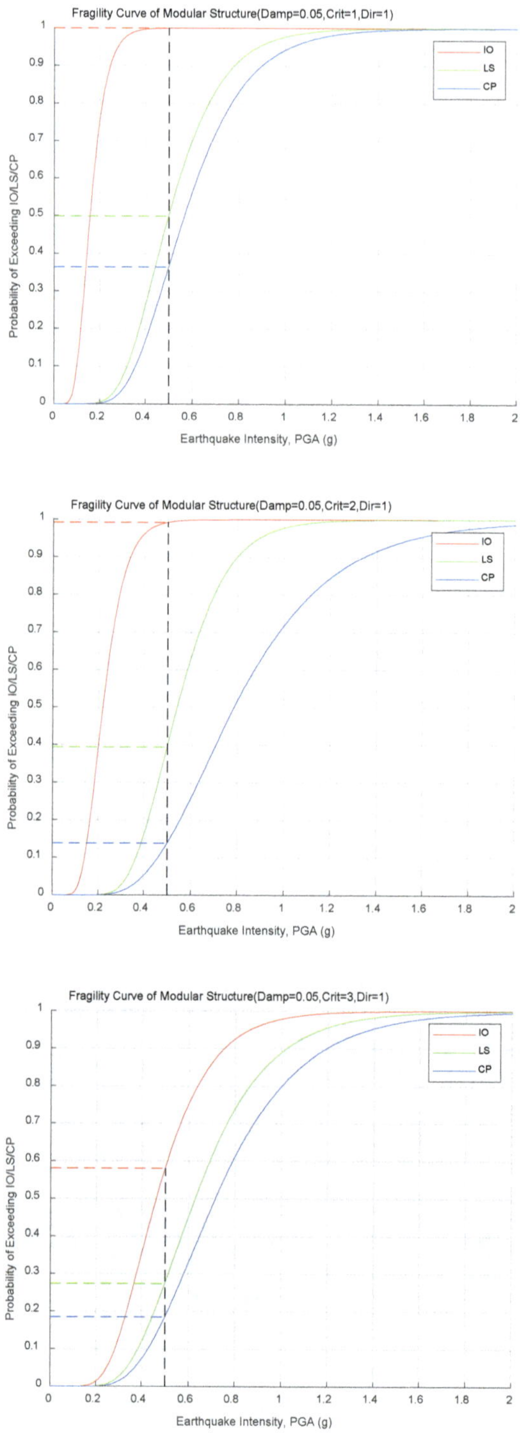

Figure 15: Fragility curves for retrofitted structure (5% damping).

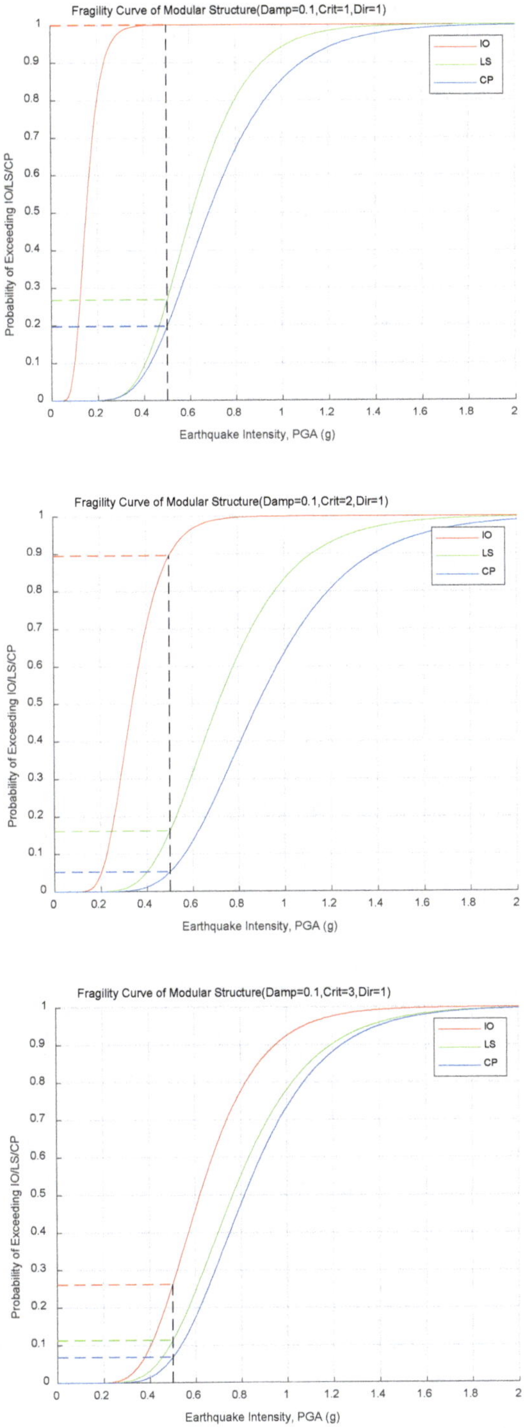

Figure 16: Fragility curves for retrofitted structure (10% damping).

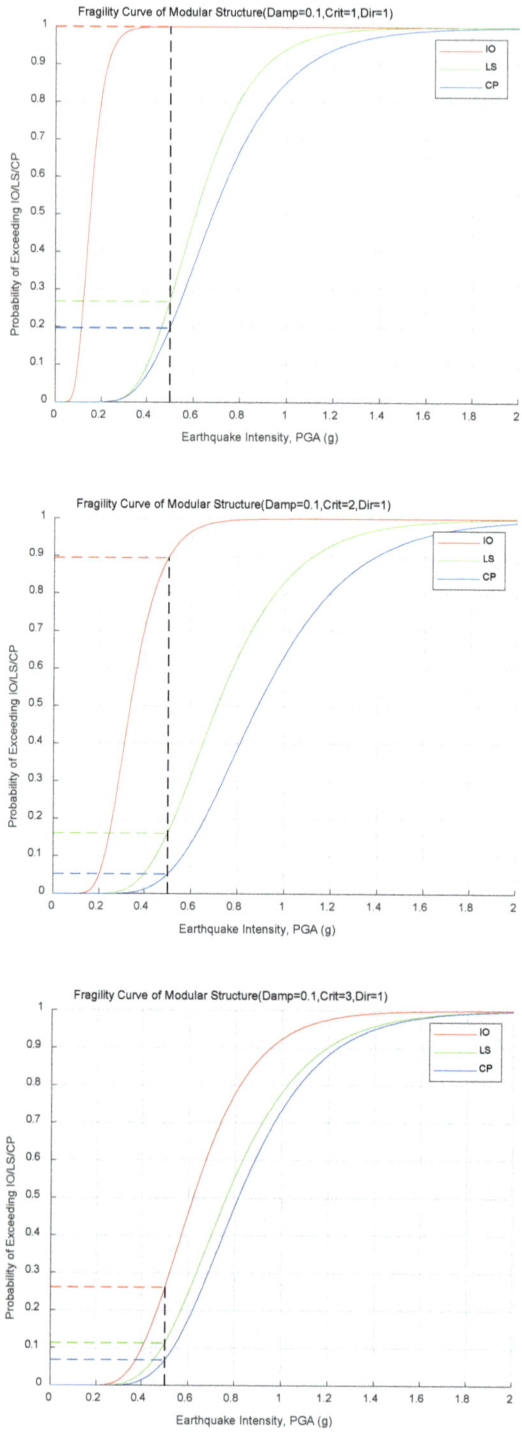

Figure 17: Fragility curves for retrofitted structure (15% damping).

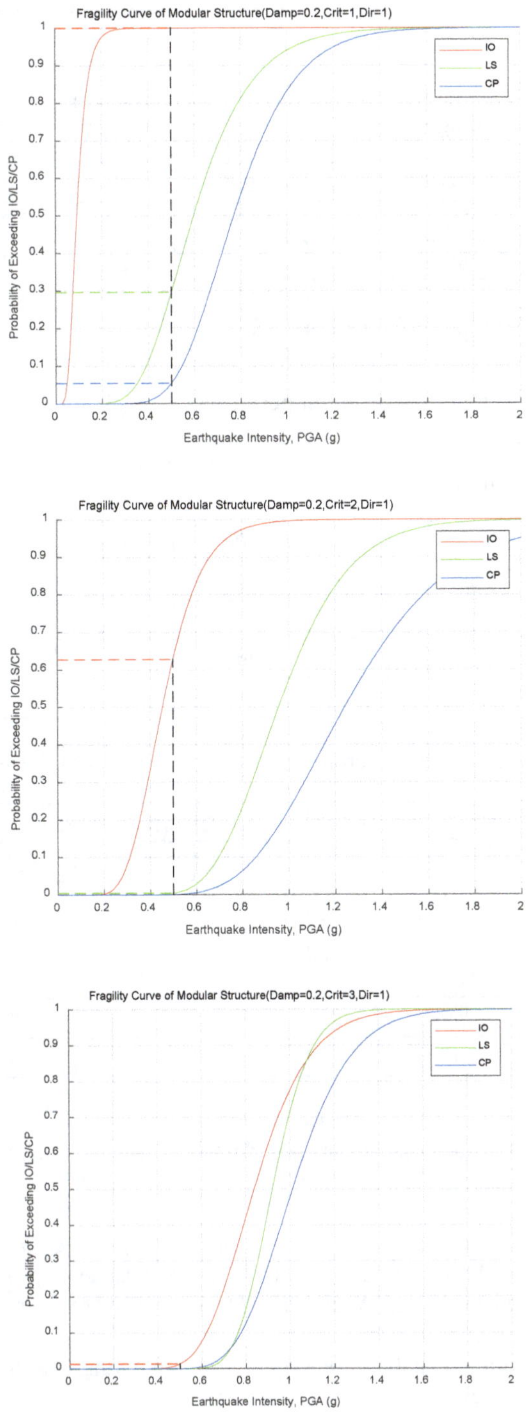

Figure 18: Fragility curves for retrofitted structure (20% damping).

Table 3: Probability of failure at different performance levels for criterion 1 (brace axial deformation).

Damping ratio	Performance level		
	IO	LS	CP
2%	100	57	40
5%	100	50	36
10%	100	26	21
15%	100	13	6
20%	100	30	9

Table 4: Probability of failure at different performance levels for criterion 2 (story drift).

Damping ratio	Performance level		
	IO	LS	CP
2%	100	67	24
5%	99	39	11
10%	89	16	6
15%	66	2	1
20%	63	1	0

Table 5: Probability of failure at different performance levels for criterion 3 (pile bending).

Damping ratio	Performance level		
	IO	LS	CP
2%	68	12	7
5%	57	27	16
10%	25	11	8
15%	4	1	8
20%	3	1	0

REFERENCES

[1] Pall, A. & Marsh, C., Response of friction damped braced frame. *Journal of Structural Engineering,* **108**(6), 1982.

[2] Quaketek, www.quaketek.com/.

[3] Hanson, R.D. & Soong, T.T., *Seismic Design with Supplemental Energy Dissipation Devices*, Earthquake Engineering Research Institute: Pasadena, p. 135, 2001.

[4] Sap2000: Integrated Structural Analysis & Design Software. Vers. 17. Berkeley, CA. *Computer and Structures*, 1997.

[5] PEER Ground Motion Database, Pacific Earthquake Engineering Research Center. https://ngawest2.berkeley.edu/.

[6] Black, G.R., Wenger, B.A. & Popov. E.P., Inelastic buckling of steel struts under cyclic load reversals. Report no. UCB/EERC-80/40. Earthquake Engineering Research Center: Berkeley, CA, 1980.
[7] FEMA. NEHRP Commentary on the guidelines for the seismic rehabilitation of buildings. FEMA-356, Federal Emergency Management Agency, Washington, DC, 2001.
[8] Karim, K.R. & Yamazaki, F.A., Simplified method of constructing fragility curves for highway bridges. *Earthquake Engineering Structural Dynamics*, **32**(10), pp. 1603–1626, 2003.

VULNERABILITY ASSESSMENT OF BUILDINGS BASED ON RAPID VISUAL SCREENING AND PUSHOVER: CASE STUDY OF GYOR, HUNGARY

ORSOLYA KEGYES-BRASSAI
Szechenyi Istvan University, Hungary

ABSTRACT

The process of carrying out an earthquake risk assessment of a town can provide important data for authorities and disaster management to better understand risks to many buildings rather than a single building. This is even more important in the case of moderate seismic areas where any mitigation measures to be taken should be justified by seismic risk determination. Moderate seismicity does not necessarily equate to moderate damage from earthquakes. Vulnerability to earthquakes even increases with extending urban areas. Seismicity in the Pannonian Basin is moderate compared to seismicity of surrounding areas, nonetheless, reports of major earthquakes in Hungary often refer to heavy building damage and liquefaction (e.g. 1763 Komarom earthquake). Gyor was chosen to be the examined area for seismic risk analysis because it is the most important city of northwest Hungary with a large number of monumental buildings and a complex geological and geographical settings. In order to make the best use of limited resources usually characteristic to moderate seismic zones, the presented method used existing soil data, rapid visual screening of buildings, a limited number of field tests and free, but sophisticated, software. This paper focuses on the results of vulnerability analysis of buildings; however, it considers the results of seismic hazard and local site effects based on response analysis with more than 6000 realizations. Vulnerability of the buildings with different structural types were evaluated based on a rapid visual screening of 5000 building. Vulnerability based on visual screening was compared to a pushover analysis of the typical constructions. As one would expect, since the hazards and vulnerabilities were not uniformly distributed in the city districts of Gyor, there were zones of higher and lower risk. These results can then serve as useful tools for decision makers and can be applied directly to risk management plans.

Keywords: vulnerability assessment, rapid visual screening, pushover.

1 INTRODUCTION

The research presented in this case study estimates the seismic risk based on seismic hazard evaluation [1] and building vulnerability functions however this paper focuses on the results of vulnerability assessment. The goals of the research were: to present the process of seismic risk evaluation of a city; to integrate local site effects to previously determined seismic hazards for Hungary (National Annex of EC8) [2]; to examine the vulnerability of the built environment focusing on residential and public buildings; and finally to combine the results and map risk scenarios to aid in the mitigation process. As one would expect, since the hazards and vulnerabilities were not uniformly distributed around the examined area, there were zones of higher and lower risk [3]. Based on these risk assessments, engineers and planners can decide where to improve buildings, whether to permit further development, and formulate emergency response plans in case of a seismic event.

The next quote clearly states the importance of the work in seismic risk assessment: "The fundamental role of the seismic risk assessment for the society is to provide all the information for each community or organizations to support the risk mitigation decision-making. These decisions are generally related to the likelihood and significance of structural collapse to the life-safety or business interruption. Hence a high risk of structural collapse is

WIT Transactions on The Built Environment, Vol 185, © 2019 WIT Press
www.witpress.com, ISSN 1743-3509 (on-line)
doi:10.2495/ERES190051

not accepted by the current standards and there are various methods available for making decisions to reduce that risk" [4].

Major cause of loss during an earthquake is due to the deficient seismic performance of the built environment. Catastrophic events with high socio-economic impact occurred also at places with moderate seismic events with a magnitude of 6 to 6.5 (e.g. Italy 2009 and 2016). New standards and regulations have been applied to new structures from the beginning of 21st Century, which represent only a small part of existing structures. In Hungary this is true also, just as it is in other parts of the world. Many existing buildings and other structures might fail to resist the earthquake and thus sustain some degree of damage that can vary from minor cracking to total collapse. Since other forms of damage and loss are generally defined as a function of primary structural damage, the starting point is to find out the expected damage due to a likely strong ground motion.

2 METHODS USED FOR VULNERABILITY ASSESSMENT

Vulnerability represents the link between building or structural damage and the earthquake effects and defines loss as a function of ground shaking. These functions can be developed statistically, analytically or based on expert opinions. From a probabilistic point of view, fragility functions can be derived defining the probability of some undesirable result as a function of ground shaking levels. Analogous vulnerability functions can provide a damage factor, given the intensity of the earthquake.

Vulnerability is the possibility of damage or loss of structures due to a seismic event; it is characteristic to the structure and it can be expressed in probabilistic or statistical terms. The methods to determine the vulnerability depends on the objective of the study, the required results and on available data, and can be grouped according to the space scale considered for analysis: e.g. urban level or building level, etc. [5]. Large-scale approaches are based on empirical methodologies and local scale assessments are performed using some detailed numerical analyses (Fig. 1). As these methods are getting more complex, they become more accurate, but also requiring higher computational and evaluation effort.

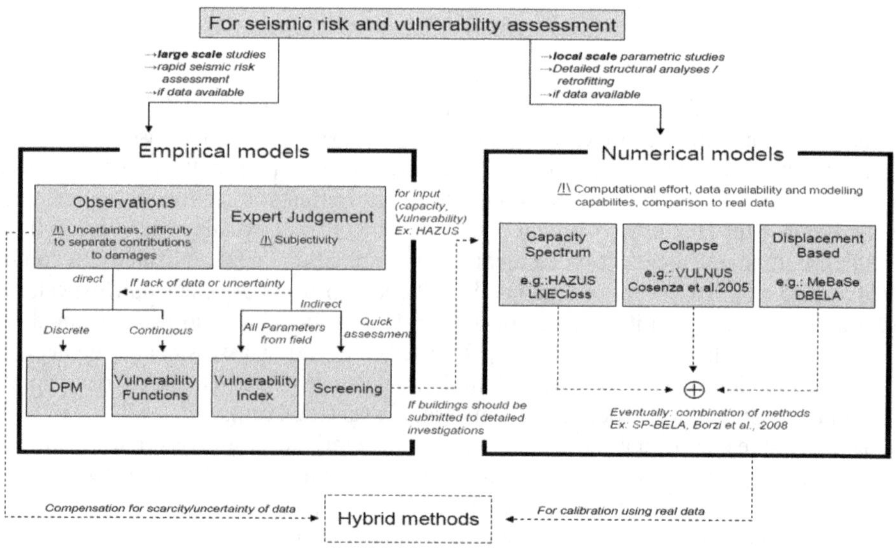

Figure 1: Schematic representation of the classical models for seismic vulnerability derived by the ENSURE project [5].

The rapid visual evaluation of buildings was performed on over 5000 structures. The evaluation method was developed from well-known approaches but modified to account for typical building designs found here. The reliability of the visual evaluations performed by trained non-experts was verified by experts over a significant percentage of all buildings.

Empirical methodologies are widely used to quickly assess the seismic risk on a large scale or to sort out structures that need more detailed analysis. Observed vulnerability refers to data about damages of past earthquake. In the case of observed vulnerability [6], [7] the damage is defined with the repair cost as a ratio of the amount of loss of all affected elements as a ratio of their value. The relation between damage and earthquake intensity is valid only for the region where it was developed. The vulnerability of structures can also be evaluated through expert opinions. Experts have to estimate the expected percentage of damage caused by a given intensity, which are implied in macroseismic scales.

Method used for this research was the score assignment procedure aiming to determine hazardous structures by identifying structural deficiencies. The screening methods gather quantitative information, by using vulnerability forms including parameters, such as quality of materials, type of foundations, number of stories, or stiffness of the structure. Depending on the parameter's value, a score is attributed to each feature to quantify the level of damage according to the severity of potential ground motion. Potential structural deficiencies are identified from observed correlations between damage and structural characteristics [8].

As an alternative to empirical methods detailed analyses can be used however these are more time-consuming assessments. Engineers may use several structural analysis methods to estimate building vulnerability to seismic loading. These methods apply equally to zones of strong, moderate, or weak seismicity. They are simply ways to quantify the ability of a structure to resist an earthquake event.

Lateral force method applies additional lateral forces to the structure and the model is analysed as a static load case. The magnitude and location of additional lateral forces are based on the intensity of the earthquake. Beams and columns are sized so that they remain in the elastic state. Modal response method assumes the structure will respond as a combination of harmonic motions. The contribution of each of these modes to the response of the structure is scaled according to its mass participation factor. The material behaviour is still elastic. Nonlinear time history analysis represents the most accurate method to evaluate structural response to earthquake loading however being the most time-consuming. The earthquake acceleration record is used as input to the base of the model and the structure's deformation, stiffness, damping, hinge formation and changes in geometry are all evaluated.

Further evaluation of building vulnerability was performed by pushover analyses. This method has become very popular to determine the non-linear capacity of a building to lateral forces. The basis of analysis is a static, deflection-controlled approach. As the structure is pushed over, its resistance increases to a peak value, then declines to a residual value. Values of deflection vs. base shear are often compared to seismic demand curves.

3 AREA OF THE CASE STUDY

Gyor was chosen to be the examined area for seismic risk analysis because it is the most important city of northwest Hungary. Being the capital of Western Transdanubia region has an important economic and cultural role. Gyor is referred to as the City of Waters as it lies at the confluence of the Moson-Danube, the Raba and the Rabca; not far from the main channel of the Danube. From a geological point of view Gyor lies in the eastern part of the macro region of Little Hungarian Plain, which is a deflational lowland and on the southern part the city extends to the Pannonhalmi Hills, which is a part of the Transdanubian Mountains.

Thus the formation of the area is related to these two large regions [9]. This setting means a high variability in the local soil conditions posing a different level of seismic hazard on the different parts of the city.

Geomorphological conditions had a significant impact on the development and evolution of the town. The elevated flood-free terraces at the confluence of rivers offered a good possibility for the very first settlement. The Downtown (Belvaros) is basically the oldest part of the city with characteristic corner-balconies and narrow lanes, churches, museums are all reminders of a historic past (Fig. 2). Gyor is Hungary's second richest town in historic buildings outside Budapest making it vulnerable to earthquakes.

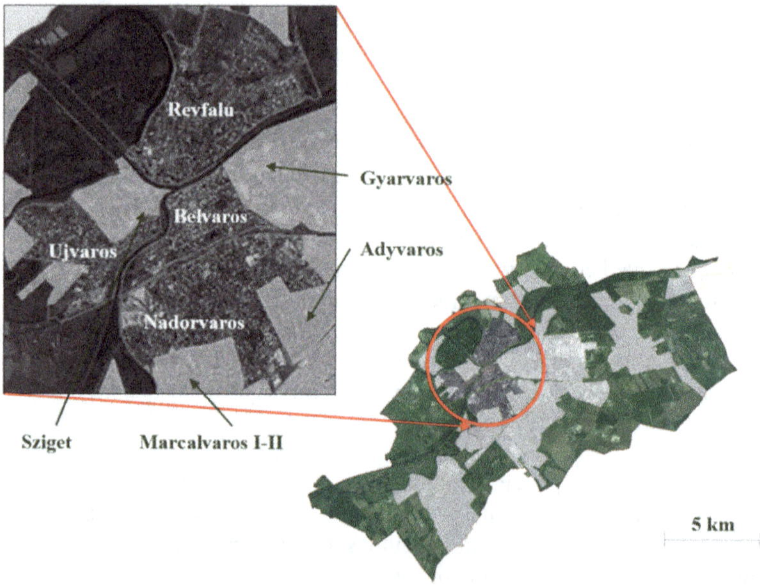

Figure 2: Historical, urban districts of Gyor.

The fact that the town is expanding to the south and east is basically due to the surface conditions. Surface water and the high-water table in every area provided adequate water supply, also for industrial purposes. The current city structure was the result of the merge of ten historical town parts. The suburb regions gradually built together with the historical city. Development of the urban areas were strongly influenced by the railway lines built in 19th Century. The industry was the key factor in the growth of several town districts such as Revfalu and Nadorvaros in that time (Fig. 2).

After the Second World War, the most striking developments took place in two areas: first, the neighbourhood villages had been attached to the city, secondly the rapid industrial development led to the creation of new industrial areas between Downtown and Gyarvaros (factory area). Accelerated urbanization enhanced the development of panel housing estates (Sziget, Gyarvaros, Adyvaros, Marcalvaros). These panel estates formed a large contiguous zone in the last fifty years of the 20th Century. They were connected with small town and suburban residential areas. Bridges built over Raba and Danube connected the districts of the city. The road network of Gyor has basically a radial structure, passed by the motorway from south. Recently, the main restructuring is the connection of the city districts, the development of the infrastructural system and the complex rehabilitation of several districts.

3.1 Seismicity of the area

The seismicity of the Pannonian region is moderate (Fig. 3). Earthquakes causing light damage to occur every year, while stronger, more damaging magnitude 5 quakes happen about every 20 years and the return period of magnitude 6 events is about 100 years [10]. The distribution of earthquakes is diffuse; however, there are certain areas where the occurrence is more concentrated, e.g. at the surroundings of Komarom-Mor-Berhida, known as Mori-trench.

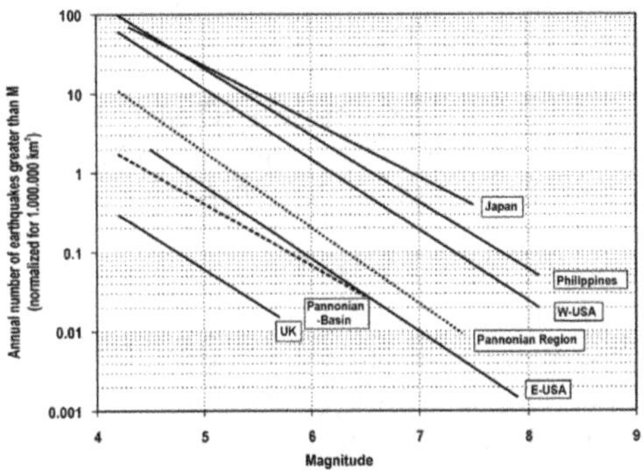

Figure 3: Annual number of earthquakes in seismically different parts of the World [10].

The largest earthquake in Hungary occurred in the city of Komárom in 1763 with an estimated magnitude of 6.3. Overall more than 100 victims were reported. The taller (2–3 story high) ecclesiastic and noble buildings with more complex structure, typical for that age, suffered serious damage, both in Komarom and Gyor, while the buildings of common people with only a ground floor made of adobe and timber survived the quakes [11]. Remedial measures were applied after the earthquake: the construction of 2–3 story high buildings were forbidden by authorities. Based on the 24% buildings destroyed, 30% seriously damaged, and 18% requiring reconstruction, the intensity can be assumed to be IX on the EMS (European Macroseismic Scale) [12].

Other significant faults contribute to the hazard concerning Gyor. One of the lineaments lies beneath the Raba River, called Raba line and meets the Diosjeno-Ogyalla or Hrubanovo fault beneath Danube River close to Gyor.

3.2 Typical building structures

The research focused on the residential buildings representing the highest number of building stock in Gyor. A total number of 21 city districts are grouped and presented according to the typical urban fabric. It can be differentiated between urban areas, housing estates, garden cities, and attached villages, recreational, industrial and commercial areas.

In case of residential buildings the most typical building structures are the unreinforced masonry loadbearing walls with only a ground floor with one or two flats. More intensively built housing areas were constructed in the second part of the 20th Century with one or two-story height apartment houses. Free standing buildings can be found in the urban city districts with a ground floor and 3–4 stories. The horizontal structure of these masonry buildings depends on the time of construction. After the sixties came the increased use of reinforced concrete slabs, before that time wooden structures were typical.

The Downtown area has a typical pattern with closely constructed buildings with different heights preserving the structure of 16–17th Century. The majority of buildings are one or two-story buildings, with a significant number of built in roof spaces turned into apartment houses. Typical are the enclosed courtyard formation. The area within the former city walls has a historic importance and is a protected area. As one could expect these are the most vulnerable structures to lateral dynamic forces.

One-third of the inhabitants live in reinforced concrete large panel housing areas with 4 or 10 story height buildings. Even though the number of these buildings are much less compared to masonry structures, they behaviour is crucial in case of a seismic event because of the high number of people living in those buildings.

4 VULNERABILITY RESULTS

More than 5000 buildings were evaluated based on score assignment method and compared to some pushover results of typical buildings on a district level.

4.1 Score assignment results

Vulnerability of buildings were investigated according to FEMA 155, however the checklist was created based on literature concerning inventories for score assignment, and the knowledge about the most-used building construction of the area. First the buildings need to be classified according to their possible performance. A simplified model often neglects some of the positive influence of structural elements because they may only be present in some buildings and not others of the same classification. Building inventory with more than 5000 buildings were established based on the screening evaluation of trained staff. The checklist consists of questions about:

- General data of each building (age and function of the buildings, regularity in plan and elevation, position of the building, changes in function, previous damages, etc.)
- Structural data of each building (construction system, quality of materials, workmanship)
- Other specific items

Based on classification and the data of each building base structural scores were assigned. Score modifiers were taken into account, such as:

- Vertical irregularity score modifier;
- Construction code score modifier;
- Score modifier concerning the height of the building;
- Soil score modifier.

The construction time determined the level of seismic design used at the construction of each building. Based on the height of buildings two categories were taken into account: low-rise building with no modifier and mid-rise buildings. Vertical irregularity was based on the questionnaire.

Resulting scores could be mapped for each individual building. However from an administrative view, this would be problematic since evaluation, retrofit, repair, and planning efforts would be applied by district, or perhaps sub-district. The city districts provide a good way to categorize building structures, in general, since each district has a predominant building type and construction era. Using a district approach is quite common in risk analysis; however some decisions are necessary about how to assign a district score. One may pick a dominant structure type, either by building count (small residential buildings would have more influence) or total resident occupancy (apartment houses would have more influence). The average score for the dominant structure would then be applied to the district. Another approach is to use an unweighted or weighted average of all the buildings in the district.

For this study, both the dominant building approach and the average score approach produced essentially the same result. Table 1 represents the data of 4 city districts and also the results of Structural Hazard Scores for two scenarios.

Table 1: FEMA structural scores in different city districts.

Town parts					Total
Function of the area	Belvaros	Nadorvaros	Ujvaros	Revfalu	
Economic, commercial, industrial	2.51%	30.23%	8.34%	1.49%	
Special (cultural, educational, ecclesiastic)	69.79%	14.72%	5.94%	9.57%	
Rural residential	-	7.69%	50.25%	36.97%	
Garden city residential	-	29.74%	15.11%	34.49%	
Urban residential	4.89%	12.46%	0.40%	0.14%	
Total area [ha]	90.49	256.37	117.39	263.81	*728*
Number of residents	10 358	20 130	4 397	6 640	*41 525*
Soil profile	C1	C1/C2/C5	C2/C5	C1/C3	
Basic structural hazard score	2.7	4.2	4.2	4.2	
Modifier of soil	-0.4	-0.6	-0.6	-0.5	
Structural hazard scores scenario I	2.3	3.6	3.6	3.7	
Structural hazard scores scenario II	0.7	2.0	2.0	2.1	

Fig. 4 shows the distribution of the Structural Hazard Scores. A score below 2.0 means a vulnerable building with a need of an individual evaluation. Since the EC8 requirements with higher demand for designing against earthquake loads is in use only from the beginning of 20th century these results are aligned with expectations.

Figs 5 and 6 represents the number of buildings and the scores of buildings for the district of Nadorvaros, respectively.

Masonry buildings dominate the final scores of the districts with two peaks: the lower score for the irregular buildings and higher for regular buildings. Lower scores represent higher vulnerability.

4.2 Pushover analysis results

Vulnerability functions were determined for typical layouts in case of masonry and reinforced concrete buildings (Fig. 7). The steps of the pushover analysis have the same order in case of both masonry and reinforced concrete building taking into account building material features respectively. With the given value of possible PGA the expectable damage can be estimated.

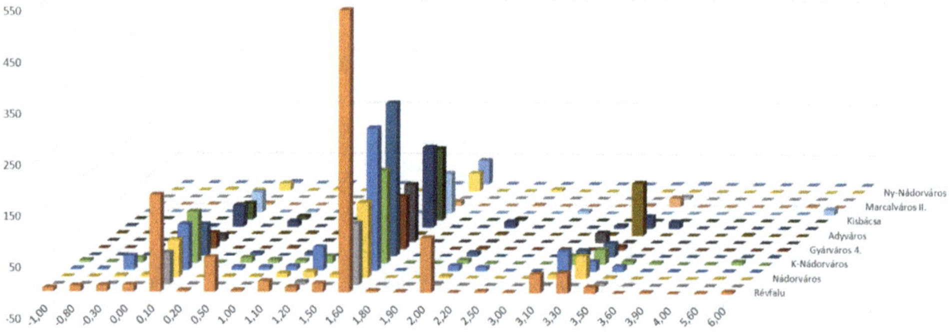

Figure 4: Distribution of structural hazard scores.

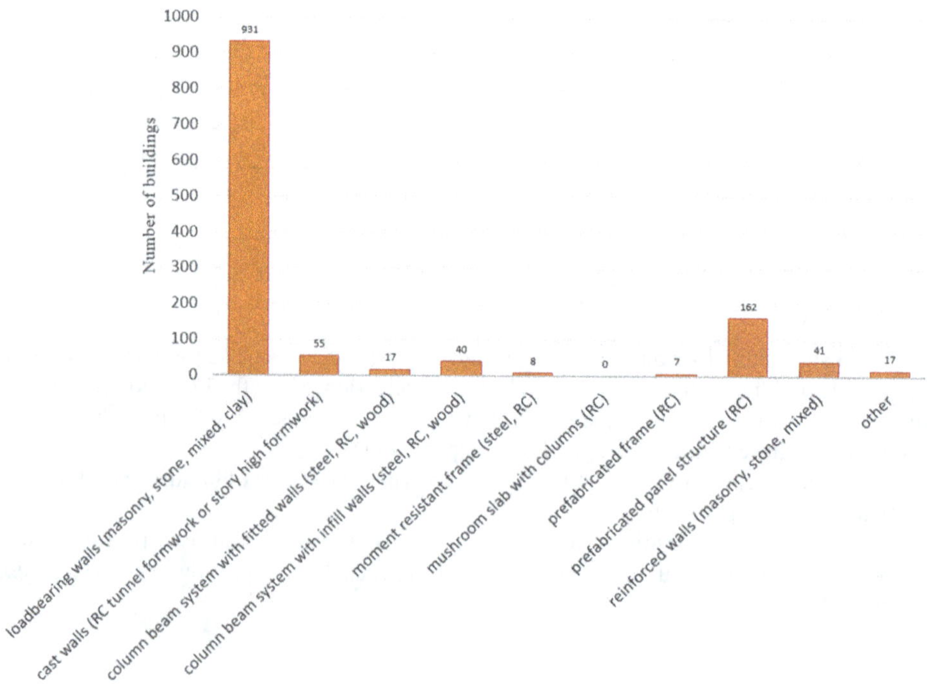

Figure 5: Number of building structures in Nadorvaros.

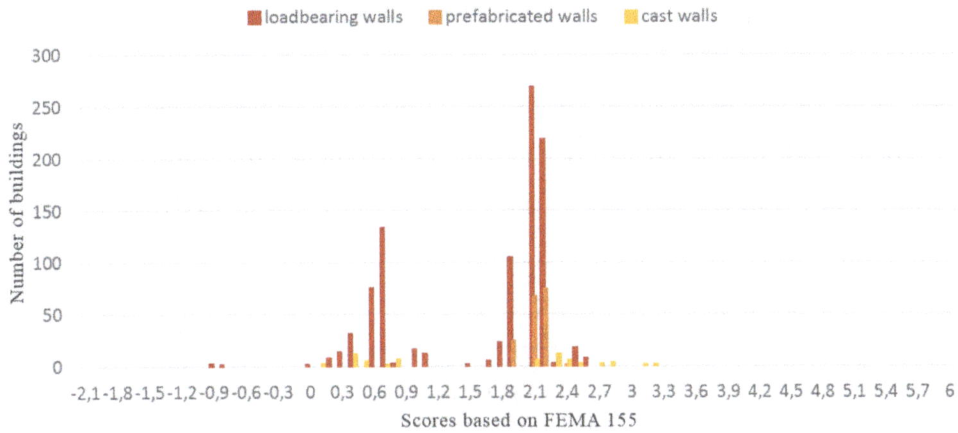

Figure 6: Structural hazard score for buildings in Nadorvaros.

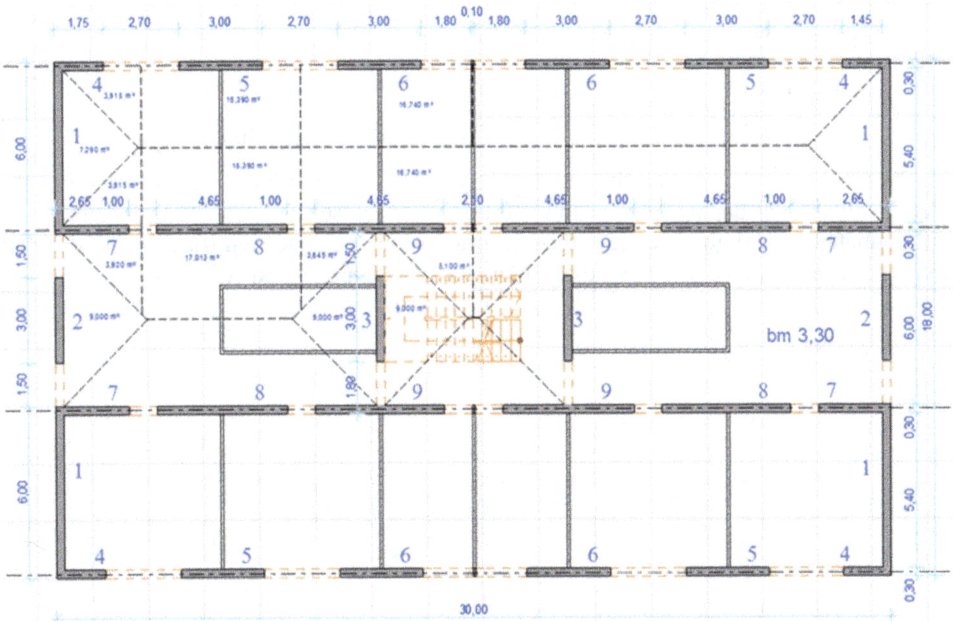

Figure 7: Layout of a typical reinforced concrete building.

All structural elements, which contribute to the loadbearing capacity of the building should be taken into account. According to the bilinear approximation the capacity curve of a wall element can be characterized by three elements: the maximum shear strength of the wall element, the nominal yield displacement (Δ_y) at the top of the wall and the ultimate displacement (Δ_u) at the top of the wall. The transition between the linear elastic and perfectly plastic region is called the yield point even if masonry structures do not yield. Summarizing the shear capacities of the separate walls the shear capacity of the whole building can be obtained (Fig. 8).

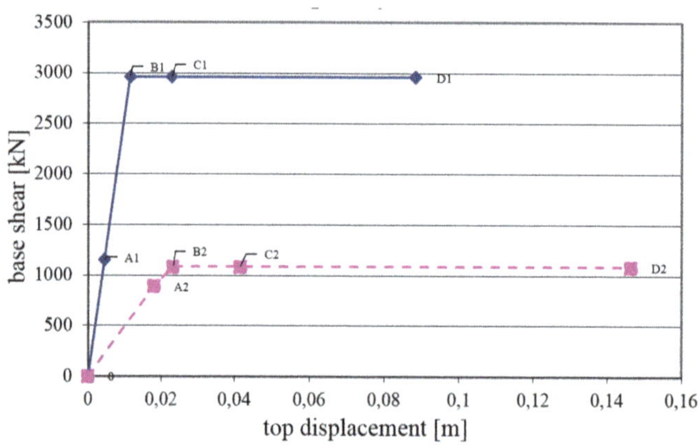

Figure 8: Capacity of the wall elements of the examined reinforced concrete building.

Material properties:

f_y [kN/m²]	500000		
$f_{c0'}$ [kN/m²]	33000	Es [kN/m²]	210000000
$f_{c0'}$ [kN/m²]	33000	E_c [kN/m²]	30000000

The main task is to obtain the vulnerability function of the building. This requires the identification of the damage grades. In the European Macroseismic Scale [12] five grades are differentiated from slight damage to total destruction allowing the identification of the points were the capacity curve of the building enters the next damage grade (Fig. 9).

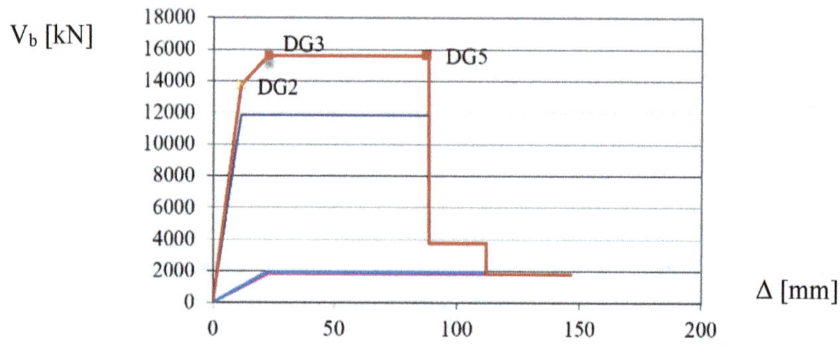

Figure 9: Capacity of the examined reinforced concrete building.

The seismic demand is determined using an elastic acceleration response spectrum (Fig. 10). It represents the maximum response of equivalent single degree of freedom system as a function of their frequencies. The spectral displacement should be compared with the displacement capacity of the building.

Based on the known PGA value the top displacement of the building can be determined. From the vulnerability function (Fig. 11) it can be read the level of damage the building would suffer in case of seismic event.

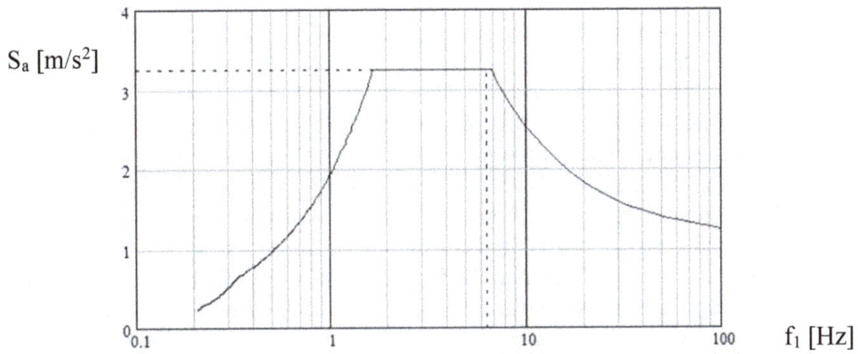

Figure 10: Spectral acceleration demand for the examined reinforced concrete building.

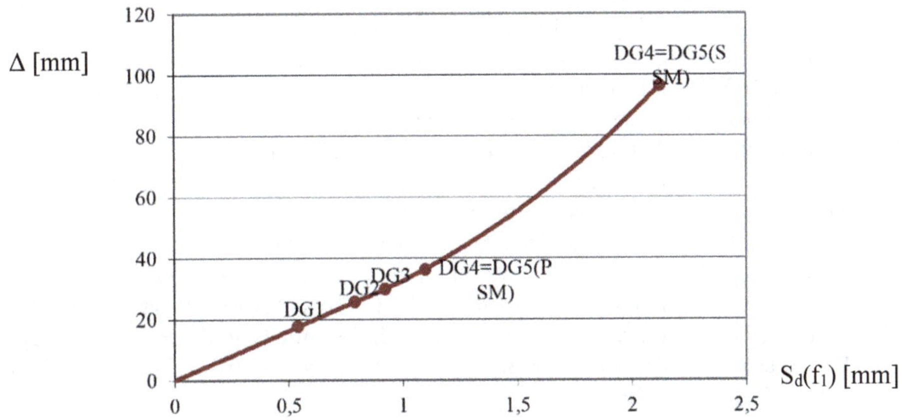

Figure 11: Vulnerability function of the examined reinforced concrete building.

5 CONCLUSION

The rapid visual evaluation of buildings was performed on over 5000 structures in Gyor. The evaluation method was developed from well-known approaches but modified to account for typical building designs found here. Further evaluation of building vulnerability was performed by pushover analyses by simplified methods. The analyses showed that the vulnerability assessments were consistent and reasonable.

ACKNOWLEDGEMENT

The research was supported by the EFOP–3.6.1–16–2016–00017 project: “Internationalization, initiatives to establish a new source of researchers and graduates, and development of knowledge and technological transfer as instruments of intelligent specializations at Szechenyi University”.

REFERENCES

[1] Kegyes-Brassai, O.K. & Ray, R.P., Comparison of the 1D response analysis results of typical Hungarian soil types and the EC8 spectra based on a case study of seismic risk

analysis in Gyor. *Earthquake Resistant Engineering Structures X,* ed. C.A. Brebbia, WIT Press: Southampton and Boston, pp. 111–122, 2015.

[2] GeoRisk Earthquake Engineering Ltd., *Seismic Hazard Map of Hungary*, 2006. http://georisk.hu/. Accessed on: 3 May 2014.

[3] Kegyes-Brassai, O., Ray, R.P. & Kuti, R., Seismic risk and disaster management perspectives in Hungary, presented on a case study performed in Gyor. *Academic and Applied Research in Military and Public Management Science,* **16**(2), pp. 5–16, 2017.

[4] ETH Zurich, European Facility for Earthquake Hazard and Risk, 2013. http://portal.share-eu.org:8080/jetspeed/portal/. Accessed on: 13 May 2014.

[5] Foerster, E., Krien, Y., Dandoulaki, M. & Priest, S.J., Del. 1.1.1: Methodologies to assess vulnerability of structural systems. *ENSURE (Enhancing Resilience of Communities and Territories Facing Natural and Na-tech Hazards) Project WP1: State-of-the Art on Vulnerability Types* s.l., European Commission Seventh Framework Programme, pp. 1–139, 2009.

[6] Castano, J.C. & Zamarbide, J.L., A seismic risk reduction program for Mendoza City, Argentina. *Tenth World Conference on Earthquake Engineering*, International Association for Earthquake Engineering, Madrid, Spain, 1992.

[7] Haddar, F.N., Urban seismic vulnerability analysis: The case of Algeria. *Proceedings of the. Tenth World Conference on Earthquake Engineering*, International Association for Earthquake Engineering, Madrid, Spain, 1992.

[8] FEMA, 2002. *FEMA 155: Second Edition, Rapid Visual Screening of Buildings for Potential Seismic Hazards: Supporting Documentation*, Federal Emergency Management Agency: Washington, DC, 2002.

[9] Göcsei, I., 2.1 Győr természeti adottságai (Natural endowments of Győr). *Győr Megyei Jogú településrendezési eszközeinek felülvizsgálata 2003–2005 (Evaluation of Settlement Assets in Győr)*, Győr Megyei Jogú Város Önkormányzata: Győr, 2005.

[10] Tóth, L., Mónus, P., Zsíros, T. & Kiszely, M., Seismicity in the Pannonian Region – Earthquake data. *European Geosciences Union Stephan Mueller Special Publication Series,* **3**, pp. 9–28, 2002.

[11] Varga, P., Az 1763. évi komáromi földrengés (The Komárom Earthquake of 1763). *Természet Világa,* **145**(2), 2014.

[12] Grünthal, G., *European Macroseismic Scale 1998,* European Seismological Commission: Luxembourg, 1998.

PREDICTING THE PEAK SEISMIC RESPONSE OF A RETROFITTED NINE-STOREY STEEL REINFORCED CONCRETE BUILDING WITH STEEL DAMPER COLUMNS

KENJI FUJII[1], HIROKI SUGIYAMA[1] & KAZUAKI MIYAGAWA[2]
[1]Chiba Institute of Technology, Japan
[2]JFE Civil Engineering and Construction Corp., Japan

ABSTRACT

A steel damper column using low-yield strength steel is an attractive choice for the seismic retrofitting of existing buildings because it does not introduce architectural problems, unlike braces and walls. In a seismic retrofit design using a steel damper column, the design of the connection joint between the damper column and the existing concrete member is very important. The designer needs to evaluate the maximum shear force and moment of the joint, in addition to the peak storey drift and member forces. In this study, the nonlinear peak response of a retrofitted nine-storey steel reinforced concrete building with steel damper columns was analytically investigated. A steel damper column was added on the side of the exterior frame by connection joint, using mortar, anchors, and studs. The peak response was predicted using nonlinear static (pushover) analysis, the peak storey drift, and the maximum moment and shear force at the connection joint, and the results were compared with the results obtained by nonlinear time-history analysis. Thus, it was revealed that the predicted peak storey drift, the maximum shear force, and moment at the connection joint are in good agreement with the time-history analysis results. The largest shear force of the anchor in the connection joint was also evaluated and compared with the time-history analysis results.

Keywords: seismic retrofit, existing steel reinforced concrete building, steel damper column, equivalent linearization technique, pushover analysis.

1 INTRODUCTION

The seismic retrofitting of existing buildings is important in most seismically active regions. In Japan, the seismic evaluation and retrofitting of existing reinforced concrete (RC) and steel reinforced concrete (SRC) buildings, and particularly that of public buildings including school buildings and government offices, has accelerated since the 1995 Hyogo-Ken Nanbu Earthquake. Although, in Japan, the seismic evaluation and retrofitting of school buildings have almost been completed, there still exist many condominiums and office buildings that are designed in accordance with the old seismic design code.

In recent years, many studies have been conducted on passive control systems for the seismic retrofitting of existing RC buildings [1]–[3] and the seismic design of new RC buildings [4]. A popular system is the buckling-restrained brace (BRB) system, which uses low-yield-strength steel. BRBs are undoubtedly effective in the seismic retrofitting of RC buildings because the structural system is clear; that is, the deformation of braces equals the shear deformation of the existing RC frame. However, the application of BRBs to the seismic retrofitting of existing buildings may introduce architectural problems. For example, in the case of the seismic retrofit of a condominium, installing BRBs in front of a balcony may obstruct the view from the living space window. Therefore, the use of a damper column with a shear panel [5] may be an attractive choice for the seismic retrofitting of existing condominiums and office buildings. In the damper column, shown in Fig. 1(a), low-yield-strength steel was used for the shear panel to absorb the hysteresis energy. One possible scheme for the seismic retrofitting of a concrete office building with a steel damper column involves a direct connection to the side of exterior frame [6], as shown in Fig. 1(b). The

WIT Transactions on The Built Environment, Vol 185, © 2019 WIT Press
www.witpress.com, ISSN 1743-3509 (on-line)
doi:10.2495/ERES190061

details of the additional steel damper column connection to the existing concrete structure are shown in Fig. 2. An additional steel damper column is connected to an existing concrete structure through the joint using mortar, anchors and studs. Another possible scheme for a condominium with a balcony involves adding a new steel frame to the side of the balcony, as has been demonstrated by a previous study [3].

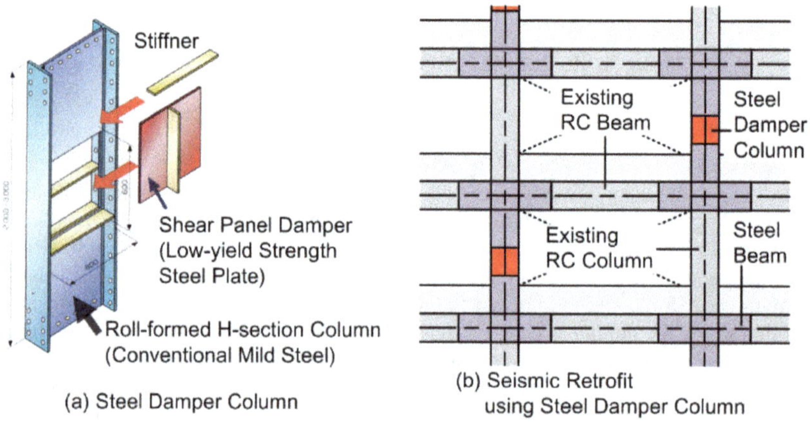

Figure 1: The seismic retrofit of an existing concrete office building using steel damper columns.

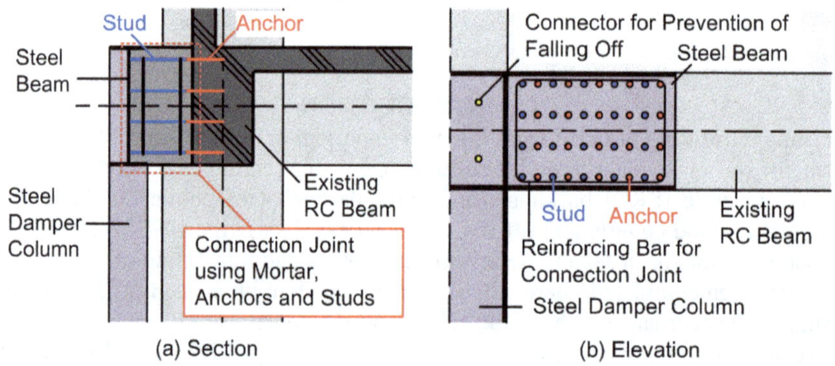

Figure 2: Detail of the connection of an additional steel damper column to an existing RC structure.

In seismic retrofit design using a steel damper column, the design of the connection joint between the damper column and the existing concrete member is very important. The designer needs to evaluate the maximum shear force and moment of the joint, in addition to the peak storey drift and member forces. In this study, the nonlinear peak response of a retrofitted nine-storey SRC building with steel damper columns was analytically investigated. The peak response of the retrofitted building was predicted through a previously proposed simplified procedure [3]. The predicted peak storey drift, maximum shear force and moment of the connection joint were compared with the time-history analysis results.

2 OUTLINE OF A SIMPLIFIED PROCEDURE

The authors have proposed a simplified procedure to predict the peak response of retrofitted buildings using steel damper columns with consideration to the strain hardening effect [3]. In the abovementioned simplified procedure, the strain hardening effect is considered by using two models; namely, the lower-bound (LB) model and the upper bound model. In the LB model, the yielding strength of the damper panel is assumed to be the LB yielding strength Q_{yDL}, which corresponds to the initial yielding strength before the strain hardening. In the UB model, the yielding strength of the damper panel is assumed to be the UB yielding strength Q_{yDU}, which corresponds to the yielding strength after an appreciable amount of cyclic loading. The predicted peak response of each component is determined from the envelope of the prediction results obtained by the LB and UB models. The simplified procedure is outlined as follows:

- Step 1: Construct two nonlinear frame structure models, namely the LB model and the UB model.
- Step 2: Carry out pushover analysis of the LB model. Then determine the following properties of the equivalent single-degree-of-freedom (SDOF) model based on the pushover analysis results: (a) the relationship between the equivalent acceleration A_1^* and the equivalent displacement D_1^* (capacity curve); (b) the relationship between the equivalent damping h_{1eq} and D_1^*.
- Step 3: Predict the peak equivalent displacement $D_1{}^*{}_{max}$ employing the equivalent linearization [7].
- Step 4: From the results obtained in STEPS 2 and 3, obtain the peak response for each component (e.g., the peak storey drift, peak shear deformation and force of the damper panel) by referring to the pushover analysis result corresponding to $D_1{}^*{}_{max}$. The maximum response occurring until the equivalent displacement reaches $D_1{}^*{}_{max}$ is the predicted peak response of each component obtained from the LB model.
- Step 5: Repeat STEPS 2 to 4 to obtain the peak response of the UB model.
- Step 6: Determine the predicted peak response of each component from the LB and UB models' envelope of predicted peak responses.

The details of the simplified procedure have been discussed in a previous paper [3].

3 BUILDING AND GROUND MOTION DATA

3.1 Building data

This study investigated a nine-storey steel reinforced concrete office building model designed in accordance with the pre-1981 seismic code of Japan. Fig. 3 shows the building model considered in this study. Because the plan of this building is symmetric, only two frames (frame Y1 and Y2) were analysed. Additional shear reinforcement through carbon fibre was assumed to have been properly provided to the existing members shown in frame Y1 to prevent shear failure. The model without an additional steel damper column is termed as Model-O, and had a base shear coefficient of 0.224. Two retrofitted building models, namely, Model-D100 and Model-D225, were considered in this study. The two models were different with respect to the steel damper column. The parameters of the steel damper column for each model are presented in Table 1. The thickness of the damper panel is denoted as t_d, the assumed initial nominal yield stress is denoted as σ_{yL}, the assumed nominal yield stress after appreciable cyclic loading, σ_{yU}, the initial shear yield strength is denoted as Q_{yDL}, and the shear yield strength after an appreciable amount of cyclic loading is denoted as Q_{yDU}. In both retrofitted building models, a steel damper column was added in each storey shown in Fig. 3(b). The ratio of Q_{yDL} to the total weight of the structure was 0.033 for Model-D100, and 0.054 for Model-D225.

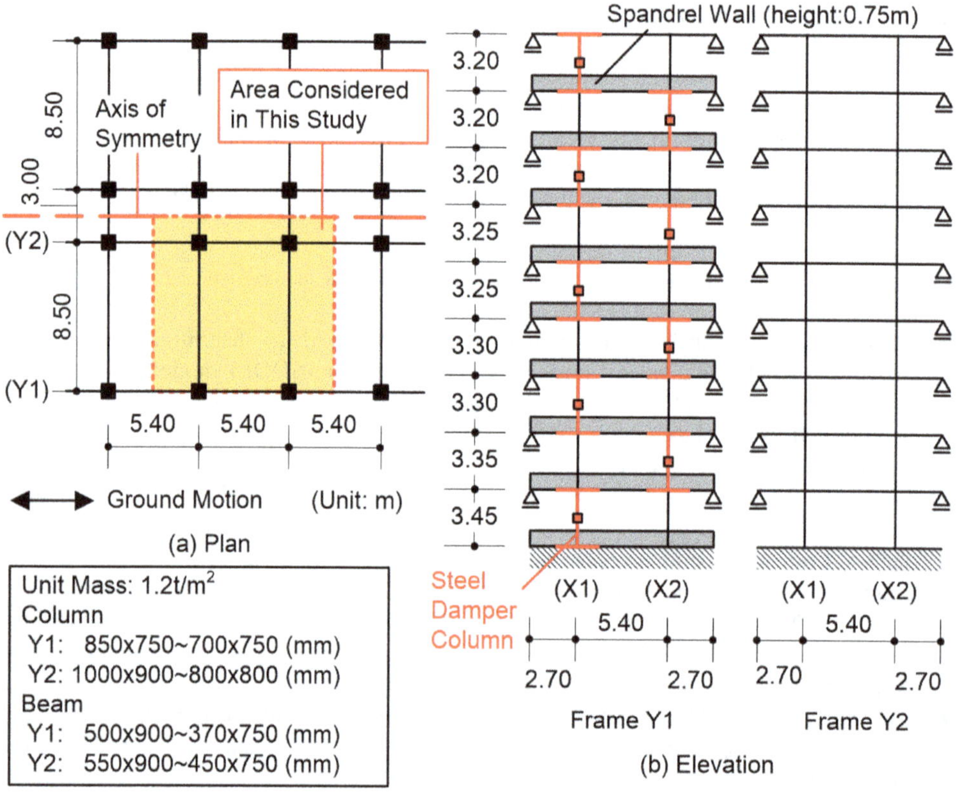

Figure 3: Building model.

Table 1: Steel damper column parameters for each model.

Model	H-Section Column	Damper Panel				
		t_d (mm)	σ_{yL} (N/mm²)	σ_{yU} (N/mm²)	Q_{yDL} (kN)	Q_{yDU} (kN)
Model-D100	H-750×250×14×28	9.0	100	200	382	764
Model-D225		6.0	225	300	618	824

Fig. 4 shows the detail of the connection joint. In this study, the size of the anchor is assumed as D16 (shear strength q_a = 64kN). The number of anchors was determined such that the shear force of the outermost anchor q did not exceed the shear strength q_a, by considering the design moment of the connection joint, M_{JD}, as follows:

$$q = \frac{\sqrt{L_X^{\ 2} + L_Y^{\ 2}}}{r_J^{\ 2}} \frac{M_{JD}}{N_a} \leq q_a = 64\text{kN}. \tag{1}$$

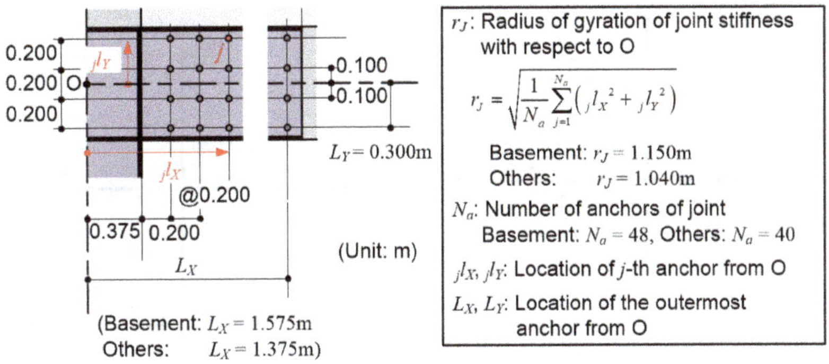

Figure 4: Detail of connection joint.

In this study, the same connection joints were assumed for both models. The design moment of joint M_{JD} was assumed 1990 kNm (= 0.7 × 824 kN × 3.45 m) for the basement, and as 1380 kNm (= 0.5 × 824 kN × 3.35 m) elsewhere.

3.2 Modelling of building structures

Fig. 5 shows the numerical modelling of the building structures, while Figs 6 and 7 show the force-deformation relationship for each member (envelopes in Fig. 6; hysteresis rules in Fig. 7).

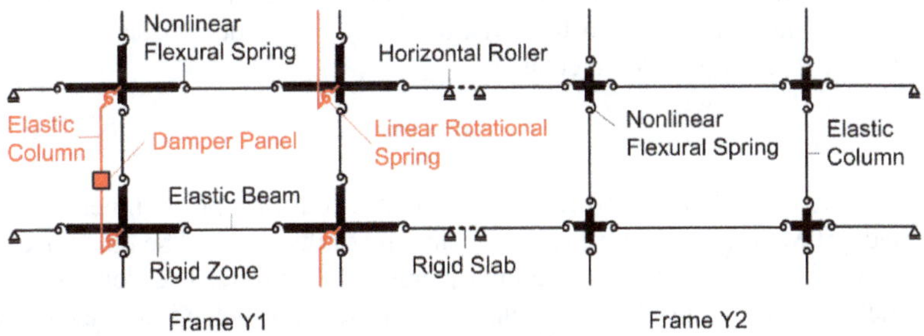

Figure 5: Numerical modelling of building structure.

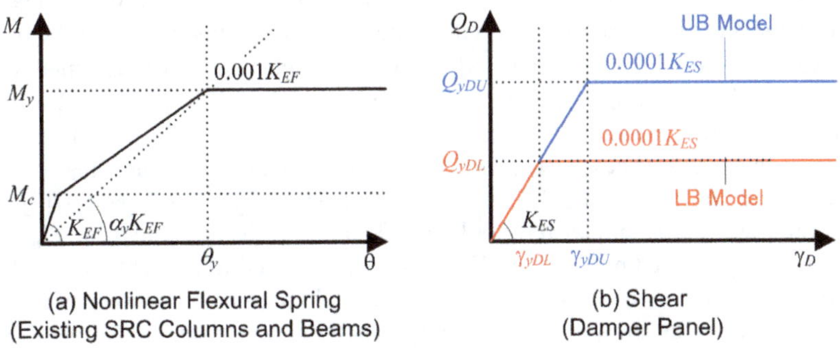

Figure 6: Envelope of force–deformation relationship.

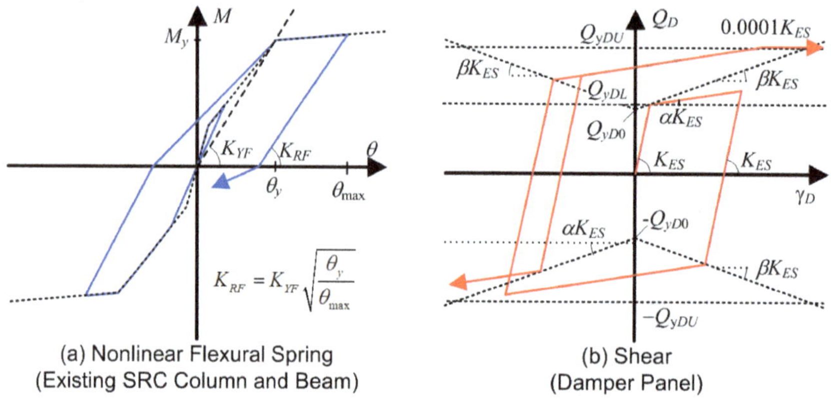

Figure 7: Hysteresis rule for each member.

In this study, the building was modelled as a plane frame model. Frames Y1 and Y2 were connected through the rigid slab. All existing beams and columns were modelled as an elastic beam/column with a nonlinear flexural spring at each end, except beam with the horizontal roller. The flexural and shear deformations were considered in the elastic beam, while the flexural, shear and axial deformations are considered in the elastic column. The rigid zone was considered in the existing members, and particularly in the case of Models D100 and D225. The rigid zone length of all beams in Frame Y1 was assumed as the same length as L_X shown in Fig. 4, with consideration to the additional steel beam. The joint of the additional steel damper column connection to the existing frame was modelled as a linear rotational spring, and its stiffness was calculated as:

$$K_J = N_a k_a r_J^{\ 2}. \tag{2}$$

Here, k_a is the elastic shear stiffness of the connection joint per number of anchors and was assumed 250MN/m, based on the experimental results in [6]. The damper column was modelled as an elastic column with a damper panel in the middle. The flexural behaviour of damper panel was assumed to be linear elastic, and the shear behaviour was assumed to be nonlinear. The height of damper panel is assumed 0.75m.

For all members, the force-deformation relationship envelopes were assumed to be symmetric in the positive and negative loading directions. For the existing SRC members, the envelope was assumed to be the tri-linear curve shown in Fig. 6(a). At yielding, the ratio of secant stiffness degradation to the initial stiffness α_y was assumed 0.10 for the beams with spandrel wall based on Takahashi et al. [8], For the other beams, α_y was assumed as 0.25. To determine the tri-linear curve for the columns, the rotational angle at yielding θ_y was assumed in the range between 1/250 and 1/150, depending on each column's clear height to depth ratio H_0/D. The bi-linear envelopes shown in Fig. 6(b) were assumed for the shear behaviour of the damper columns (in the LB/UB models). The Muto hysteresis model [9] with one modification was used to model the nonlinear flexural spring, as shown in Fig. 7(a). The hysteresis model proposed by Ono and Kaneko [10] was used to model the shear behaviour of the damper columns with consideration to the strain–hardening behaviour for the nonlinear time-history analysis, as shown in Fig. 7(b). Parameter α was assumed as 0.005 for Model-D100 and as 0.022 for Model-D225, while parameter β is assumed 0.013 for both models.

Note that, for the nonlinear time-history analysis, the initial yield strength of the damper panel shown in Fig. 7(b) was assumed to be Q_{yDL} while the UB yield strength was assumed to be Q_{yDU}.

The damping matrix was assumed to be proportional to the instant stiffness matrix without a damper column. The damping ratio of the elastic first mode of the model without a damper column was assumed to as 0.05.

3.3 Ground motion data

Fig. 8 shows the elastic response spectra of the artificial ground motions used in this study. Twelve artificial ground motions generated in a previous study (Art-L00 to 11 in [3]) were used to conduct the nonlinear time-history analysis. The target elastic spectrum was the design spectrum determined from the Building Standard Law of Japan [11] with consideration to the type-1 soil condition. The horizontal major component of Sendai Government Office building #2, which was recorded during the 2011 earthquake off the Pacific coast of Tohoku [12], was used as the source of phase angle of the artificial ground motions.

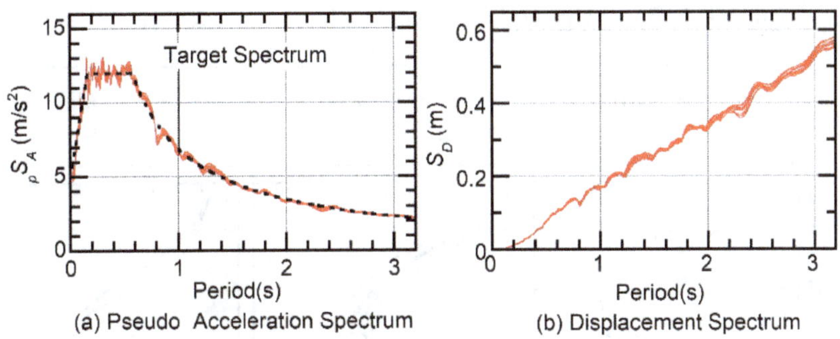

Figure 8: Elastic response spectra of artificial ground motions

4 ANALYSIS RESULTS

4.1 Prediction of peak response in simplified procedure

Fig. 9 shows the prediction of the peak equivalent displacement $D_1{}^*{}_{max}$ for all models. As expected, the comparison results revealed that the predicted peak response of the model building was effectively reduced with the addition of dampers. The predicted $D_1{}^*{}_{max}$ for Model-O was 0.220 m, while the values for the models with dampers were within the range of 0.144–0.151 m.

4.2 Comparison results obtained from the nonlinear time-history analysis and simplified procedure

4.2.1 Peak storey drift

Fig. 10 shows the comparisons of the peak storey drift R_{max}, which were obtained from the nonlinear time-history analysis and simplified procedure, respectively. As can be seen, the predicted R_{max} is in good agreement with the time-history analysis results. Fig. 10(b) and (c)

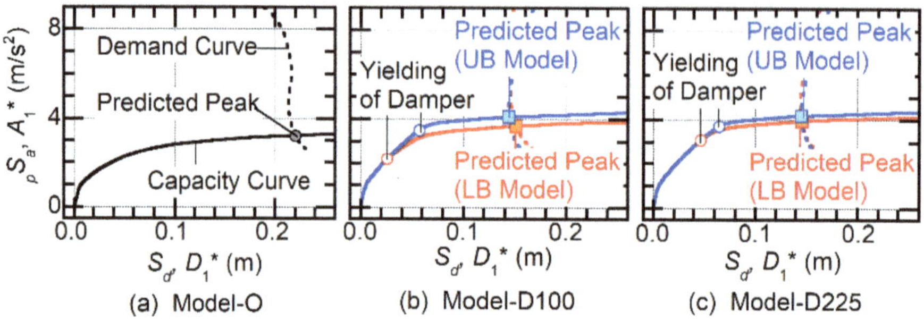

Figure 9: Prediction of the peak equivalent displacement.

also show that, for the models with dampers, the predicted (envelope) R_{max} for the part of the structure between the fifth to seventh stories was determined from the result obtained by the LB model. Moreover, the predicted (envelope) R_{max} for the other part of the structure was determined from the result obtained from the UB model. This implies that it was reasonable to use of two models (LB and UB models) to predict the peak response.

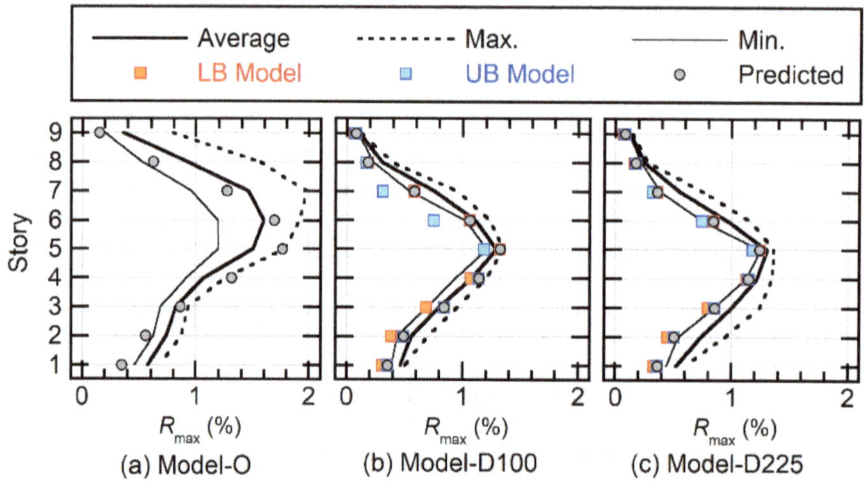

Figure 10: Comparisons of peak storey drift.

4.2.2 Maximum shear force and moment of connection joint

Fig. 11 shows the comparison results for the maximum shear force at the connection joint Q_{Jmax}, which were obtained from the nonlinear time-history analysis and simplified procedure. In this study, Q_{Jmax} was obtained as the maximum shear force of each steel damper column. As shown in Fig. 11(a), the Q_{Jmax} prediction was conservative for Model-D100, because the results obtained from the time-history analysis lie between the results obtained from the LB and UB models, except for the eighth and ninth floor of column X1. However, for Model-D225 shown in Fig. 11(b), the predicted Q_{Jmax} agrees with the nonlinear time-history analysis results, with the exception of the upper floors.

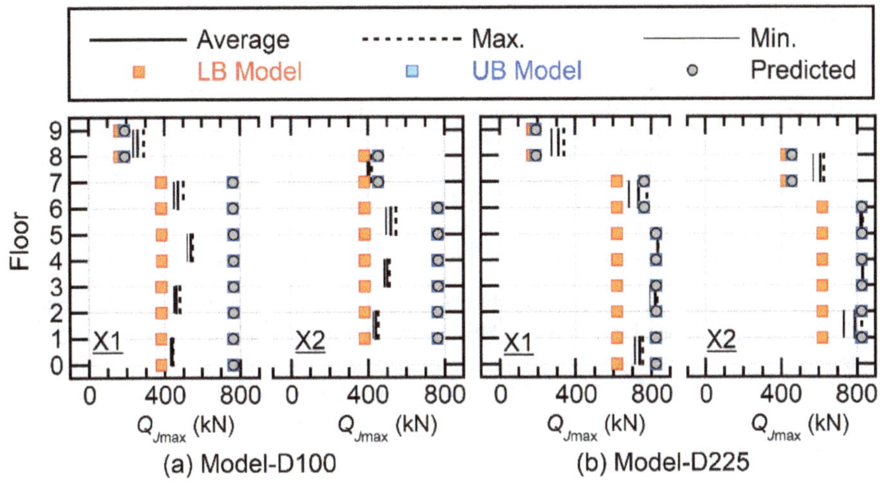

Figure 11: Comparisons of maximum shear force at connection joint.

Fig. 12 shows the comparison results for the maximum moment at the connection joint M_{Jmax}, which were obtained by the nonlinear time-history analysis and simplified procedure. As shown in Fig. 12, the M_{Jmax} prediction is conservative for Model-D100 (Fig. 12(a)). However, for Model-D225 shown in Fig. 12(b), the predicted M_{Jmax} agrees with the nonlinear time-history analysis results.

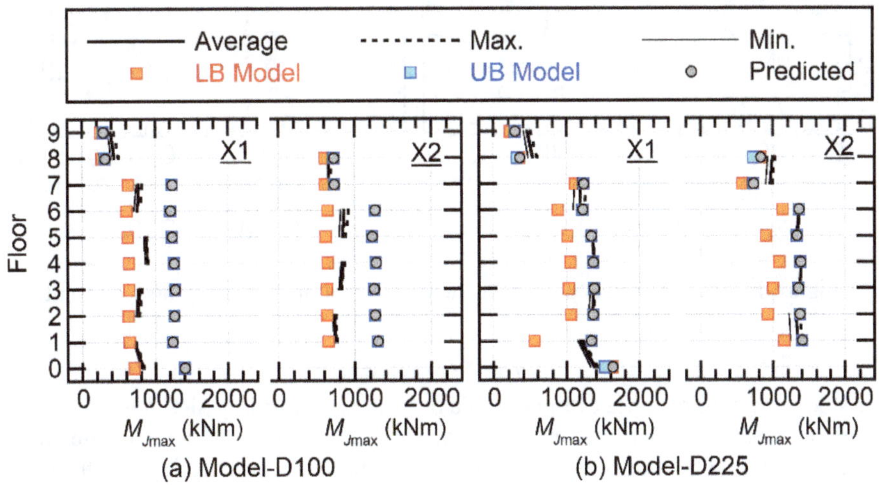

Figure 12: Comparisons of maximum moment at connection joint.

For most of the connection joint, the Q_{Jmax} and M_{Jmax} were predicted from the results of the UB model. However, the M_{Jmax} at the basement of column X1, top of column X2 of the eighth storey, and bottom of column X1 of the ninth storey were predicted from the result obtained by the LB model.

4.2.3 Largest shear force of anchor in connection joint

Fig. 13 shows the comparison results of obtained by the nonlinear time-history analysis and simplified procedure for the largest shear force of anchor in connection joint q_{max}. In this figure, the shear strength of the D16 anchor q_a (= 64kN) is also shown. Moreover, in this study, q_{max} is calculated from Q_{Jmax} and M_{Jmax}, as follows:

$$q_{max} = \frac{1}{N_a}\sqrt{\left(Q_{J\,max} + \frac{L_Y}{r_J^{\,2}} M_{J\,max}\right)^2 + \left(\frac{L_X}{r_J^{\,2}} M_{J\,max}\right)^2} \quad . \tag{3}$$

As shown in Fig. 13(a), the q_{max} prediction is conservative for Model-D100, except for the eighth and ninth floor of column X1. However, for Model-D225 shown in Fig. 13(b), the predicted q_{max} agrees with the nonlinear time-history analysis results, with exception of the upper floors. This figure also shows that, in all connection joints, q_{max} was smaller than the shear strength. Therefore, it was expected that all steel damper columns would be properly connected to the existing structure by the joint shown in Fig. 4.

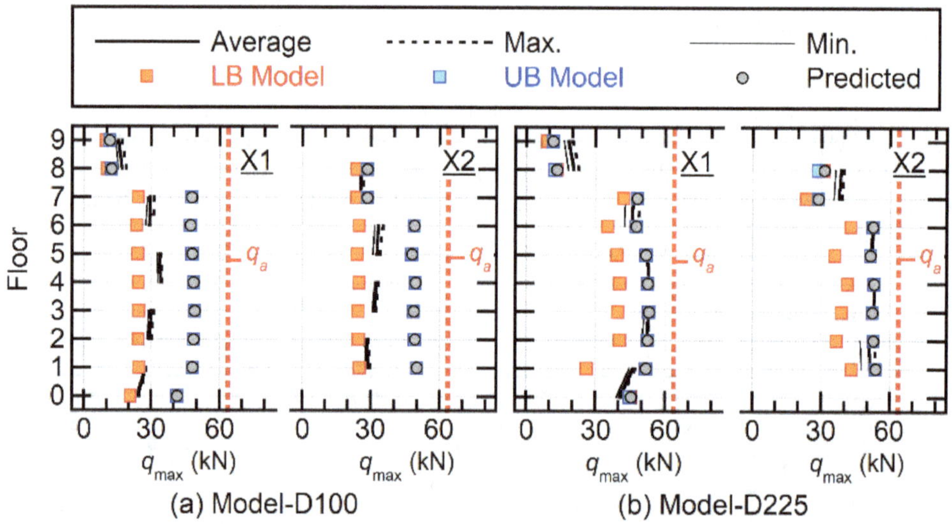

Figure 13: Comparisons of largest shear force for anchor in connection joint.

5 CONCLUSIONS

In this study, the nonlinear peak response of a retrofitted nine-storey SRC building with steel damper columns was analytically investigated. The peak response of the retrofitted building was predicted through a previously proposed the simplified procedure [3], and the obtained results were compared with the results obtained by nonlinear time-history analysis.

The comparison results revealed that the simplified procedure could satisfactorily estimate the peak response. The largest shear force of the anchor in the connection joint was also properly estimated through the simplified procedure.

ACKNOWLEDGEMENT

We thank Edanz Group (www.edanzediting.com/ac) for editing a draft of this manuscript.

REFERENCES

[1] Ishii, T., Mukai, T., Kitamura, H., Shimizu, T., Fujisawa, K. & Ishida, Y., Seismic retrofit for existing R/C building using energy dissipative braces. *Proceedings of the 13th World Conference on Earthquake Engineering*, paper no. 1209, 2004.

[2] Mazza, F. & Vulcano, A., Displacement-based design procedure of damped braces for the seismic retrofitting of RC framed buildings. *Bulletin of Earthquake Engineering*, **13**, pp. 2121–2143, 2015.

[3] Fujii, K. & Miyagawa, K., Nonlinear seismic response of a seven-storey steel reinforced concrete condominium retrofitted with low-yield-strength-steel damper columns. *Proceedings of the 16th European Conference on Earthquake Engineering*, paper no. 10736, 2018.

[4] Izumi, M., Chiba, O., Takahashi, K. & Iizuka, S., Earthquake resistant performance of reinforced concrete frame with energy dissipation devices. *Proceedings of the 13th World Conference on Earthquake Engineering*, paper no. 2498, 2004.

[5] Katayama, T., Ito, S., Kamura, H., Ueki, T. & Okamoto, H., Experimental study on hysteretic damper with low yield strength steel under dynamic loading. *Proceedings of the 12th World Conference on Earthquake Engineering*, paper no. 1020, 2000.

[6] Kitamura, K., Fujinaga, T., Miyagawa, K., Kitamura, W. & Sun, Y., Study on cyclic behavior of indirect connection between existing R/C beam and retrofitting steel frame. *Proceedings of the 10th Symposium on Research and Application of Hybrid and Composite Structures*, paper no. 60, 2013. (In Japanese.)

[7] Otani, S., New seismic design provision in Japan. *Proceedings of the Second US–Japan Workshop on Performance-Based Earthquake Engineering Methodology for Reinforced Concrete Structures*, PEER Report 2000/10, pp. 3–14, 2000.

[8] Takahashi, S., Ueda, H., Sumi, A., Ichinose, T., Sobue, M. & Tanaka, H., A flexural-design model of reinforced concrete members with spandrel and wing wall. *Journal of Structural and Construction Engineering* (*Transaction of AIJ*), **74**(641), pp.1321–1326, 2009. (In Japanese.)

[9] Muto, K., Hisada, T., Tsugawa, T. & Bessho, S., Earthquake resistant design of a 20 storey reinforced concrete buildings. *Proceedings of the Fifth World Conference on Earthquake Engineering*, pp. 1960–1969, 1974.

[10] Ono, Y. & Kaneko, H., Constitutive rules of the steel damper and source code for the analysis program. *Passive control symposium 2001,* Structural Engineering Research Center, Tokyo Institute of Technology, pp. 163–170, 2001. (In Japanese.)

[11] BCJ, *The Building Standard Law of Japan on CD-ROM*, The Building Center of Japan, 2016.

[12] Kashima, T., Koyama S. & Okawa I., *Strong Motion Records in Buildings from the 2011 off the Pacific Coast of Tohoku Earthquake*, Building Research Institute, 2012. (In Japanese.)

EARTHQUAKE RESISTANCE OF REINFORCED CONCRETE SHELLS OF DIVERSE FORMS IN THE PLANE AND WITH DIFFERENT CONTOURS

MOSHE DANIELI[1] & ARCADY ARONCHIK[2]
[1]Department of Civil Engineering, Ariel University, Israel
[2]Anatoly Borushansky Engineering, Jerusalem, Israel

ABSTRACT

In this study, we present the earthquake resistance of three types of reinforced concrete shells, having three different plans and contours. In the Memorial Museum near the city of Karachaevsk, Israel, a reinforced concrete monolithic shell serves as the building's roof and fence. The middle surface of the building is a connected shell in the form of a revolution and truncated cone, with a spherical segment. The shell on the perimeter rests on the foundation beam. In the market in the city of Sukhumi, Republic of Georgia, a precast monolithic spherical reinforced concrete shell square overlaps the Central Hall. Contour diaphragms are made of steel trusses transmitting the load to monolithic reinforced concrete columns. In Borjomi, Israel, the roof of a pump room is made of a monolithic triangular reinforced concrete spherical shell whose corners are supported in the corners on points. Three bowstring arches, connected to a support, border the shell's contour. The results of theoretical and full-scale experimental analyses of the static and dynamic characteristics of these shells are presented. Theoretical analysis was performed by means of the Finite Element Method (FEM). We show that all three constructions have the required degree of seismic resistance, yet shells with stiffer and more rounded contours exhibit better behavior under seismic conditions.

Keywords: concrete shell, construction, contours, diaphragm shapes, earthquake resistance, finite element method, modal linear analysis, reinforced concrete, shells, steel trusses.

1 INTRODUCTION

Various forms of shells are widely used in the construction of civil and industrial buildings. The first large roof known is a stone dome. Currently there are many stone domes all over the world that are now considered architectural works of art, and whose preservation is a societal responsibility. With the appearance of concrete in the late nineteenth century, the use of more lightweight concrete domes replaced heavy stone domes. Reinforced concrete shells of various forms were consequently developed, which became an effective kind of spatial coverage in the twentieth century.

The widespread use of concrete shells as roofs began in Germany and gradually spread to other countries in Europe, the US and Japan. Originally, reinforced concrete shells typically had a maximum span of 20 m and an effective thickness of a minimum of 10–15 cm: These were implemented of monolithic reinforced concrete on solid, mostly wooden, framework. Although today there are reinforced concrete shells with a span of over 200 m, the most common coverings worldwide have a span between 20 m and 60 m.

The first concrete dome known to the author, built in the Republic of Georgia, is a spherical monolithic concrete dome atop the hall of the synagogue of Ashkenazi Jews in Tbilisi (now the D. Baazow historical-ethnographic museum of Georgian Jews). The dome was built in 1915 (span of dome, 15 m; rise f = 7.0 m; thickness t = 0.15–0.2 m) [1]. Since 1940, monolithic reinforced concrete shells with relatively small spans have been widely constructed in the Republic of Georgia. Since the 1950s, more than 40 precast shells have been constructed there [1].

WIT Transactions on The Built Environment, Vol 185, © 2019 WIT Press
www.witpress.com, ISSN 1743-3509 (on-line)
doi:10.2495/ERES190071

One of this paper's authors participated in full-scale experiments on a number of monolithic concrete shells. In this paper, we present the results of experiments involving the following [2]–[5]:

1. The reinforced concrete monolithic shell which serves as a roof and fence of the Memorial Museum adjacent to the city of Karachaevsk, in Caucasus, Russia, which was built in 1967–1968. The middle surface of the shell is a connected shell of a revolution in the form of a truncated cone with a spherical segment. The shell on the perimeter leans on the foundation beam. The shell was designed by the Giprogorstroi Design Institute of the Republic of Georgia and the Institute of Structural Mechanics and Earthquake of the Georgian Academy of Sciences [2], [3].
2. A square precast monolithic spherical reinforced concrete shell overlapping the central hall of the new market erected in the city of Sukhumi, Republic of Georgia in 1978–1980. The shell was designed by the Institute of Structural Mechanics and Earthquake of the Georgian Academy of Sciences and the Department of Giprotorg of the Republic of Georgia, as well as the Design Institute No. 1 of Leningrad, Russia [1], [4].
3. The roof of a pump room, in Borjomi, Republic of Georgia, built in 1965–1966. The roof is made of monolithic reinforced concrete. The triangular spherical shell is supported in the corners. Three bowstring arches, connected among themselves in the support, border the shell as a contour. The shell was designed by the Giprogorstroi Design Institute and the Institute of Structural Mechanics and Earthquake of the Georgian Academy of Sciences, Republic of Georgia [5].

The shells are located in different seismic regions of the Caucasus. The expected horizontal acceleration of the ground for these regions is $a_{h,max} = 0.75 \div 2.00$ m/s^2. It is important to mention that in 1970 the Borjomi shell successfully overcame the seismic effects of a ground horizontal acceleration of $a_{h,max} = 1.0$ m/s^2 (1970 Borjomi earthquake) [2].

The shells differ in plan and contour. A series of studies was conducted on site to estimate the static and dynamic characteristics of shells. The main results of these studies and investigations are presented in [1]–[5]. Some results of these studies were used in this article. The aim of the present work was to examine these three types shells, using static and dynamic modal linear analysis. Data for the structural models (geometry, load, etc.) were taken from structural sketches and drawings, presented in other references [1]–[5]. According to the numerical analysis of seismic actions, displacement ratios and stresses are lower than the maximum permitted levels, for all three shells; therefore, we conclude that all three constructions have sufficient seismic resistance. However, shells with stiffer and more rounded contours exhibit better seismic behavioral responses.

A series of static and dynamic modal linear analyses on all three models was performed using the FEM (Finite Element Method) on LIRA and STR (Ukraine and France) software certified by CONCEIL (June, 1992). All three structural models consisted of several groups of three-dimensional (3D) finite elements: shells, bars and springs. The damping ratio used in all three numeral models was $\zeta = 0.10$.

Information on the dynamic characteristics of shells is presented in many referenced and unreferenced sources [6]–[9].

2 MONOLITIC REINFORCED CONCRETE SHELL OF REVOLUTION

2.1 Structure of shell of revolution

The Memorial Museum for the Heroes and Defenders of the Caucasus Passes, near the city of Karachaevsk, Russia, is a reinforced concrete monolithic shell, built in 1967–1968.

Fig. 1 shows that the shell serves as the building's roof and fence [2], [3]. The middle surface of the building is a connected shell of revolution in the form of a truncated cone with a spherical segment. The diameter of the resting contour is 21.74 m, the spherical segment is 19.65 m, the thickness of the conical section is 10 cm and the thickness of the spherical section is 8 cm. The building is 6.9 m high, the height of its walls is 4.2 m, and the rise of the spherical segment is 1.6 m. On the perimeter, the shell leans on the foundation beam. The main geometrical dimensions are presented in Fig. 2. Where the spherical and conical sections meet, the shell thickness increases to 20 cm.

(a) (b)

Figure 1: The Memorial Museum near Karachaevsk, Russia. (a) A general view; and (b) View while under construction.

2.2 Structural analysis

2.2.1 Static study

The museum structure was designed using the membrane theory of shells, taking into account the boundary effect, and the theory of limit analysis on the basis of the three-parameter axisymmetric mechanism of plastic collapse [3], [10]. The design grade of concrete is M 30.

2.2.2 Study of the dynamic characteristics

The dynamic characteristics of reinforced concrete structures were studied by analytic and numerical calculations. The dynamic characteristics of natural oscillations were determined theoretically, according to other's publications [11]. The results were a frequency f = 18.86Hz and a period T = 0.053s, referring exclusively to the spherical segments without the conical-shaped surface. The main results of structural dynamic numerical analysis in the horizontal earthquake direction are provided in another publication [12].

For the simulations, we selected three modes (1, 47, 88) of natural vibrations for these forms that demonstrate characteristic mode shapes. The results of our simulations for the Museum building are shown in Fig. 3. The displacements, contour of normal stress and bending moments for these forms of the Museum are shown in Table 1.

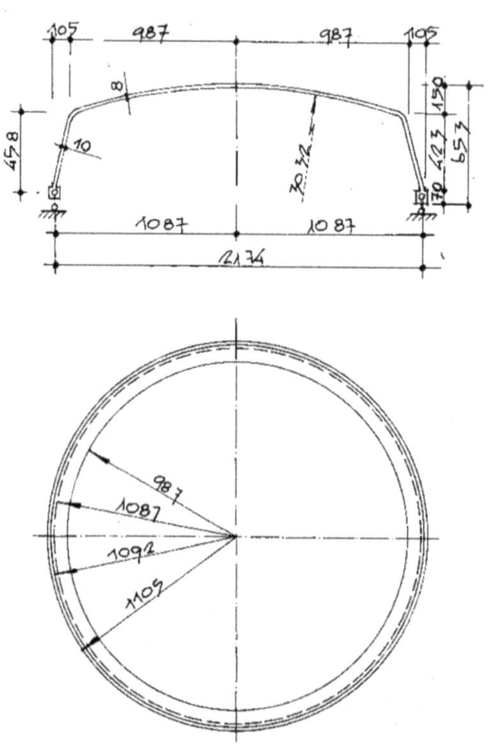

Figure 2: The Memorial Museum near Karachaevsk, Russia. Top image: Meridian section; bottom image: Plan (Unit: cm).

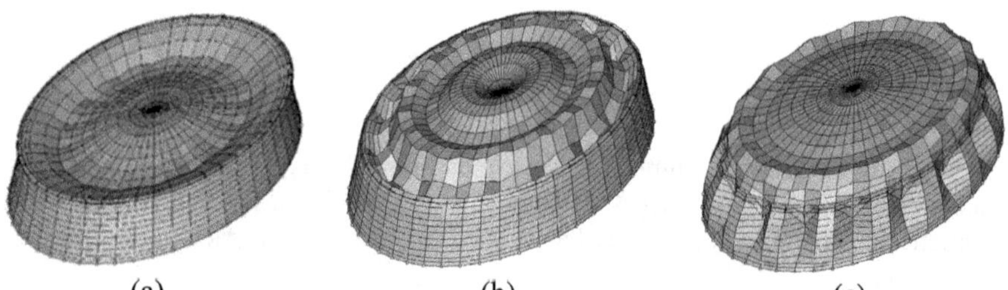

(a) (b) (c)

Figure 3: Mode shapes of natural vibrations of the museum structure. Earthquake direction: vertical. (a) Mode shape 1; (b) Mode shape 47; and (c) Mode shape 88.

Frequencies for the first six modes of natural vibration (mode shapes) according to known studies are [13]: f_1 = 8.82Hz; f_2 = 10.82Hz; f_3 = 10. 32Hz; f_4 = 12.03Hz; f_5 = 12.62Hz; and f_6 = 13.97Hz. These frequencies are for spherical segments with the cone surface.

Notably, the calibration of numerical models was not the aim of these numerical analyses. The structure itself clearly includes several groups of elements that were not presented in the numerical model given (which focused exclusively on masses). The geometry, characteristics of materials, and loads were based on information taken from the actual structures.

Table 1: Main results of structural analysis of Museum: Maximum values.

Parameters	Seismic loads			Experimental shape
	Shape 1	Shape 47	Shape 88	
Periods of natural vibrations, s: $T_1 = 0.1688$; $T_2 = 0.1416$; $T_3 = 0.1388$;	0.1688	0.0598	0.0208	0.0699*
Base force F(h, z), t	9.49	1.78	6.71	
Max. displacement, mm Vertical, w Horizontal, u	0.559 0.085	0.043 0.011	0.014 0.088	
Maximum stress, MPa σ_{Ny} (+)/ σ_{Ny} (–) σ_{My} (+)/ σ_{My} (–)	0.210/0.160 0.083/0.072			
Earthquake direction: vertical; Seismic zone factor – Z = 0.2; Site coefficient – S = 1; Seismic importance factor – I = 1; Reduction factor – K = 1; Damping ratio – $\zeta = 0.1$ *Value T of Experiment shape is closest to Shape 47.				

2.3 Full-scale experiment

2.3.1 Static data study

Before stripping, the strength of concrete was found to be ≈ 30MPa. In the course of stripping, the deflection of the spherical part of the shell was measured (Chizanishvili took part in this experiment) [3]. The curve of vertical deflection corresponding to the loading from self-weight, based on the numerical calculations of the general theory of shells, is shown (Fig. 4). The curve of deflection, obtained by experimental methods, is represented by a dotted line. The experimental deflection values are marked by a bar.

Fig. 4 shows that the directions of both experimental and theoretical deflection coincide qualitatively. The obtained relative experimental and theoretical maximum values of the deflection from the diameter of the lower base ($W_e/2R = 1.41/2 * 10.87 = 1/15400$ and $W_t/2R = 1.73/2 * 10.87 = 1/12000$) show that there is high shell rigidity, which may contribute to structures' high earthquake resistance.

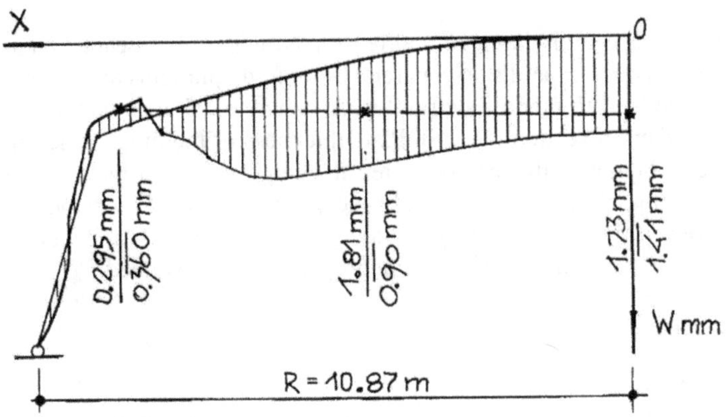

Figure 4: Schematic meridian section. The curve of vertical theoretical (" – " continuous line) and experimental deflections (" -- -- " dotted line).

2.3.2 Study of dynamic characteristics

The dynamic characteristics of the shell were examined experimentally (Zaalishvili took part in the experiment) [13]. Oscillations were excited by vertical pulses concentrated in the shell center (each pulse of 3.5Kn/m^2). In doing so, vertical free oscillations of the shell were recorded with the aid of a H338-6П graph plotter and VEGIK vibration detectors installed on the shell surface. As a result of analyzing the oscillograms, a three-half-wave form of low-tone free oscillations across the shell diagonal section and frequency f = 14.28Hz, period T = 0.07sec and logarithmic decrement λ = 0.22 were found [14].

3 PRECAST-MONOLITHIC DOUBLE CURVATURE SHELL COVER IN SUKHUMI

3.1 Structure of shell

A monolithic spherical precast reinforced concrete shell, covering a rectangular area, overlaps the central hall of the new market, erected in the city of Sukhumi (Fig. 5) [4]. In the plan, shell dimensions are 40 x 40 m, the surface is spherical and the rise is 7.6 m (Fig. 5). The main part of the roof, confined in the plan by a circle of 40 m in diameter, comprises two rows of precast reinforced concrete ribbed plates with inflexions near the ring ribs and subsequent embedment of joints between these ribs.

(a)

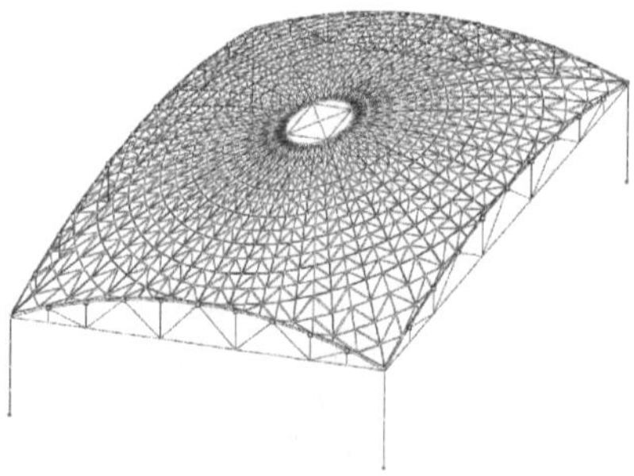

(b)

Figure 5: Building for the market. (a) General view; and (b) View of the design model.

The dome drum is 6 m in diameter and reinforced with radial beams; angular zones and contour band are made of monolithic concrete. Contour diaphragms are made of steel trusses which transmit the load to monolithic reinforced concrete columns that are 9.5 m in height and 90 x 90 cm in cross-section (Fig. 5). Plate thickness of the monolithic portion is 10 cm near the band and 8 cm near the precast plates; the height of plate ribs is 30 cm. The contour band, whose height is 35 cm, is reinforced by a continuous three-dimensional (3D) cage over the entire contour of the shell, forming a closed band for the shell and welded to the upper band of the contour steel trusses, whose support parts are also welded.

3.2 Structural analysis

3.2.1 Static study

The design grade of concrete of the precast plates, monolithic sections, and the concrete for filling joints between the plates and columns, is M30. The roof design is based on the membrane theory of shells with a boundary effect for design load of 4 kN/m^2 [15]. The critical load for stability, according to the Russian publication *Instruction Manual on Design of the Reinforced Concrete Space Coverings and Floor Structures* [15], is 6.0 kN/m^2, and the limit load 5.4 kN/m^2, in compliance with the theory of limit analysis according to Akhvlediani [16]. The building is designed for a seismic effect of horizontal acceleration of $\boldsymbol{a}_{h,max} = 0.75$ m/s^2 for the ground. Vertical displacement of the structure due to applied static loads (weight 300 kg/m2) reached 27.3 mm for the trusses, 25.2 mm for the apex of the shell, and 0.54 mm for the columns; and the corresponding horizontal displacement of trusses at zones near the columns reached +/ - 6.24 mm. Maximum horizontal displacement at the top of columns was similar in both main horizontal directions, +/ - 5.85 mm.

3.2.2 Study of the dynamic characteristics

The dynamic characteristics were estimated both by analytical and numerical analysis. Dynamic characteristics of natural oscillations, determined by the analytic method [11], were as follows: frequency $f = 18.86$Hz, period $T = 0.053$s. The first modes of vibrations for the reviewing shell were caused by shell horizontal displacement (mainly due to bending of the columns): $T_{1(hor)} = 0.717$s; $T_{2(hor)} = 0.305$s. Vibrations in a vertical direction were obtained for periods of 0.304s and 0.21s, and smaller. Apparently, just higher modes of shell vertical vibrations were measured during the dynamic loading of the shell [4]: (f_1, test = 8Hz; or T, test = 0.125s) (Table 2). The three modes (## 1, 9, 45) of natural vibrations for these forms of the market building are shown in Fig. 6.

Table 2: Main results of the structural analysis on the market building: Maximum values. Earthquake direction: horizontal and vertical.

Parameters	Seismic loads. Earthquake direction: vertical			Experiment vertical
	Mode 1 horizontal	Mode 9 (1) vertical	Mode 45 vertical	
Periods of natural vibrations, s: $T_1 = 0.1688$; $T_2 = 0.1416$; $T_3 = 0.1388$;	0.717	0.304 (0.213)	0.139	0.125
Base force F (h, z), *t*	27.1	27.1	0.62	
Max. displacement, mm Vertical, w Horizontal, u	1.940 0.434	1.94 0.434	0.044 0.011	
Maximum stress, MPa $\sigma_{Ny}(+)/\sigma_{Ny}(-)$ $\sigma_{My}(+)/\sigma_{My}(-)$	0.210/0.160 0.083/0.072			
Earthquake direction: vertical; Seismic zone factor – Z = 0.2; Site coefficient – S = 1; Seismic importance factor – I = 1; Reduction factor – K = 1; Damping ratio – $\zeta = 0.1$				

3.3 Full-scale experiment

3.3.1 Static study

When stripping the shell from the scaffolds, the strength of the concrete of the shell roof elements was determined using a cube crushing test and reference hammer within 30 to 40 MPa. In the process of striking of the shell, deflection of the shell and contour diaphragm

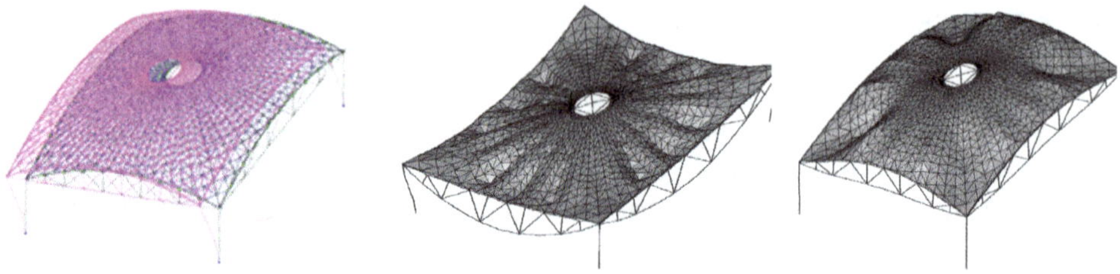

(a) (b) (c)

Figure 6: The modes of the market building's natural vibrations. (a) Mode shape 1 (Ed: horizontal); (b) Mode shape 1 (Ed: vertical); and (c) Mode shape 25 (Ed: vertical).

trusses' flex were measured (Fig. 5), as well as the horizontal and vertical movements of the support joints of the contour diaphragm trusses [15]. The location of deflection meters made it possible to control the uniform activation of all parts of the shell. Maximum deflection from self-weight in the shell center was equal to 27mm, with 20 mm at the middle of the truss span. Relative values of these deflections ($W_{sh}/ L = 2.7/4000 = 1/1480$ and $W_{tr}/L = 2.0/4000 = 1/2000$) bear record to sufficient rigidity of the shell and contour diaphragm trusses. Vertical movements of the support joints of the diaphragm trusses reached 1 mm and the divergence of the support joints in the plane of a diaphragm truss was 13.65mm, which corresponds to the tensile strength in the lower band: 72MPa. This value is approximate to the design stress from the shell's self-weight as determined according to a previous study [15], which is 94MPa. No cracks or other local defects were detected after the removal of the supporting scaffolding.

3.3.2 Study of the dynamic characteristics

The dynamic characteristics of the shell were examined experimentally [4]. Oscillations were excited by making vertical and horizontal pulses concentrated in the shell center (by sudden removal of a load of 500 kg). Vertical free oscillations of the shell were recorded with the aid of a H338-6П graph plotter and VEGIK vibration detectors installed on the shell surface.

As a result of analyzing the oscillograms, we obtained a three-half-wave of the form of low-tone free oscillations across the shell's diagonal section, with a frequency $f = 8$Hz and logarithmic decrement $\lambda = 0.15$ [4], [14]. The value of the frequency obtained approximates the theoretical value ($f = 9.8$ Hz) for monolithic shells as determined according to Oniashvili [11], yet it exceeds the experimental values of frequency f given in Danieli [1] for similar precast shells, by 30% on the average. As a result of the tests, it was established that the degree of concrete embedding of a precast-monolithic shell is sufficiently high, which is indicative of its seismic stability.

4 TRIANGULAR REINFORCED CONCRETE MONOLITIC SHELL

4.1 Structure of the shell

The roof of a pump room in Borjomi, Georgia, was built in 1967 (Fig. 7(a)) [5]. The roof was made of monolithic reinforced concrete, the spherical shell is triangular in shape, and it is supported in the corners on points. The shell is bordered by three bowstring arches on the contour, connected among themselves in the support (Fig. 7(b)). The distance between the supports is 24.4 m, the rise of the shell is $f = 8.0$ m and of the arch is $f_0 = 6.8$ m. Shell thickness t varies: In the center, shell thickness is 6 cm, yet it reaches 10 cm near the contour elements. The dimensions of the cross-section of the arch are 30 × 40 cm. The pump room building was designed based on the membrane theory of shells and the theory of the limit analysis on the basis of the three-parameter mechanism of plastic collapse [5], [10]. The design grade of the concrete is M 30.

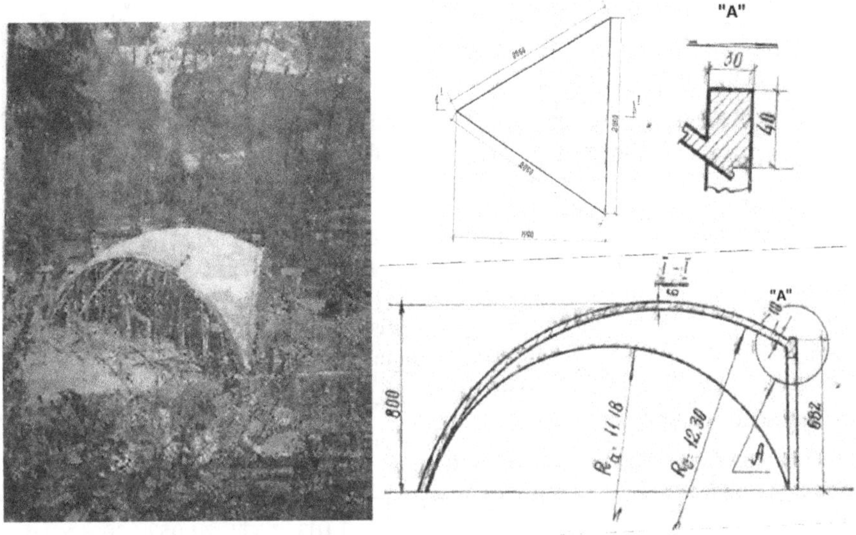

(a) (b)

Figure 7: The roof of the pump room in the Borjomi. (a) General view; and (b) Design model.

4.2 Numerical structural analyses

Loads and masses were taken from past literature [5].

4.2.1 Static analysis

Maximum vertical displacement of the structure due to applied 300 kg/m^2 static loads reached 9.29 mm for the apex of the shell. The corresponding maximum horizontal displacement is 7.14 mm for the *X* direction and 5.21 mm for the *Y* direction. Maximum internal forces, normal and bending stress in the reinforced concrete shell elements were as follows: at the body of the shell: normal stresses $\sigma_x = -4.0 \div 6.0$kg/cm^2, $\sigma_y = -5.0 - 6.5$ kg/cm^2 (by most area: 80%), $\sigma_{xmax} = -12.2$ kg/cm^2 in local corner areas: $\sigma_x = +1.2 \div 1.5$ kg/cm^2,$\sigma_y = 2.0$ t/m^2. Bending moments in the corner areas were: $M_{xmax} = \pm 0.38$ t*m/m, $M_y = \div 0.32 \div 0.40$ t*m/m, t/m^2 $\sigma_{mxmax} = \pm 3.0$ t/m^2; and $\sigma_{mymax} = \pm 2.52 \div 3.75$ t/m^2. Maximum internal forces in the contour bordered arches: axial forces: $N_{max} = -22.3$t; normal stress $\sigma = 18.58$ kg/cm^2. In the corner zones - $M_{ymax} = \pm 3.7$; - 2.7t*m; $M_{zmax} = \pm 2.7$ t*m; $\sigma_{ymax} = \pm 0.462$ kg/cm^2. Axial forces in the steel ties - bars (profile - cross- section UPB N140) N = 6.7 ÷ 6.9t. The three modes (3, 7, 9) of natural vibrations for these forms of the pump room roof are shown in Fig. 8. The displacement, contour of normal stress and bending moments for these forms of the pump room roof are shown Table 3.

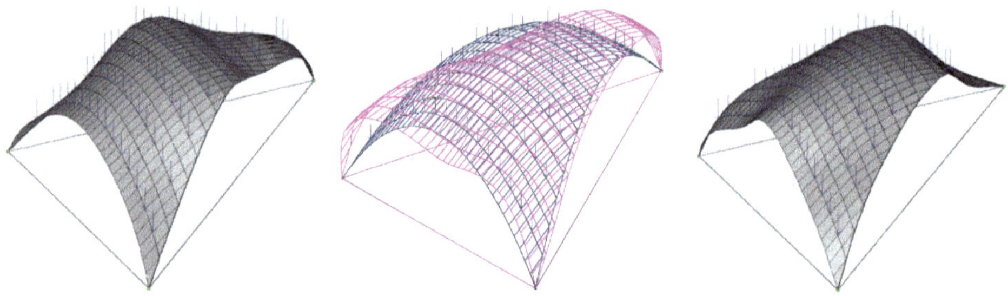

Figure 8: The mode shapes of the cover of pump room. Ed: (a) Mode shape 3 (Ed: vertical); (b) Mode shape 7 (Ed: vertical); and (c) Mode shape 9 (Ed: horizontal).

Table 3: Main results of structural analysis for the Pump room roof. Maximum values. Earthquake directions: horizontal and vertical.

Parameters	Seismic loads. Earthquake direction: vertical			Experiment vertical
	Mode 3	Mode 7	Mode 9	
Periods of natural vibrations (s)	0. 175	0.067	0.049	0.2
Base force F(h, z), *t*	1.15	1.13	0.96	
Max. displacement, mm Vertical, w Horizontal, u_{x-x} u_{y-y}	0.703 0.515 0.396	0.103 0.069 0.099	0.052 0.048 0.022	
Maximum stress (Mpa) σ_{Ny} (+)/ σ_{Ny} (–) σ_{My} (+)/ σ_{My} (–)	0.210/ 0.160			
Earthquake direction: vertical; Seismic zone factor Z = 0.2; Site coefficient S = 1; Seismic importance factor I = 1; Reduction factor K = 1; Damping ratio ζ = 0.1				

4.3 Full-scale experiment

4.3.1 Static study

In the process of striking the shells, vertical deflections in the planes of median of the triangle and contour bordered arch from the self-weight of the shell, formwork and support scaffolding of shell were measured ($q = 2.5kN/m^2$). In Fig. 9, the vertical experimental is represented by a continuous line and the theoretical deflections are represented by a dotted line. Maximum deflection in the shell in the planes of the median of the triangle and contour-bordered arches were $W_m = 16.81$ mm and $W_a = 6.91$ mm, respectively. The relative values of these deflections, $W_m/R_c = 1.681/1200 = 1/715$ and $W_a/R_c = 0.691/1200 = 1/1740$, indicate there is sufficient rigidity of the shell and contour bordered arches. No cracks or other local defects were detected after the removal of the supporting scaffolding.

4.3.2 Study of the dynamic characteristics

The dynamic characteristics of the shell were examined experimentally [2], [14]. Oscillations were excited by a horizontal acting load concentrated in the shell corner. In doing so, the vertical free oscillations of the shell were recorded with the aid of a H338-6П graph plotter and VEGIK vibration detectors that were installed on the shell surface. Based on an analysis of the oscillograms, we found a frequency $f = 5$ Hz and logarithmic decrement of $\lambda = 0.1$.

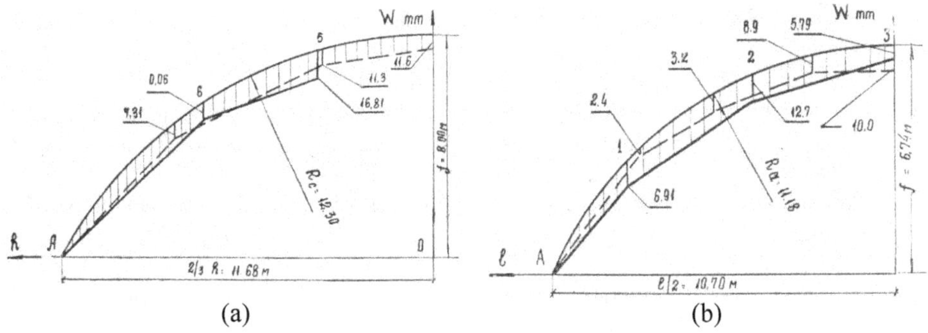

Figure 9: The curve of the vertical experimental (continuous line) and theoretical deflections (dotted line) in the planes of (a) the median of the triangle; and (b) the contour bordered arch.

5 CONCLUSIONS

1. Shells of double positive curvature are very rigid. Therefore, in approximate calculations it is possible for horizontal vibrations to simulate a hard disk. Durations of vibrations for such structures are durations of vibrations for their supporting structures (see the shell in Sukhumi, for example). Durations for vertical vibrations for this type of shells are low: 0.15–0.20 s.
2. The major part of the shell (80–85%) is in compression in both directions. Corresponding reinforcement can be minimal, as normal stress due to bending moments is low. Local edge zones (for example, especially the shell in Borjomi) and local zones with significant geometric changes (for example, see the shell in Karatchaevsk), normal stress of bending moments is significant. Consequently, corresponding dimensions of contour elements and their reinforcement and additional reinforcement of local zones of the shell body is necessary.
3. Similar problems are typical of the shells of ancient architecture in the East, although shell buckling does not typically occur as the spans are not so large. Massive walls are able to carry tension and bending due to significant weight and section dimension; however, attempts to open holes in these walls leads to a reduction in the wall load-carrying capacity, which might become dangerous even for existing load levels.
4. All three constructions were shown to have the required seismic resistance; however, shells with stiffer and more rounded contours were found to exhibit better seismic behavior.
5. Alternative solutions for supporting shell areas may include the use of ties (see the shell in Borjomi) or bottom joists of the trusses (see the shell in Sukhumi).

REFERENCES

[1] Danieli, M., Construction of precast concrete shells in Georgia. *Earthquake Resistant Engineering Structures XI*, **97**, pp. 149–160, 2017.

[2] Akhvlediani, N. et al., Investigation, design, and construction of aseismic spatial structures based on the theory of limit equilibrium. *Proceedings of the International*

Congress of IASS Theoretical and Experimental Studies of Spatial Structures: The Use of Shells in Civil Engineering, pp. 190–201, 1985.

[3] Danieli (Danielashvili), M., Limit analysis of reinforced concrete shells of revolution and its application. *Proceedings of the Ninth International Conference on Civil and Structural Engineering Computing, B.H.V.*, pp. 1–20, 2003.

[4] Akhvlediani, N. et al., Shell roof for market in Sukhumi-town. *Concrete and Reinforced Concrete*, **10**, pp. 6–7, 1983 [in Russian].

[5] Akevlediani, N. et al., Investigation of the carrying capacity of hoisting concrete shells of double curvature on three supporting points. *Structural Mechanics of Spatial Structures*, **4**, pp. 39–49, Metsniereba, Tbilisi [in Russian].

[6] Akhvlediani, N. & Danielashvili, M., Limit analysis of reinforced concrete shells of revolution. *Archiwum Inzynerii Ladowey, Polska Akademia Nauk*, XXXVI, Z-3, pp. 187–205, 1990.

[7] Richard, N. et al., International symposium on ultra-high performance fiber-reinforced concrete. *RILEM Symposium*, 2013.

[8] Fragonara, L.Z., *Dynamic Models for Ancient Heritage Structures*. Doctoral Thesis, Politecnico di Torino: Italy, 2012.

[9] ter Maten, R.N., *Ultra-High Performance Concrete in Large Span Shell Structures*. Master's Thesis, Delft University of Technology, Structural and Building Engineering, Delft: The Netherlands, p. 313, 2012.

[10] Hosseini, M., Hajnasrollah, S. & Herischian, M., A comparative study on the seismic behavior of ribbed, Schwedler and diamatic space domes by using dynamic analysis. *Proceedings of the 15th World Conference of Earthquake Engineering*, pp. 30222–30231, 2012.

[11] Oniashvili, O., *Some Dynamic Problems of the Theory of Shells*, USSR Academy of Sciences: Moscow, p. 320, 1957.

[12] Danieli, M., Bloch, J. & Halperin, I., Analysis of seismic stability of shells of revolution using probabilistic methods. *Earthquake Resistant Engineering Structures VIII*, WIT Press: Southampton and Boston, pp. 249–260, 2011.

[13] Danielashvili, M., Katsapchuk, K. & Melnik-Melnikov, P., Probabilistic analysis of the seismic stability of concrete shells of revolution. *Spatial Structures of Buildings and Edifices*, Belgorod: Moscow, pp. 5–12 [in Russian].

[14] Danielashvili, M. & Zhorzholadze, T., Experimental and theoretical studies of the dynamic characteristics of shells. *Proceedings of the International Conference on Applied Mechanics*, pp. 1733–1738, 1989.

[15] Instruction manual on design of the reinforced concrete space coverings and floors structures, Stroiizdat: Moscow, p. 320, 1979 [in Russian].

[16] Akhvlediani, N., On calculation the load carrying capacity of shallow reinforced concrete shells of dual curvature. *Studies of the Theory of Structures*, **2**, pp. 78–86, 1962 [in Russian].

SEISMIC PERFORMANCE COMPARISON BETWEEN STRUCTURE-IMPROVEMENT TECHNIQUES AND GROUND-IMPROVEMENT TECHNIQUES: APPLICATION TO A REINFORCED CONCRETE SCHOOL BUILDING

MARIA-VICTORIA REQUENA-GARCIA-CRUZ[1], ANTONIO MORALES-ESTEBAN[2], MARIA-LUISA SEGOVIA-VERJEL[1], EMILIO ROMERO-SÁNCHEZ[1], JAIME DE MIGUEL-RODRÍGUEZ[1] & JOÃO MIGUEL CARVALHO ESTÊVÃO[2]
[1]Department of Building Structures and Geotechnical Engineering, University of Seville, Spain
[2]Department of Civil Engineering, ISE, University of Algarve, Portugal

ABSTRACT

The seismic retrofitting of reinforced concrete (RC) buildings has been widely analysed. Most of the solutions proposed are focused on the building's structure improvement. However, the effects of ground-improvement techniques combined with the building's structure-improvement techniques have not been usually analysed. Therefore, this paper aims to assess the seismic performance of a building by adding different seismic retrofitting techniques in the structure and the ground. A RC school building is proposed in this work. This has been selected because it was constructed prior to the current seismic code. Schools are some of the buildings most vulnerable to earthquakes. This is due to the low adult/child ratio. This paper is framed within the PERSITAH project (*Projetos de Escolas Resilientes aos SISmos no Território do Algarve e de Huelva*, in Portuguese). The main goal of the project is to analyse the seismic vulnerability of schools' buildings located in the Algarve-Huelva region. This area is characterized by earthquakes of long-return period and large magnitude. Therefore, the population is not aware of the seismic hazard of the area. Different seismic retrofitting techniques have been added to the building and they have been compared and analysed. The techniques have consisted of the addition of X-bracings within the buildings' bays, steel jackets in columns and soil injection grouting. These solutions have been added both individually and combined to generate hybrid models. Nonlinear static analyses have been carried out to determine the seismic performance of the building including each technique. The N2-method has been considered to obtain the performance displacement. Moreover, the damage level probability and the mean damage index have been determined for each retrofitting technique. Results have shown that the addition of X-bracings is the most efficient solution. However, this solution causes a great architectural impact. Therefore, the solution of steel jackets and/or injection grouting emerges as an interesting alternative.

Keywords: seismic vulnerability, seismic retrofitting, reinforced concrete, buildings, soil grouting, ground considerations.

1 INTRODUCTION

Among existing natural disasters, earthquakes have historically caused the most outstanding damage and human losses in Europe. Moreover, earthquakes occurred in Europe during the early 20th century have cost around 29 billion euros and have caused 19,000 casualties [1]. The destructive potential of an earthquake depends on the damage produced, the preparation and the resilience of the population.

The seismic buildings' performance plays a key role in the destructive potential of an earthquake. Existing buildings vulnerability has been the focus of European interests over the past years. This is due to the catastrophic consequences produced by recent earthquakes such as the 2009 L'Aquila (Italy), the 2011 Lorca (Spain) and the 2016 Amatrice (Italy) earthquakes [2]–[4]. A major part of these cities' building stock was severely damaged during

WIT Transactions on The Built Environment, Vol 185, © 2019 WIT Press
www.witpress.com, ISSN 1743-3509 (on-line)
doi:10.2495/ERES190081

these earthquakes. Therefore, the improvement of the seismic performance of buildings has become a major concern [5], which can be achieved by the implementation of seismic retrofitting techniques.

In this context, the European project named PERSISTAH (*Projetos de Escolas Resilientes aos SISmos no Território do Algarve e de Huelva*, in Portuguese) is under development. The main goal of the project is to analyse and reduce the seismic vulnerability of schools' buildings located in the Algarve-Huelva region. This area is located near the Eurasian and the African tectonic plates boundary [6]. This is characterized by earthquakes of long-return period and large magnitude such as the famous 1755 Lisbon earthquake (Mw = 8.5) and the 1969 earthquake (Mw = 8) [7]. Consequently, the population of the area is not aware of this seismic hazard.

Focusing on schools, these are some of the buildings most vulnerable to earthquakes. The capacity to evacuate the building, the low ratio adult/child and the trauma that they can suffer after the event [8] make children the least resilient part of the society. In the case of the Algarve-Huelva region, a major part of the schools was constructed prior to seismic considerations or, if considered, the requirements were not very restrictive. Half of them were designed with reinforced concrete (RC) wide beams, short columns and two to three floors. Buildings of these characteristics were considerably damaged during the 2011 Lorca earthquake [2]. Furthermore, the area is characterized by the presence of soft superficial soil strata, which can amplify the effects of earthquakes as reflected by every seismic code [9], [10].

The seismic retrofitting of RC buildings has been widely analysed. Two of the most implemented solutions have been the addition of X-bracings within the buildings' bays and steel jackets in columns. In Valente and Milani [11], different steel braces were experimentally assessed. The authors concluded that X-bracings had the potential to perform suitably against earthquakes. This is due to the fact that this system's cracks and failure type were much better than other samples. In Requena-García-Cruz et al. [12], the authors compared different seismic retrofitting techniques based on the efficiency, the construction costs and the architectural impact. They concluded that the most efficient solution was the addition of X-bracings while adding steel jackets was the least expensive solution and had the least architectural impact.

Most of the solutions proposed are focused on the building's structure-improvement. Nevertheless, the effects of ground-improvement techniques combined with the building's structure-improvement techniques have not been extensively analysed. Soil injection grouting solutions are commonly used such as the jet-grouting system. This system can help to reduce the structural damage of buildings due to an earthquake [13]. However, as proved in Ozener et al. [14], the assessment of its behaviour is complex. The authors concluded that experimental tests should be carry out to better understand the response of jet grouting columns during the earthquake.

In this context, this paper aims to assess the seismic performance of a building by adding different seismic retrofitting techniques in the structure and the ground. A RC school building is proposed in this work. Different seismic retrofitting techniques have been added to the building and they have been compared and analysed. The techniques have consisted in the addition of X-bracings within the buildings' bays, steel jackets in columns and soil injection grouting. These solutions have been added individually and combined to generate hybrid models. Moreover, the damage level probability and the mean damage index have been determined for each retrofitting technique.

2 METHOD

2.1 Case study

A typical two-storey school has been selected as case study (Fig. 1). This is located in Huelva (Spain). The school was constructed during the seventies, therefore no restrictive seismic requirements were considered during its design. The building is composed of RC wide-beams, columns and ribbed slabs.

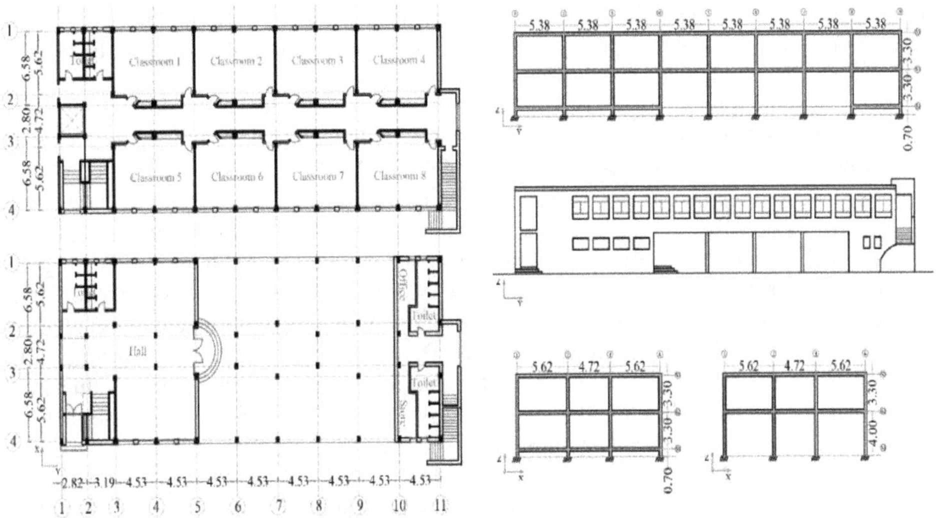

Figure 1: School's distribution in plan and elevation.

Regarding the RC mechanical characteristics, the RC compressive strength (f_{ck}) is 17.5 MPa and the modulus of elasticity (E_c) is 25,000 MPa. For the reinforcing steel, the steel yield stress (F_y) is 420 MPa and the E_c 200,000 MPa. In Table 1, the RC frames characteristics are listed. The thickness of the ribbed slabs is 30 cm. The load bearing direction is the X.

Table 1: Characteristics of the RC frames.

Characteristic	Columns	Load beams	Tied beams
Dimensions	40 x 30 cm	40 x 30 cm	30 x 30 cm
Longitudinal rebar	4Ø12 mm	Top: 2Ø12	Top: 2Ø12
		Lower: 4Ø16	Lower 2Ø12
Transversal rebar	Ø6 mm/20	Ø6 mm/20 cm	Ø6 mm/20 cm

The determination of the loads considered in the analyses has been according to the Spanish Building Code-Structural Security [15]. The value of the gravitational loads has been 5.5 kPa, including the self-weight of the elements: the RC frames and slabs, the internal partitions, the ceiling and the ceramic flooring. The variable load has been 3 kPa for public spaces.

2.2 Seismic performance assessment

The seismic performance of buildings can be analysed by means of the so-called performance-based methods [16]. They are focused on the fact that maximum storey drifts can efficiently represent the seismic response of buildings. There are numerous methods to obtain the performance of buildings. One of them is the capacity spectrum method presented in Freeman [16]. This procedure is implemented in the American ATC-40 [17]. In this work, the N2-method has been considered as this is the procedure established in the Eurocode-8 Part 1 (EC8-1) [10]. This was developed by Fajfar [18] and is based on the intersection of the capacity curve of the building and the inelastic response spectrum. The result of this intersection is the performance point (PP). This represents the displacement that the building will have for the seismic action considered.

Nonlinear static (pushover) analyses have been carried out to determine the capacity curve of the building including each retrofitting technique. These analyses have been performed in the two orthogonal directions of the building (X and Y) by means of the SAP2000 v.19 software [19]. As indicated in the EC8-1, two load patterns have been considered: one based on the mass of each storey and one based on the first vibration mode displacements. The nonlinear behaviour of the RC frames has been considered by adding plastic hinges at the end of the elements as recommended by the EC8-1. In this case, PM2M3 plastic hinges have been added to columns whereas M3 hinges have been included in the beams as in Lodi and Mohammad [20]. The rigid diaphragm effect of slabs has been considered. The contribution of infills has been obviated to obtain unfavourable capacity curves values.

The inelastic spectrum has been defined according to the EC8-1 elastic response spectrum procedure. A PGA of 0.1g has been obtained from the 2012 Spanish update of the ground acceleration values [21]. This must be multiplied by the importance factor (γ_I) of value 1.3. This has been selected since schools are considered as especially important buildings. The product of the seismic action and the γ_I-factor represents the EC8-1 design ground acceleration (a_g). The soil type has been defined as D according to a nearby geotechnical study. The area is characterized by soft superficial soil strata such as silt-sand and clays until 2.40 m of depth. The N_{SPT} determined has been four and a medium value of eight has been obtained for the N_{DPSH}.

2.3 Damage assessment

Once the capacity curves and the PP's displacements have been obtained, the building's damage has been analysed. In this work, this assessment has been carried out according to the fragility curves and the mean damage index (DI). Fragility curves represent the probability of reaching or exceeding a damage state for a given seismic action. They have been obtained according to the RISK-UE lognormal distribution [22]. They depend on the capacity curve of the building and the PP's displacement. The DI determines the most probable damage level that the building will suffer for a given seismic action. It is obtained from the damage probability percentages determined from the fragility curves and the corresponding PP's displacements [23]. This ranges from 0 to 4. Values higher than 3 represent severe damage and, therefore, those models will not comply with the ultimate life safety requirements established by the EC8-1.

Furthermore, the non-collapse safety condition required $d_u^*/d_t^*>1$ according to the EC8-1 has been obtained for each model analysed. d_u^* represents the ultimate displacement for the single-degree-of-freedom (SDOF) and d_t^* the SDOF inelastic target displacement.

2.4 Retrofitting techniques

In this work, the retrofitting techniques considered have been the addition of X-bracings (X) within the buildings' bays, steel jackets (SJ) in columns and soil injection grouting (IG) techniques such as the jet-grouting system (Fig. 2). These techniques have been experimentally tested as previously mentioned. However, they have not been extensively compared nor analysed in conjunction.

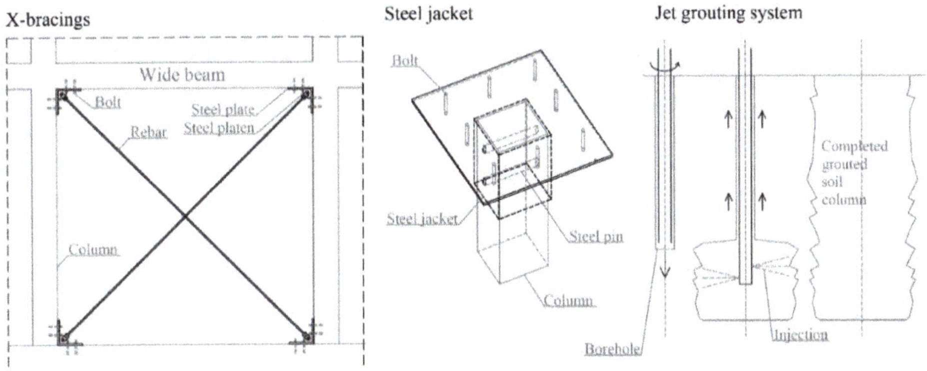

Figure 2: Details of the seismic retrofitting techniques considered in the analyses.

These solutions have been added to the original building according to Fig. 3. The structural steel determined has been S275, the E_c 210,000 MPa and the F_y 275 MPa. The X-bracings have been of Ø16 mm. The steel jackets have been simulated by adding steel plates of 30x30x0.5 cm. In addition, the soil injection grouting [24] effects have been taken into account by considering the Mohr-Coulomb criterion [25]. The model X1x2yF12 includes X-bracings (X) in one bay in the X (1x) and two in the Y (2y) for both storeys or floors (F12). The model SJmF12 adds steel jackets (SJ) in the middle (m) of the building and for both storeys (F12).

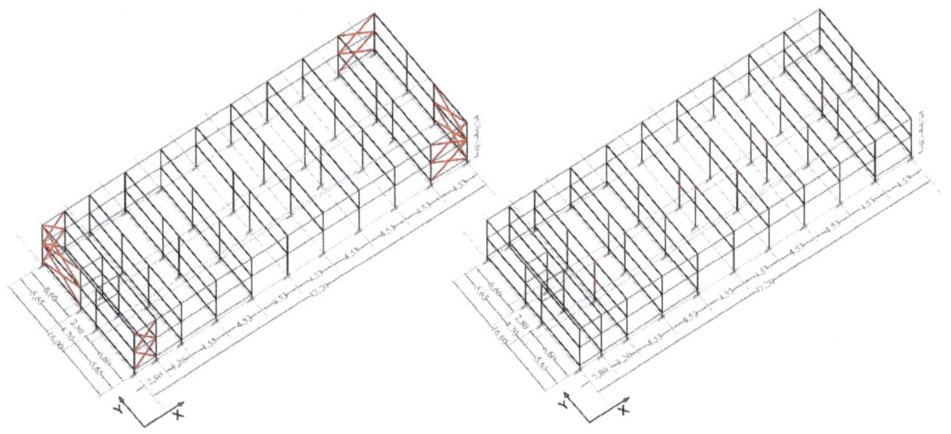

Figure 3: Models including the retrofitting techniques: X-bracings (left) and steel jackets (right).

3 RESULTS

The results of the addition of the retrofitting techniques are shown below. Pushover analyses have been carried out for three models: the original building and adding X-bracings and steel jackets to this first configuration. In total, 6 capacity curves have been obtained for both orthogonal directions of the building. Here, only the results concerning the modal load pattern have been shown since this has produced more unfavourable results. Then, the N2-method has been applied to obtain the target displacements considering the soil improvement produced by the jet grouting system. In Fig. 4, the fragility curves for each model analysed are depicted. Also, the target displacements have been highlighted for each case.

In Fig. 5, the DI for each model analysed is depicted considering both orthogonal directions of the building.

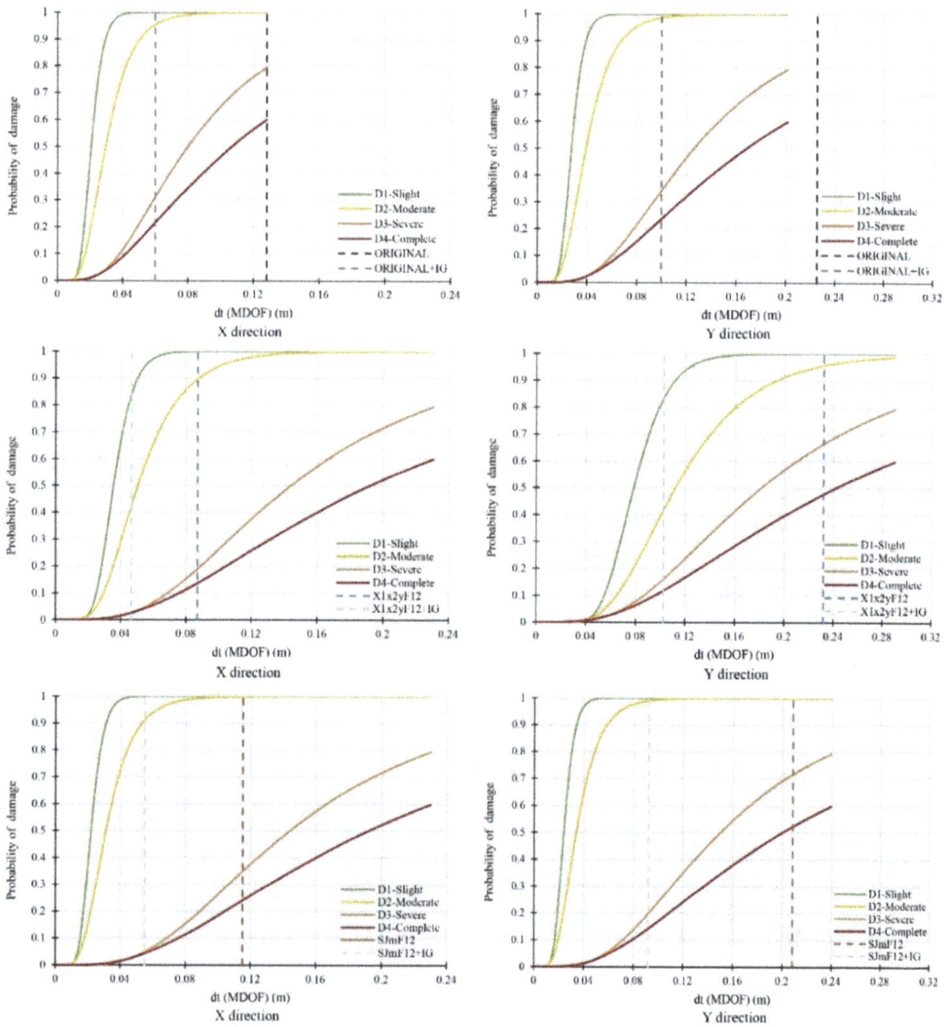

Figure 4: Fragility curves obtained for the three models analysed including the dt obtained for each case and for both orthogonal directions.

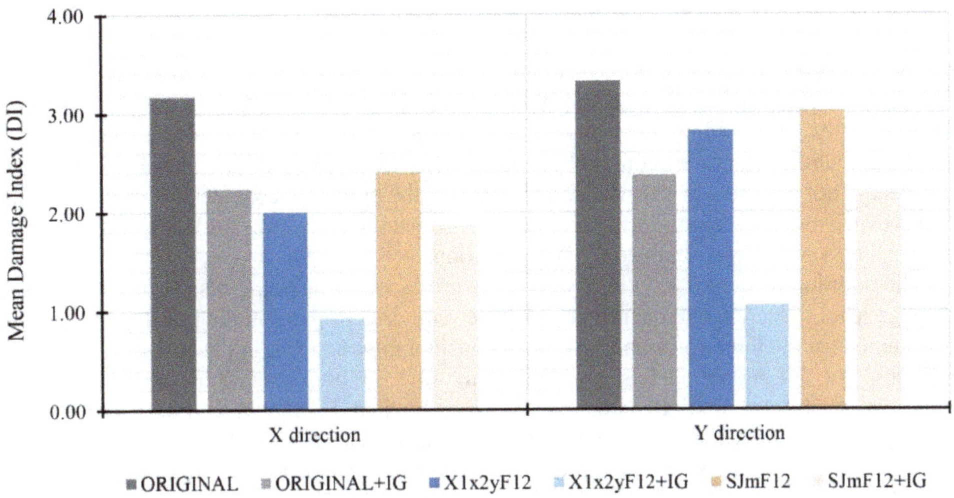

Figure 5: DI for each model analysed in the X and Y directions.

In addition, the EC8 non-collapse safety condition $d_u^*/d_t^*>1$ has been obtained for each model analysed. The results have been listed in SDOF ultimate and target displacements for each capacity curve and corresponding ratios of d_u^*/d_t^* (Table 2).

Table 2: The SDOF ultimate and target displacements for each capacity curve and corresponding ratios of d_u^*/d_t^*.

	X direction			Y direction		
	d_t^*	d_u^*	d_u^*/d_t^*	d_t^*	d_u^*	d_u^*/d_t^*
ORIGINAL	0.1053	0.1068	1.0143	0.189	0.1669	0.8833
X1x2yF12	0.0726	0.1918	2.6423	0.1931	0.2330	1.2068
SJmF12	0.0961	0.1918	1.9961	0.1733	0.2000	1.1519
ORIGINAL+IG	0.0494	0.1068	2.1621	0.0838	0.1669	1.9921
X1x2yF12+IG	0.0388	0.1918	4.9441	0.0854	0.2330	2.7288
SJmF12+IG	0.0461	0.1918	4.1612	0.0766	0.2000	2.6062

4 ANALYSIS OF RESULTS

Results have shown that the original building has not complied with the life safety condition $d_u^*/d_t^*>1$ in the Y direction, being the ratio 0.88. In the case of the X, the ratio is close to 1. The damage expected in each direction has been severe, resulting a DI of 3.17 and 3.33 in the X and Y, respectively.

The addition of X-bracings and steel jackets has produced higher values of displacements, as shown in the fragility curves plot. This is due to the improvement of the building's ductility. Owing to this, the expected damage has been reduced. In the case of X-bracings, this has been 2 and 2.82 in the X and Y respectively. Expressed as percentage, the damage has been reduced in a 58% and 17% in each direction. Nevertheless, adding steel jackets has not produced a considerable improvement. The DI has been 2.41 and 3.02 in the X and Y direction respectively. Hence, in the Y direction, the expected damage has been severe. Consequently, only adding steel jackets will not be enough to reduce the damage. These results are similar to those obtained by Requena-García-Cruz et al. [12]. In this work, steel jackets did not enhance the seismic performance of the buildings considerably in spite of not being expensive and having lower architectural impact. Concerning the d_u^*/d_t^* ratio, values higher than 1 have been obtained for both solutions.

The addition of the jet grouting system has produced outstanding results in terms of the damage level reduction and the ratio $d_u^*/d_t^*>1$ complying. Regarding the fragility curves, the d_t obtained for the three solutions have been considerably reduced. In the case of the original building, the DI reduction has been close to 40% in both directions. Combined with X-bracings, values near slight damage have been obtained, being 0.93 and 1.05 in the X and Y, respectively. In the case of steel jackets, the DI has been 1.87 and 2.19 for each direction.

Regarding the $d_u^*/d_t^*>1$, higher values of 1 have been obtained for all the solutions adding the jet grouting system. In the case of the original building, the ratio has been enhanced a 100%. Concerning the hybrid models, the ratio has been considerably higher than the values obtained for the models adding only the structure's improvement solutions.

5 CONCLUSIONS

This paper has analysed and compared the seismic performance of structure-improvement and ground-improvement seismic retrofitting techniques. These solutions have consisted in the addition of X-bracings within the buildings' bays, steel jackets in columns and soil injection grouting techniques such as the jet-grouting system. These solutions have been added individually and combined to generate hybrid models. A RC school building has been proposed. These buildings are one of the most vulnerable typologies to earthquakes. In addition, the school is located in the Algarve-Huelva region. This area is characterized by earthquakes of long-return period and large magnitude. Consequently, the population is unaware of the hazard.

Nonlinear static analyses have been carried out to determine the seismic performance of the building including each technique. The N2-method has been considered to obtain the performance displacement. The fragility curves have revealed that the addition of X-bracings and steel jackets produce higher values of displacements due to the building's ductility enhancement.

Results have shown that the original building has not complied with the life safety condition. Also, the expected damage has been determined as severe in both directions. Therefore, this all leads to the fact that solutions must be provided in order to reduce the expected damage. Contrariwise, the implementation of the jet grouting system in the original building has reduced the expected damage to moderate.

The addition of X-bracings has been the most efficient technique. This solution has presented a higher reduction percentage than the addition of steel jackets. In fact, this has reduced the expected damage to slight and moderate in the X and Y respectively. Nevertheless, this solution causes a great architectural impact. Combined with the jet grouting, the reduction has been outstanding, resulting in slight damage for both directions.

In the case of the steel jackets, this technique has not considerably reduced the expected damage, being moderate and severe in the X and Y respectively. However, combined with the jet grouting system, the expected damage and the ultimate life safety ratio have been considerably improved.

In the light of the results, the solution of steel jackets and/or injection grouting emerges as an interesting alternative. This hybrid technique is of low architectural impact and can be easily implemented. Also, further research should be carried out on the determination of the seismic effects of adding ground-improvement retrofitting techniques.

ACKNOWLEDGEMENTS

This work has been supported by the INTERREG-POCTEP Spain-Portugal programme and the European Regional Development Fund through the 0313_PERSISTAH_5_P project and the VI-PPI of the University of Seville by the granting of a scholarship. The grant provided by the *Instituto Universitario de Arquitectura y Ciencias de la Construcción* is acknowledged.

REFERENCES

[1] Battarra, M., Balcik, B. & Xu, H., Disaster preparedness using risk-assessment methods from earthquake engineering. *European Journal of Operational Research*, **269**(2), pp. 423–435, 2018.

[2] Ruiz-Pinilla, J.G., Adam, J.M., Pérez-Cárcel, R., Yuste, J. & Moragues, J.J., Learning from RC building structures damaged by the earthquake in Lorca, Spain, in 2011. *Engineering Failure Analysis*, **68**, pp. 76–86, 2016.

[3] Del Gaudio, C., Ricci, P., Verderame, G.M. & Manfredi, G., Urban-scale seismic fragility assessment of RC buildings subjected to L'Aquila earthquake. *Soil Dynamics and Earthquake Engineering*, **96**(July 2016), pp. 49–63, 2017.

[4] Fiorentino, G. et al., Damage patterns in the town of Amatrice after August 24th 2016 central Italy earthquakes. *Bulletin of Earthquake Engineering*, **16**, pp. 1399–1423, 2018.

[5] Mazzoni, S. et al., 2016–2017 Central Italy earthquake sequence: Seismic retrofit policy and effectiveness. *Earthquake Spectra*, **34**(4), pp. 1671–1691, 2018.

[6] Amaro-Mellado, J.L., Morales-Esteban, A., Asencio-Cortés, G. & Martínez-Álvarez, F., Comparing seismic parameters for different source zone models in the Iberian Peninsula. *Tectonophysics*, **717**(September), pp. 449–472, 2017.

[7] Sá, L., Morales-Esteban, A. & Durand Neyra, P., The 1531 earthquake revisited : loss estimation in a historical perspective. *Bulletin of Earthquake Engineering*, **16**(10), pp. 4533–4559, 2018.

[8] Bal, A. & Jensen, B., Post-traumatic stress disorder symptom clusters in Turkish child and adolescent trauma survivors. *European Child & Adolescent Psychiatry*, **16**(7), pp. 449–457, 2007.

[9] Spanish Ministry of Public Works [Ministerio de Fomento de España], *Spanish Seismic Code* [*Norma de Construcción Sismorresistente: Parte general y edificación (NCSE-02)*], Spain, 2002.

[10] European Union, Eurocode 8: Design of structures for earthquake resistance. Part 1: General rules, seismic actions and rules for buildings. *Journal of Constructional Steel Research*, **54**(2), pp. 18–20, 2004.

[11] Valente, M. & Milani, G., Alternative retrofitting strategies to prevent the failure of an under-designed reinforced concrete frame. *Engineering Failure Analysis*, **89**(January 2017), pp. 271–285, 2018.

[12] Requena-García-Cruz, M.-V., Morales-Esteban, A., Durand-Neyra, P. & Estêvão, J.M.C., An index-based method for evaluating seismic retrofitting techniques. Application to a reinforced concrete primary school in Huelva. *PLOS ONE*, **14**(4), pp. e0215120, 2019.

[13] Durgunoglu, H.T. et al., Soil improvement with jet-grouting columns: A case study from the Kocaeli earthquake. *5th International Conference on Case Histories in Geotechnical Engineering*, New York, 2004.
[14] Ozener, P., Dulger, M. & Berilgen, M., Numerical study on effectiveness of Jet-grout columns in liquefaction mitigation. *6th International Conference on Earthquake Geotechnical Engineering*, Christchurch, 2015.
[15] Spanish Ministry of Public Works [Ministerio de Fomento de España], *Spanish Technical Code of Buildings. Structural Security. Actions* [*Código Técnico de la Edificación (DB-SE-AE)*], Spain, 2009.
[16] Freeman, S.A., Review of the development of the capacity spectrum method. *ISET Journal of Earthquake Technology*, **41**(1), pp. 113, 2004.
[17] Applied Technology Council (ATC), ATC-40: Seismic evaluation and retrofit of concrete buildings, United States, 1996.
[18] Fajfar, P., Capacity spectrum method based on inelastic demand spectra. *Earthquake Engineering and Structural Dynamics*, **28**, pp. 979–993, 1999.
[19] Computers and Structures INC, SAP2000 v.19, 2019.
[20] Lodi, S.H. & Mohammad, A.F., Nonlinear static analysis of an infill framed reinforced concrete building. *15th World Conference on Earthquake Engineering*, (2006), 2012.
[21] Spanish Ministry of Public Works [Ministerio de Fomento de España], *Update of the Seismic Hazard Maps* [Actualización de mapas de peligrosidad sísmica de España], Spain, 2012.
[22] Milutinovic, Z.V. & Trendafiloski, G.S., *An Advanced Approach to Earthquake Risk Scenarios with Applications to Different European Towns. WP4: Vulnerability Assessment of Lifelines and Essential Facilities*, 2003.
[23] Calvi, G.M. & Pinho, R., Development of seismic vulnerability assessment methodologies over the past 30 years. *ISET Journal of Earthquake Technology*, **43**(472), pp. 75–104, 2006.
[24] Kazemian, S. & Huat, B.B.K., Assesment and comparison of grouting and injection method in geotechnical engineering. *European Journal Of Scientific Research*, **27**(2), pp. 237–247, 2009.
[25] Abbo, A.J. & Sloan, S.W., A smooth hyperbolic approximation to the Mohr-Coulomb yield criterion. *Computers & Structures*, **54**(3), pp. 427–441, 1995.

FEASIBILITY OF THE UNDERGROUND EARTHQUAKE BRACELET

AMAL GERGES & NABIL FARES
Department of Civil Engineering, University of Balamand, Lebanon

ABSTRACT

Earthquakes are sudden phenomena that occur worldwide and have the potential to cause significant devastation. A number of methods have emerged over the years in an effort to decrease their destructive potential, including the emergence of seismic design, dampening mechanisms and shock absorbents. The earthquake bracelet is a new solution which entails digging a gap around a building and then engineering the fill as well as the basement floors to provide more desirable behavior under earthquake load. A dynamic analysis study was completed to explore the response of the structure and assess the feasibility of the idea. Many cases were explored to identify the change in the response spectrum with and without the earthquake bracelet. The results of the preliminary study have been encouraging and indicate that the method is feasible and can be conceptualized as utilizing underground stories and their surroundings as a sophisticated base-isolation system. In order to accomplish this, the system must have: (i) a flexible bracelet, (ii) relatively flexible underground stories and (iii) preferably, a higher relative mass of understructure over superstructure is preferred.

Keywords: earthquake bracelet, dynamics of structure, base isolation, earthquake, understructure, preservation, preliminary study.

1 INTRODUCTION

Lebanon, a Mediterranean country situated astride the Dead Sea Transform, has experienced a number of major earthquakes over its history. The most recent major event occurred on the 25th of November 1759 impacting Beirut and Damascus and resulting in more than 40,000 fatalities. Prior to this, earthquakes had not been considered in building design. More recently, earthquake design has become mandatory prerequisite for construction permits, and buildings have been designed as situated in a 2c seismic zone such that the seismic zone factor is Z=0.25 (which represents the peak ground acceleration in studied site as a percentage of gravity acceleration g (9.81 m/s^2)).

If an earthquake having a magnitude close to that of the 1759 event were to happen today, it would be a national catastrophe. A few minutes of earthquake could cause damage equivalent to 16 years of war. Thus, it may be necessary to retrofit existing buildings, prioritizing essential buildings such as schools and hospitals due to the probable expense. Hopefully, this study will show that a decrease in the earthquake effect is achievable and can be reasonably implemented on some critical existing structures.

Try the following: Obtain a mug, a plastic water bottle and cotton wool. Place the water bottle on top and then put a layer of cotton wool around the bottle between the mug and the bottle. Hold the mug by its handle and vibrate it. Now compare by vibrating the bottle directly. Do you notice the difference? The bottle vibrates much more when you directly move it. This is what our study resembles.

1.1 Background on isolation systems of buildings

The first application of seismic isolation took place in Japan and the United States during the early 1980s. As a demonstration project in 1982 Japan built an isolated structure, which was a small house. Likewise, in 1985 in the US, the first isolated building was a county

WIT Transactions on The Built Environment, Vol 185, © 2019 WIT Press
www.witpress.com, ISSN 1743-3509 (on-line)
doi:10.2495/ERES190091

administration building in Southern California [1]. Cooperation between scientists in Japan, New Zealand and the US took place throughout the 1980s, resulting in advancement in both the technology of seismic isolation and passive energy dissipation. As Japan's economy boomed, New Zealand and the US couldn't keep up, and funding for such projects came via government-mandated research investment for construction companies [1], [3].

Recently base isolators and damping devices have been more commonly used. Base isolation is used to decouple the structure from the foundation lying on the ground, which enhances the structure's earthquake resistance.

There are many types of base isolation systems such as elastomeric bearings, friction pendulum bearings, rollers and ball bearings.

Now damping devices are able to absorb energy in order to reduce the structure's response to earthquake. They can be combined with the base isolators or placed in diagonal braces along the height of the structure. They can also be applied for tall buildings that can't be properly base-isolated [2], [4].

1.2 Scope of work

The response spectrum is often used in building design to mitigate the impact of potential harmonic ground motion or earthquakes. With this approach, the periods of the first few modes of vibration of the building are determined and the corresponding maximum displacements are then obtained from the response spectrum. These maximum displacements with their corresponding mode shapes are used to identify the maximum loading cases and their impacts are combined appropriately. By having an underground mass hugged by an isolating polymer with the mass of superstructure above it, the response spectrum is altered which modifies the maximum loading conditions. There are parameters associated with this isolation system and with the variation of these parameters we get a response spectrum family a.k.a. response spectra. In this study we are investigating whether there are some combinations of parameters that lead to a significant beneficial improvement in the effective response spectrum.

2 MODEL AND REVIEW OF RESPONSE SPECTRA

2.1 Model

First, we isolate the underground mass with a material of a certain thickness and depth (Fig. 1). This eliminates the side forces in an earthquake and thus creates an isolated underground mass with a certain stiffness; the superstructure above it also has a certain stiffness and mass [5].

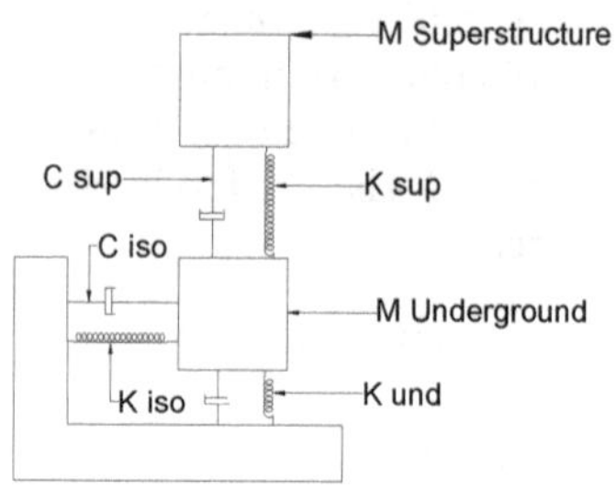

Figure 1: Basic model.

From the model depicted in Fig. 1, we obtain the following equations of motion:

$$\begin{bmatrix} M_{und} & 0 \\ 0 & M_{sup} \end{bmatrix}\begin{bmatrix} U''_{und} \\ U''_{sup} \end{bmatrix} + \begin{bmatrix} K_{iso} + K_{sup} + K_{und} & -K_{sup} \\ -K_{sup} & K_{sup} \end{bmatrix}\begin{bmatrix} U_{und} \\ U_{sup} \end{bmatrix} +$$

$$\begin{bmatrix} C_{iso} + C_{sup} + C_{und} & -C_{sup} \\ -C_{sup} & C_{sup} \end{bmatrix}\begin{bmatrix} U'_{und} \\ U'_{sup} \end{bmatrix}$$

$$= \begin{bmatrix} K_{iso} \times U_g + K_{und} \times U_g + C_{iso} \times U'_g + C_{und} \times U'_g \\ 0 \end{bmatrix}, \quad (1)$$

such that:

- M_{und}: Underground mass in kg; M_{sup}: Mass of superstructure in kg.
- K_{und}: Stiffness of the underground structure in KN/m; K_{sup}: Stiffness of the superstructure in KN/m; K_{iso}: Stiffness of isolation in KN/m.
- C_{und}: Damping of the underground structure in KN×s/m; C_{sup}: Damping of the superstructure in KN×s/m; C_{iso}: Damping of the isolation in KN×s/m.
- U_{und}: Displacement of the underground in m; U_{sup}: Displacement of the superstructure in m; U'_{und}: Velocity of the underground in m/s; U'_{sup}: Velocity of the superstructure in m/s; U''_{und}: Amplitude of the underground in m/s^2; U''_{sup}: Amplitude of the superstructure in m/s^2.
- U_g: Ground motion displacement in m; U'_g: Ground motion velocity in m/s; U''_g: Ground motion amplitude in m/s^2.

To simplify the equations, let:

$$U_{n1} = U_{und} - U_g \text{ and } U_{n2} = U_{sup} - U_g,$$

having:

$$\text{Mund} \times \text{U''und} + (\text{Kiso}+\text{Ksup}+\text{Kund}) \times \text{Uund} - \text{Ksup} \times \text{Usup} = \text{Kiso} \times U_g + \text{Kund} \times U_g$$

$$\text{Msup} \times \text{Usup} - \text{Ksup} \times \text{Uund} + \text{Ksup} \times \text{Usup} = 0. \quad (2)$$

The equations become the following:

$$\text{Mund} \times U''_{n1} + \text{Mund} \times U''_g + (\text{Kiso} + \text{Ksup} + \text{Kund}) \times U_{n1} - \text{Ksup} \times U_{n2} + 2\text{Ksup} \times U_g = 0$$

$$\text{Msup} \times U''_{n2} - \text{Msup} \times U''_g + \text{Ksup}\,(U_{n2} - U_{n1}) = 0. \quad (3)$$

2.2 What are response spectra and what is their role in design

In earthquake engineering the response spectra are very useful in quantifying the ground motion and the building's ability to resist earthquakes. The design response spectrum (RS) plots the displacement of a single degree of freedom system, with varying natural frequencies, resulting from a forced motion (base vibration). We expand this definition to allow Multiple Degrees Of Freedom (MDOF) where we consider the RS of a main Degree Of Freedom (DOF) with other DOFs considered as internal variables or parameters. More clearly, instruments distributed and located in several locations sense the initial ground motion of an earthquake and data about the ground motion are thus collected over time. The first data reported is the peak ground acceleration data (PGA). The PGA provides information over the intensity of the ground motion, but it doesn't provide information concerning the response of buildings with different periods. Now if the peak acceleration, displacement or velocity is calculated when an earthquake hits for several systems of different stiffness K and mass M, the locus of point plotted will form the response spectrum. Therefore, if the period of vibration is known, the elements of design required can be deduced [6].

2.3 Estimating mass, stiffness and damping of various classes of buildings

For the sake of accuracy and in order to obtain the proper ratio of masses and stiffness for each mode of movement during earthquake of the building, a model must be analyzed. The effective masses (percentage of modal mass) are needed in order to identify which mode can be readily excited by base excitation and to see whether the masses are to be used as they are, or as a certain percentage only. The same goes for the stiffness. To that end and as an example, we have collected information concerning hospitals.

Using the data in Table 1, a six-story building with two underground basements, having the masses equivalent to the average of masses collected in the hospital's survey was modeled on Etabs software where the results are presented in Tables 2 and 3. Two Etabs files were created: one with only the superstructure, and the other including only the underground basements. After running the models, the ratio of modal masses was extracted from the Etabs reports. For the superstructure, see Table 2. For the underground, see Table 3.

Table 1: Hospital data.

Hospital/ country	Built up area m^2	Area of sup	# of sup	Area of und	# of und	Sup weight	Und weight
CMC-Dubai	45,000	2,800	5	5,300	4	37,824	60,844
		1,900	3				
CMC-Riyadh	62,000	2,200	6	6,000	5	62,784	86,100
		1,900	8				
		1,700	2				
		900	1				
CMC-Lebanon		1,600	8	1,600	6	36,736	27,552
CMC-Amman	80,000	4,000	11	4,000	4	144,960	45,920
		900	35				

Table 2: Table of modal participating mass ratios part 1.

Case	Mode	Period sec	Ux	Uy	Uz	Sum Ux	Sum Uy	Sum Uz
Modal	1	0.763	0.016	0.4963	0	0.0016	0.4963	0
Modal	2	0.597	0.0184	0.2375	0	0.02	0.7338	0
Modal	3	0.527	0.6638	0.0022	0	0.6838	0.7361	0
Modal	4	0.212	0.0003	0.126	0	0.6841	0.8621	0
Modal	5	0.159	0.0016	0.0349	0	0.6856	0.897	0
Modal	6	0.117	0.2023	0	0	0.8881	0.942	0
Modal	7	0.106	0.0001	0.0451	0	0.8885	0.9443	0
Modal	8	0.075	0.0004	0.0023	0	0.8885	0.9443	0

Table 3: Table of modal participating mass ratios part 2.

Case	Mode	Rx	Ry	Rz	SUM Rx	SUM Ry	SUM Rz
Modal	1	0.11942	0.0006	0.2376	0.1942	0.0006	0.2376
Modal	2	0.0913	0.0085	0.461	0.2854	0.0092	0.6987
Modal	3	0.0008	0.3286	0.0186	0.2862	0.3378	0.7173
Modal	4	0.3491	0.0006	0.0362	0.6353	0.3384	0.7534
Modal	5	0.0883	0.0027	0.1378	0.7236	0.341	0.8912
Modal	6	5.462E-06	0.3618	0.0018	0.7236	0.7028	0.893
Modal	7	0.1067	0.0002	0.009	0.8303	0.7031	0.9021
Modal	8	0.0046	0.001	0.0558	0.8349	0.7041	0.9579

It was concluded that no reduction of mass or stiffness is necessary and therefore examples of the study can proceed using the whole masses without using contribution factors.

3 RESPONSE SPECTRA IN THE EARTHQUAKE BRACELET AND NON-DIMENSIONALISATION

3.1 The 1-DOF response spectra and how it changes in the earthquake bracelet model

In a 1-DOF function where ug[t] is a sinusoidal function, such that $u_g[t]= amp\times\sin(wt)$, if ug is replaced in the equation of motion:

$$f(u)= m\times u''[t]+c\times u'[t]+k\times u[t]=-k\times u_g[t], \quad (4)$$

such that t varies from 0 to tmax, the parameters of the equation become: m, c, K, amp, w, u_0, v_0, t_{max}.

The period T of the structure is then defined in function of both parameters m and K, afterwards K is expressed as a function of m and T. Now we are able to plot the displacement in function of t for a specific period and the maximum value of the displacement u after steady state is extracted.

We are then able to plot the maximum of u→U_{max} for a range of T. Having the plot of U_{max} in function of the period, we have thus the response spectrum. Now how does this apply to the model of the Earthquake Bracelet? Our model is based on a 2-DOF equation where we have U_{und} and U_{sup}, for the understructure mass and superstructure respectively, the superstructure mass has a stiffness and damping which we associate with the first mode of the superstructure. The understructure mass has two sets of stiffnesses and damping coefficients, the first set corresponds to the bracelet while the second to the foundation below the understructure mass. The parameters to be manipulated are: M_{und}, M_{sup}, K_{und}, K_{sup}, C_{und}, C_{sup}, K_{iso}, C_{iso}. Thus, the response spectra will be manipulated as functions of a range of all the parameters. The effect of the initial conditions U_{01}, V_{01}, U_{02} and V_{02} are removed by considering only the long time response and neglecting transitions. Finally, the earthquake period T_{eq} is used as a normalization factor.

This allows us to know for a structure, having a specific ratio of mass, what is the K_{iso}, among other design parameters, needed to retrofit the structure, by fixing the masses and manipulating K_{iso} to get the response spectrum with lowest peak (→ lowest displacement amplification).

3.2 The non-dimensionalization of the equations

Usually the response spectrum is a family of curves that depends on the natural damping of the building, but in our case it also depends on additional parameters. We want to reduce the number of parameters involved in order to more efficiently search the parameter space for beneficial behavior. To that end we resort to non-dimensionalization and we propose non-dimensionalized parameters that are more easily understood.

Since the response spectrum is our stepping stone, a program using Mathematica was written. An data inputs, the ratio of mass of the braceleted ground over that of the superstructure ($mR = M_{und}/M_{sup}$), the ratio of periods where it is the period of the underground mass over that of the earthquake ($TR_{und} = T_{und}/T_{eq}$, $TR_{iso} = T_{iso}/T_{eq}$) and damping ratios of the superstructure, understructure and isolation ($\zeta_{sup} = c_{sup} / (2\ m_{sup}\ (2\pi/T_{sup}))$, $\zeta_{und} = c_{und} / (2\ m_{und}\ (2\pi/T_{und}))$, $\zeta_{iso} = c_{iso} / (2\ m_{und}\ (2\pi/T_{iso}))$) shall be entered to generate the desirable response spectrum output. This will allow us to identify and manipulate the parameters while seeing the behavior of the response spectrum allowing us to draw conclusions concerning each manipulation of parameter, or group of parameters.

Forcing function: $u_g = A \times \sin[2\ \pi \times t/\ T_{eq}]$, all displacements are non-dimensionalized by the amplitude of the earthquake: A that we chose as one unit of length.

First equation:

$$-4\pi^2\mathrm{Usup}[\tau] + 4\pi^2\mathrm{Uund}[\tau] + \frac{4\mathrm{mR}\pi^2\mathrm{TRsup}^2\mathrm{Uund}[\tau]}{\mathrm{TRiso}^2} + \frac{4\mathrm{mR}\pi^2\mathrm{TRsup}^2\mathrm{Uund}[\tau]}{\mathrm{TRund}^2} - 4\pi\mathrm{TRsup}\zeta\mathrm{supUsup}'[\tau] + \frac{4\mathrm{mR}\pi\mathrm{TRsup}^2\zeta\mathrm{isoUund}'[\tau]}{\mathrm{TRiso}} + 4\pi\mathrm{TRsup}\zeta\mathrm{supUund}'[\tau] + \frac{4\mathrm{mR}\pi\mathrm{TRsup}^2\zeta\mathrm{undUund}'[\tau]}{\mathrm{TRund}} + \mathrm{mRTRsup}^2\mathrm{Uund}''[\tau] = \frac{8\mathrm{mR}\pi^2\mathrm{TRsup}^2\zeta\mathrm{isoCos}[2\pi\tau]}{\mathrm{TRiso}} + \frac{8\mathrm{mR}\pi^2\mathrm{TRsup}^2\zeta\mathrm{undCos}[2\pi\tau]}{\mathrm{TRund}} + \frac{4\mathrm{mR}\pi^2\mathrm{TRsup}^2\mathrm{Sin}[2\pi\tau]}{\mathrm{TRiso}^2} + \frac{4\mathrm{mR}\pi^2\mathrm{TRsup}^2\mathrm{Sin}[2\pi\tau]}{\mathrm{TRund}^2}. \quad (5)$$

Second equation:

$$4\pi^2\mathrm{Usup}[\tau] - 4\pi^2\mathrm{Uund}[\tau] + 4\pi\mathrm{TRsup}\zeta\mathrm{supUsup}'[\tau] - 4\pi\mathrm{TRsup}\zeta\mathrm{supUund}'[\tau] + \mathrm{TRsup}^2\mathrm{Usup}''[\tau] = 0. \quad (6)$$

4 COMPUTER METHOD AND IMPLEMENTATION

4.1 Solution procedure

During the study of the solution of the non-dimensionalized earthquake bracelet equations, an obstacle was encountered. While exploring the graphs presented in Fig. 2 and while manipulating the masses and period of the equation, we couldn't have clear oversight over the system's behavior, for when a change positively affected the superstructure it at times negatively affected the under structure and vice versa.

Therefore, in order to be more precise and to understand clearly the behavior of the system we proceeded to find the response spectrum and explore it.

4.2 Implementation (mathematica file description)

After non-dimensionalizing the equations as previously shown, we proceeded to manipulate the plot of the solutions with various ranges for parameters such as TR_{sup}, mR, TR_{und} and TR_{iso} between 0.1 and 10, while damping ratios ζ_{sup}, ζ_{und} and ζ_{iso} are at 0.05 varying between 0 and 1, the earthquake period ranges from 0.1 to 5 sec and u_{max} at 2 varying as a power of 2.

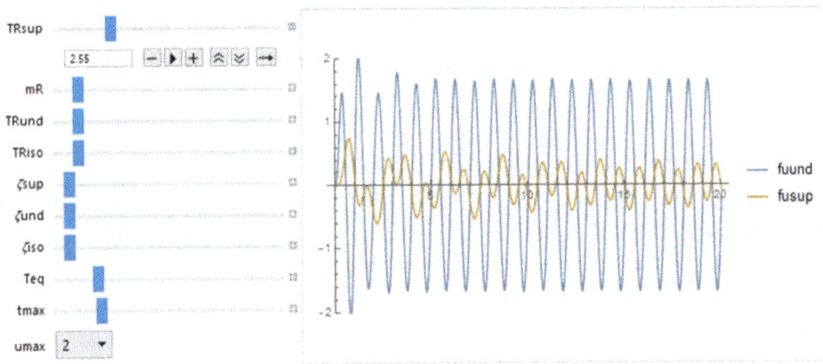

Figure 2: Solution of the equation of motion.

As mentioned in Section 4.1, the manipulation of this plot in Fig. 2 didn't lead to clear conclusions and it was hard to correlate between the two graphs. Thus, the response spectrum of the earthquake bracelet was plotted with the same previous ranges with the sole modification of dividing the earthquake period by the period of the superstructure. Afterwards a combination between the system's original response spectrum and one with an earthquake bracelet present was written with parameters varying as the following mR, TR_{und} and TR_{iso} from 0.1 to a max of 5, the damping ratios of the isolation, superstructure and understructure vary from 0 to 1, and then the following plots in Fig. 3 were generated. (Green RS: building without EQ bracelet; red RS: building with EQ bracelet).

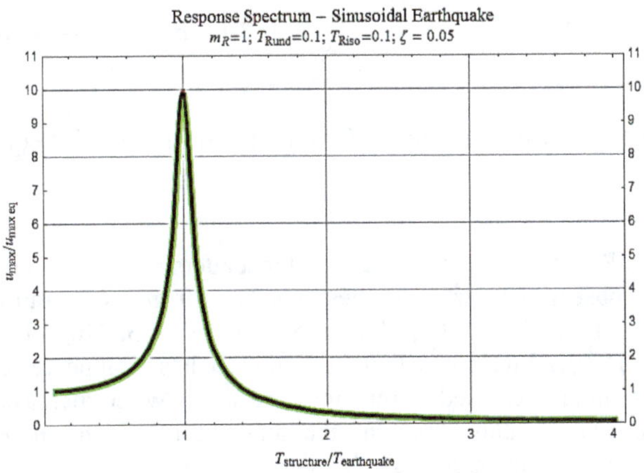

Figure 3: Combination of Response Spectra ($U_{max}/U_{max\ eq}$ in function of $T_{structure}/T_{earthquake}$).

5 PARAMETRIC STUDIES

5.1 Range of systems considered

In our trials and exploration of the graph implemented on Mathematica, we initially have the response spectrum of the regular Single Degree Of Freedom (SDOF) matching that of the earthquake bracelet since we have TR_{und} and TR_{iso} at 0.1 both.

Many cases were observed, varying, of the many cases a few will be discussed, and conclusions will follow. The cases vary between the very bad and the very good. Firstly, we studied the effect of mR on the system by fixing the periods of the superstructure and the understructure at 2.

We managed to observe in the graphs generated, for which Fig. 4 is an example, that for mR=1 as well as mR=2, mR=3 the bracelet has hugely decreased the amplitude of the response spectra. The application of the bracelet is encouraging and we could better understand its characteristics through manipulating the parameters and exploring the graphs. Another approach that could help better understand the observation is when the mass ratio is fixed and the periods manipulated to be able to compare more easily.

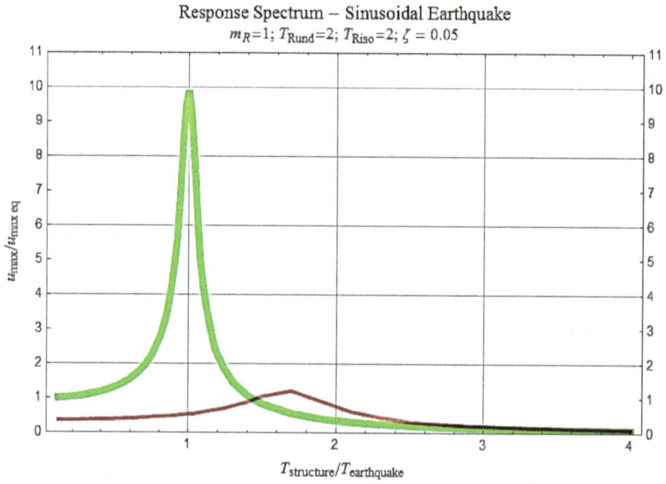

Figure 4: Response Spectra for mR=1 and period ratios equal 2.

5.1.1 For mR=0.5

The meaning of mR=0.5 is that the mass of the understructure is half the mass of the superstructure; we observed some of the best matches for this mass ratio such as, TR_{und}=1 and TR_{iso} =2, TR_{und}=1 and TR_{iso} =2.5, TR_{und}=0.85 and TR_{iso}= 3, TR_{und}=0.9 and TRiso=3.

In Fig. 5 where TR_{iso} is 2 we notice that the resonance has disappeared completely and the low displacements either decreased or remained the same. When increasing TR_{iso} 2.5 and 3 respectively we notice that it almost remains the same as for 2. Furthermore, increasing TR_{und} to 0.9 still produces a significant decrease in the resonance.

5.1.2 For mR=1

For a high period of isolation such that TR_{iso} fixed to 3s, TR_{und} varied between 1, 2, 2.5 and 3 (Fig. 6).

5.1.3 For mR=1.5

For a fixed value of TR_{und}=3 such as TR_{und}=3 and TR_{iso} =2, TR_{und}=3 and TR_{iso} =2.5, TR_{und}=3 and TRiso =3. As all previous cases the response spectrum is enhanced when the isolation's flexibility increases, as Fig. 7 illustrates.

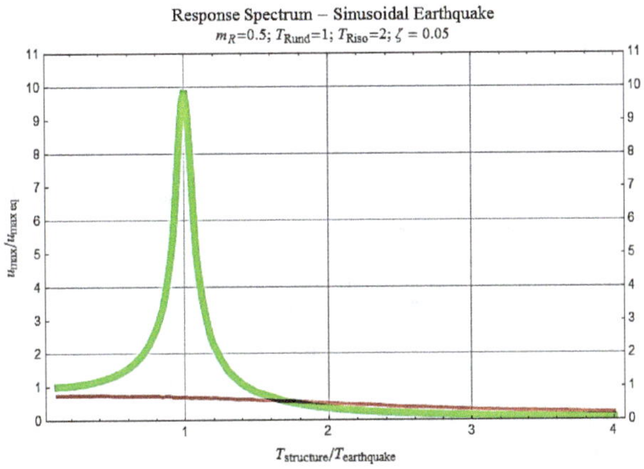

Figure 5: Response Spectra for fixed mR= 0.5, TR_{und}=1 and TR_{iso} =2.

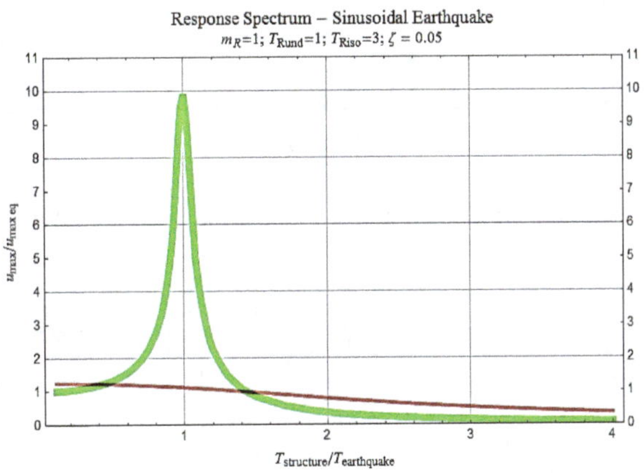

Figure 6: Response spectra for fixed mR= 1, TR_{und}=1 and TR_{iso} =3.

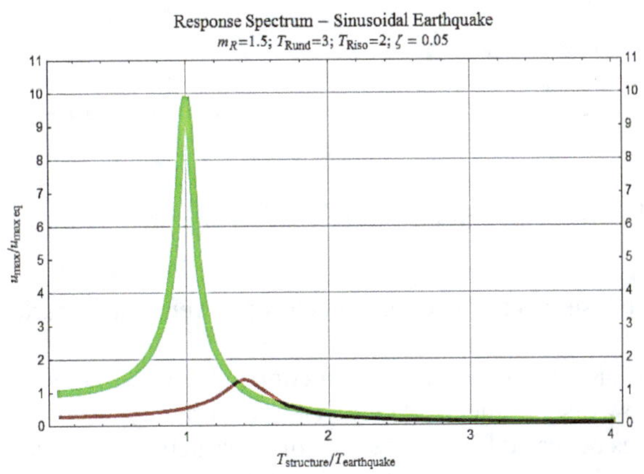

Figure 7: Response Spectra for mR= 1.5 fixed, TR_{und}=3 and TR_{iso} =2.

5.1.4 For mR=2

We observed bad matches except for high periods such as, TR_{und}=0.8→3 and TR_{iso} =0.8→3 (Fig. 8).

5.1.5 For mR=3

The bad cases are the following: TR_{und}=1 and TR_{iso} =2, TR_{und}=1.5 and TR_{iso}=1 the uncertain case is the following: TR_{und}=3 and TR_{iso} =1.5.

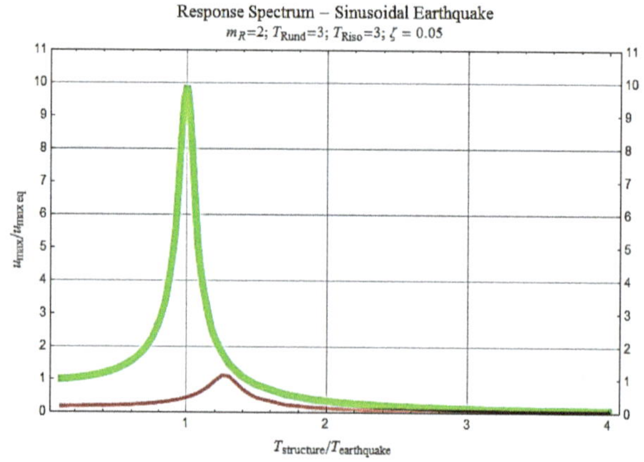

Figure 8: Response Spectra for mR=2 fixed, TRund=3 and TRiso =3.

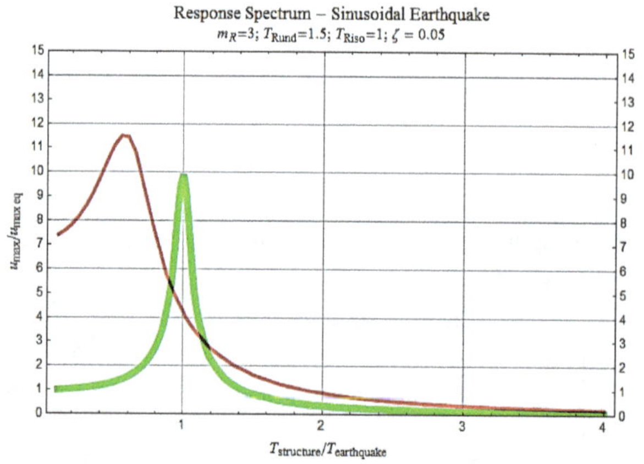

Figure 9: Response Spectra for mR=3 fixed, TRund=1 and TRiso =2.

In Fig. 9, the response spectrum became much worse versus a scenario with no earthquake bracelet.

For TR_{und}=3 and TR_{iso} =1.5 there was a reduction in the value of the displacement but also an increase in the low displacements which creates uncertainty.

The good matches occurred for high periods of substructure and isolation for mR=3 fixed, TR_{und}=2.5 and TR_{iso} =3, TR_{und}= 3 and TR_{iso} = 2, TR_{und}= 3 and TR_{iso} = 3. The graphs generated all resemble the one in Fig. 9.

5.2 Interpretation of results

After close interpretation of the above graphs, the enhancement of the RS occurred for a high period of isolation and a high period of understructure, meaning the more flexible the isolation and underground structure is, the more the displacements are decreased (as can be seen in Figs 6–8). Moreover, we observe that a high mass ratio (mass of understructure larger than that of the superstructure) with a flexible underground mass enhances greatly the results. We can also deduce that results are enhanced when the understructure is half the mass of the superstructure and has a period relatively close to that of the earthquake (TR_{und} = 1 which means $T_{und}/T_{eq}=1$).

On the other hand, keeping a larger mass of understructure relatively flexible is tricky. More thought needs to be given to achieve the desired decrease in the response spectrum in cases with a larger mass of understructure while maintaining flexibility or rendering it even more flexible (high T_{und}) needs to be given more thought to come up with ways that help in order to get the required decrease in the response spectrum.

An overall interpretation is that by using the earthquake bracelet we are actually transforming the underground stories into a complex base isolation system. This would be consistent with our results indicating that the understructure must be as heavy and as flexible as possible, as this is precisely the profile of a base isolation layer.

6 CONCLUSIONS

The hypothesis and exploration in this report is a preliminary study. In order to proceed further in the study and to provide a pathway for further development, some conclusions and recommendations are presented:

- The preliminary study is encouraging and indicates that the method is feasible.
- The concept can be conceptualized as utilizing underground stories and their surroundings as a sophisticated base-isolation system.
- To have the understructure behave as a base-isolation system, we require (i) a flexible bracelet, (ii) relatively flexible underground stories and (iii) preferably, higher relative mass of understructure over superstructure.

Our recommendations for further study are as follows:

- This study looked at the superstructure and substructure as each being a single mass. This would not capture possible mode interactions between multiple superstructure modes and multiple substructure modes. In future, we propose segmenting both the superstructure and substructure to investigate this further.
- We need to investigate the response spectra under multiple types of earthquakes, both real and simulated, and not just due to a sinusoidal applied ground motion.
- Find methods to implement the design implications on the bracelet and the substructure. Specifically, how to make a well-supporting bracelet while being highly flexible and energy absorbent perhaps by studying several types of elastomers (their dampening characteristics and flexibility) and how to make a relatively heavy yet flexible underground, such a system may be implemented.
- Designing example buildings to verify the calculations in a full setting.
- Experimental explorations using shake tables and model buildings.
- The Earthquake Bracelet should also be considered in regularly designing buildings against earthquake loads and not just for retrofitting. In particular, this approach may be very significantly cost effective and simple relative to other options.

REFERENCES

[1] Kelly, J., *Base Isolation in Japan: 1988* (No. UCB/EERC-88/20), Earthquake Engineering Research Center, University of California at Berkeley, 1998.
[2] Clark, P., Aiken, D., Nakashima, M., Miyazaki, M. & Midorikawa, M., The 1995 Kobe earthquake as a trigger for implementing new seismic design technologies in Japan. *Lessons Learned Over Time, Learning From Earthquakes, vol. III*, Earthquake Engineering Research Institute, 1999. www.siecorp.com/lfe/index.html.
[3] Buchanan, A.H., Bull, D., Dhakal, R., MacRae, G., Palermo, A. & Pampanin, S., A report written for the royal commission of inquiry into building failure caused by the Canterbury earthquakes (No. 2011-02), University of Canterbury, 2011.
[4] Seismic Isolation Engineering, Inc., Database of U.S. Seismically Isolated Buildings. Unpublished manuscript, 1999.
[5] De la Cruz, S.T., Rodríguez, M.A. & Hernández, V., *Using Spring-Mass Models to Determine the Dynamic Response of Two-Story Buildings Subjected to Lateral Loads*, Autonomous University of Ciudad Juárez, 2012.
[6] Earthquake-response spectra, Fundamental of Vibration (pp. 53). Lecture based on Chopra, A.K. (ed)., Earthquake response of linear systems. *Dynamics of structures, Theory and Applications to Earthquake Engineering*, 3rd ed., Pearson, p. 197, 2006.

TSUNAMI SHOALING THEORY

KALLE M. LAMPELA
Lamprotek Oy, Finland

ABSTRACT

When a kilometers-long tsunami wave approaches the coast, the amplitude of the tsunami wave will increase according to Green's law, depending on the depth of the seawater. Near the shoreline, the increasing tsunami wave sucks from the front seawater and it will withdraw. The sea bottom will open, creating a drawback phenomenon. The drawback flow and the propagating wave front will collide and will form a wave with a sharp drop downwards. The tsunami break wave starts flowing rapidly to the shoreline. To simulate the situation, we can imagine a virtual dam break (theory by Ritter). As a result, we can use the rules for tsunami wave break theory to calculate an emerging massive flow. The velocity of the break wave can be obtained according to the "dam wall break flow" theory. By these rules, the flow velocity, the parabolic form of flowing water surface, the depth of the sea water at shoreline, the duration, the level of the maximum run-up height onshore and flow loading for buildings can all be calculated. The whole aforementioned shoaling process is presented by easily understandable figures. A simplified calculation example is presented based on the Asian tsunami of 2004 in Xaaphuun in Somalia, from which we have a lot of input data from 0.25-degree-shallow shore. The third and strongest drawback of the tsunami was 1300 m long and 6 m deep. Knowing the shoreline velocity (11 m/s) and estimating the flow friction, we can calculate the depth of water at shoreline, the volume flow, the timing and the average run-up height onshore. The results were in accordance with real data. Also, an explanation will be formulated for the reason why the city of Faro was spared in the Lisbon earthquake and tsunami in 1755. The tsunami shoaling process includes shoreline approaching increasing tsunami wave, drawback, collision of flows, virtual dam break, massive flow to the shoreline and run-up onshore.

Keywords: tsunami, shoaling, drawback, dam break wave, kinetic energy, run-up.

1 INTRODUCTION

A tsunami consists of three phases: tsunami generation, propagation and shoaling. The velocity of the propagation of the long-distance tsunami in the constant sea depth is:

$$U = \sqrt{Dg}, \tag{1}$$

where D is seawater depth and acceleration due to gravity is $g = 9.81$ m/s^2. The propagation velocity will be, in seawater depth of 4 km, 713 km/h and in seawater depth of 10 m, theoretically only 36 km/h.

The tsunami propagation theory for constant sea depth is linear and not proper for shoaling of the rising seafloor, where the theory has been said to be nonlinear [1], [3], [11], [13]. Nonlinear theories lead to mathematical difficulties without practical results. When a tsunami wave front is approaching a shallow shore, the wave height will increase with rising seabed according to Green's law [2]:

$$\frac{H1}{h1} = \left(\frac{d1}{D1}\right)^{0.25}, \tag{2}$$

where H_1 and h_1 are wave heights and D_1 and d_1 are corresponding water depths (Fig. 1). Waves get slower, shorter and higher. The general mistake is to assume wave's continuity at the coastline.

WIT Transactions on The Built Environment, Vol 185, © 2019 K. M. Lampela published under CCBY4.0
www.witpress.com, ISSN 1743-3509 (on-line)
doi:10.2495/ERES190101

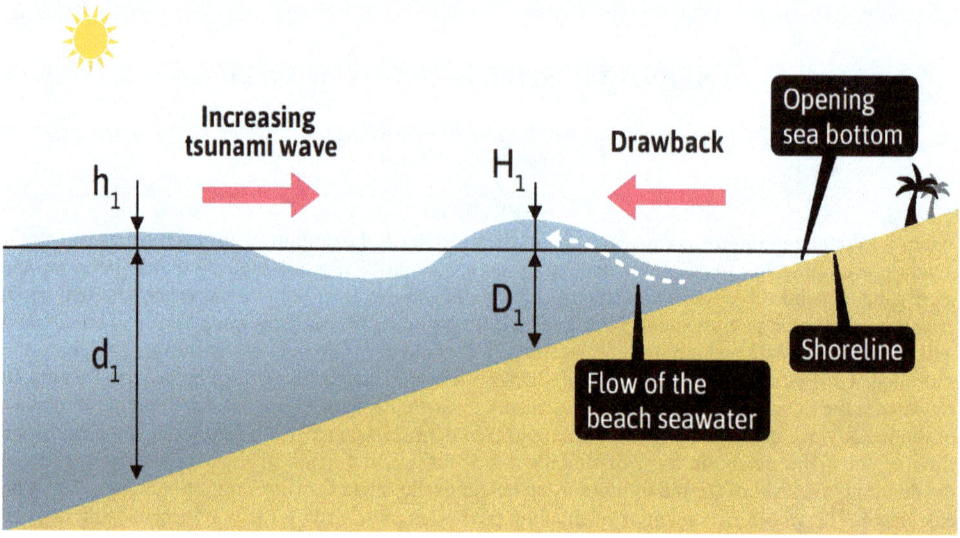

Figure 1: Tsunami wave rises with drawback.

In the shoaling theory it is very important to know the details of flow for saving human lives and preventing damages to property. A good theory also helps to find correct rules and dimensions of the shoreline buildings and structures.

The novelty in this article is that the tsunami waves coming from the ocean cannot reach the shallow shore because of the water drawback and tsunami waves' break according to dam break theory. After breaking the water masses are flooding to the shore causing great disaster.

First, a shoaling theory for the shallow shore, where the seabed slope will be some degrees, will be presented, then for the deep shore and finally for a very deep shore. In the following, the principles of the "Tsunami Shoaling Theory" will be demonstrated with figures and some general equations and calculation examples.

2 SHOALING OF THE VERY SHALLOW SHORE

Shoaling can be divided into two phases: (1) the tsunami wave's water rises on the offshore due to the drawback, and (2) the tsunami wave's water breaks resulting in a huge water flow on the dry seabed to shoreline.

When tsunami waves approach ocean shoreline, seawater first withdraws and water flows seaward forming the drawback. The volume of drawback water could be more than 1 million m^3, which means 1 million tons of seawater. The amount of seawater depends on how much energy the tsunami wave has.

The first and partly second wave lose tsunami wave's energy by lifting seawater from the sea level while approaching the shore [16]. The first wave has a low drawback and it will be detected only in very shallow shore or beach [4], [12]. The second wave loses also wave energy and generally the third wave is the strongest one. Often only two tsunami waves can be detected.

In tsunami drawback, seawater will flow to the open ocean and will hit to the edge of the kilometers-long tsunami wave. In the collision there will be a sharp drop of the seawater as in the waterfall (Fig. 2).

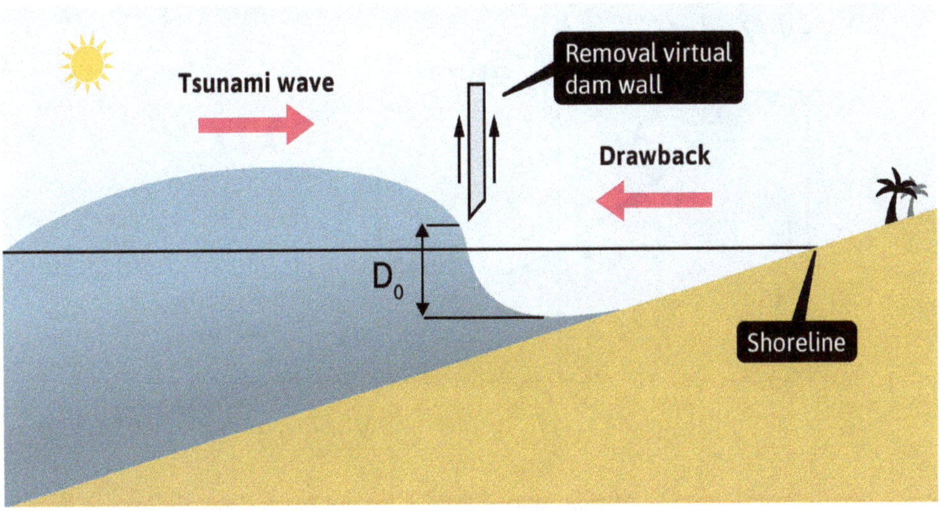

Figure 2: Tsunami wave breaks as a flow from dam wall break.

In this situation the water will approximately behave like water of a dam wall break. The water flows as a flood wave to the shore at velocity V according to the dam break theory [5], [6]. According to dam break theory, flood velocity to plain shore is given by:

$$V = 2\sqrt{D_0 g}, \tag{3}$$

where V is the flood flow velocity in m/s, D_0 is total water drop height in m and g acceleration due to gravity in m/s^2. If the total drop including the height of drawback and tsunami is altogether $D_0 = 5$ m, the shoreward velocity on the sea in the breaking point is $V_1 = 14$ m/s (50 km/h). The velocity of the negative wave is half of the positive wave $V_2 = V_1/2 = 7$ m/s (25 km/h).

The water flow of the rising seabed will be similar to an open channel flow upwards. The frictionless ideal dam break flow equation (Fig. 3) is:

$$\frac{x}{t\sqrt{gD_0}} = 2 - 3\sqrt{d/D_0}, \tag{4}$$

where x is the horizontal positive distance from breakpoint, D_0 the height of waterfall drop, d the water depth at time t. The dam break wave has also a negative wave. The breaking wave profile is parabolic. The wave tip region [7], [8] can be noticed in many tsunami videos as front turbulence edge (Fig. 4). The velocity of the open channel flow upwards is given by:

$$\frac{V}{\sqrt{gD_0}} = 2 + (S_0 - S_f)\sqrt{\frac{g}{D_0}}t, \tag{5}$$

where $S_0 = \sin \Theta$, positive downwards, negative upwards, and S_f is friction.

Point C in Fig. 3 has, during flow, constant height h_C, where flow velocity is:

$$V_C = (2/3)\sqrt{gD_0}, \tag{6}$$

and volume flow:

$$Q = 8/27 D_0\sqrt{gD_0}. \tag{7}$$

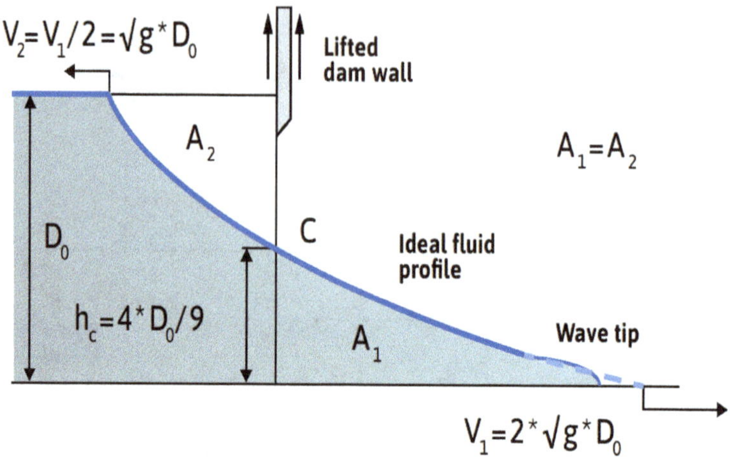

Figure 3: Visualization of the dam wall break flow.

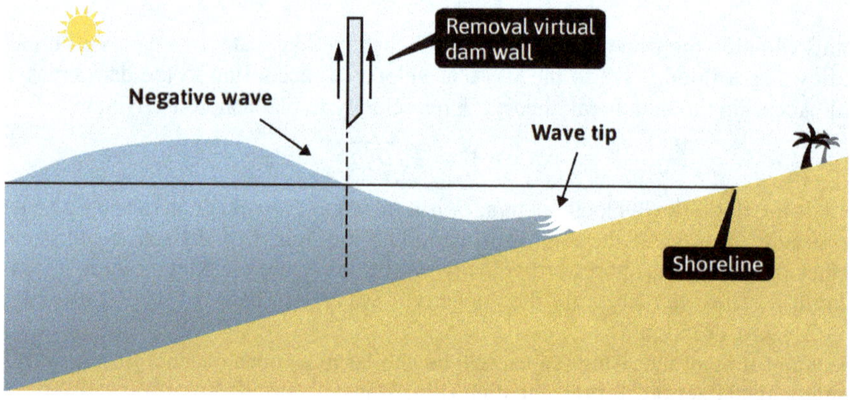

Figure 4: Tsunami break wave flow to the shoreline.

3 EXAMPLES

3.1 Asian tsunami 2004 in Africa

As a first example, the 2004 Asian tsunami in Xaafuun, a Somalian city in the Horn of Africa, is considered [4]. The initial conditions were: drawback of the third wave was 1300 m, depth 6 m, shoreline water velocity 11 m/s and slope of rising seafloor 0.25 degrees.

If we assume the height of 1.7 m for a kilometer-long tsunami wave on the coast area, we can simulate tsunami shoaling. When the drop of drawback and wave height together is $D_0 = 6{+}1.7 = 7.7$ m, the dam break velocity is $V = 17.4$ m/s.

Seawater flow slows down in about 2 minutes to the shoreline at the velocity of 11 m/s. The time to flow stopping on shore is about 4 minutes, the run-up height is 4 m about 0.8 km from shoreline. In simulation it is assumed a 0.25-degree slope onshore and a friction factor of $S_f = 0.0025$ (Fig. 5). The calculation result depends strongly on seabed friction [10].

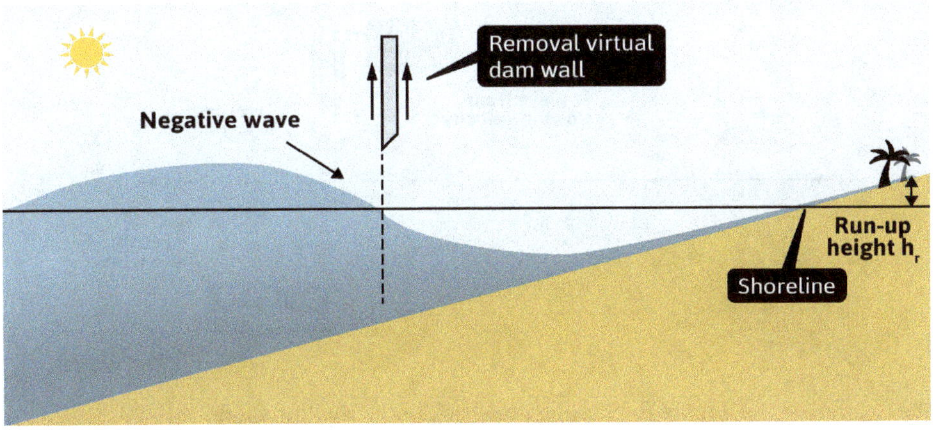

Figure 5: Tsunami break wave flow to run-up height.

The depth of the water flow at shoreline during run-up was 0.85 m, which at a water velocity of 11 m/s means 9.35 m^3/s per 1-m shoreline. For the half-kilometer-long shoreline in Xaafuun, this means nearly 20 million m^3/h and tons of water. It is understandable, that the walls of houses along the shoreline break due to the fast-moving seawater. According to the simulation the seawater flow on shoreline was supercritical with Froude number $F_r = 3.8$. The developed kinetic energy can be dissipated, for example, by "a hydraulic jump".

3.2 Portugal south coast tsunami 1755

As a second example we consider the earthquake of 1755 which hit the city of Faro in southern Portugal [9]. While the neighboring cities of Olhao and Tavira experienced the destructive tsunami after the earthquake, the city Faro was unaffected by the tsunami. Faro, in fact, was protected by wetland known nowadays as the Ria Formosa nature park with sand banks, barrier islands and marches, which caused a big dissipation of the kinetic tsunami energy. Faro survived the tsunami, which was a miracle at that time of history.

4 SHOALING OF THE SHALLOW SHORE

When the seawater depth is from 10 m to 30 m, the velocity of the approaching tsunami wave is lower and the wave starts to suck water from the front lifting the height of the tsunami wave (Fig. 6). Depending on the tsunami wave's lower velocity, the phenomenon of the negative wave drawback starts to grow. The flow of the virtual dam break directs to the open ocean. The drawback velocity must be faster than the wave velocity ($V_d > V_{wave}$) and the drawback phenomenon starts (Fig. 7). In the case of deep shore, no seawater withdraws at the shoreline.

The flow process continues until the drawback touches the seabed or the shoreline. The flow of the virtual dam break will start at the velocity (Fig. 8):

$$V = 2\sqrt{gD_t}, \tag{8}$$

where $D_t = H + h$.

The drawback phenomenon changes the vertical movement of the seawater to a horizontal flood wave. This is the fundamental phenomenon in the tsunami shoaling due to the horizontal flow.

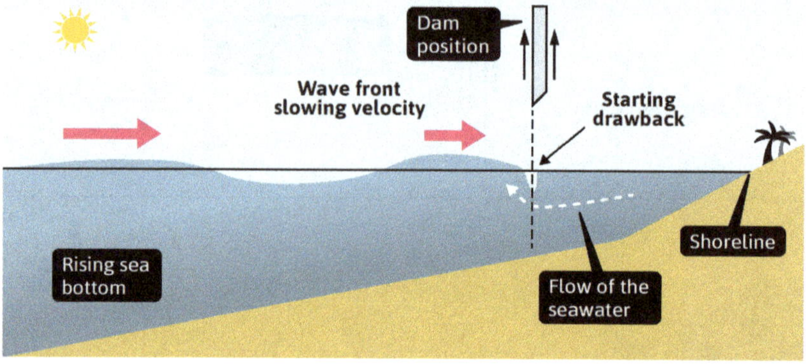

Figure 6: Starting drawback of the shallow shore.

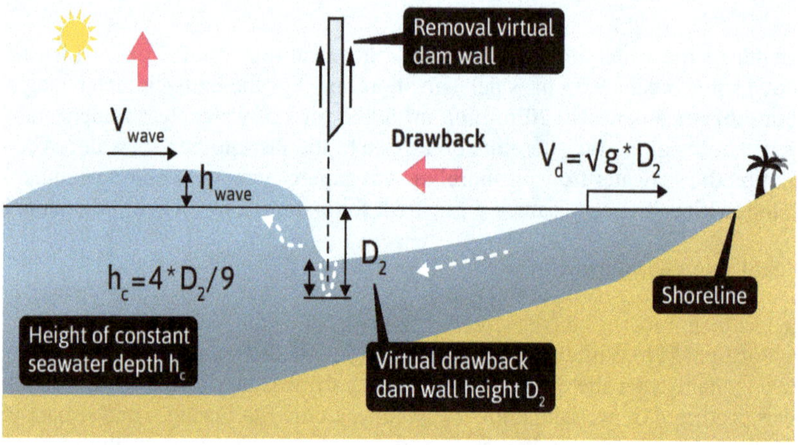

Figure 7: Drawback of the virtual dam break flow to the open ocean.

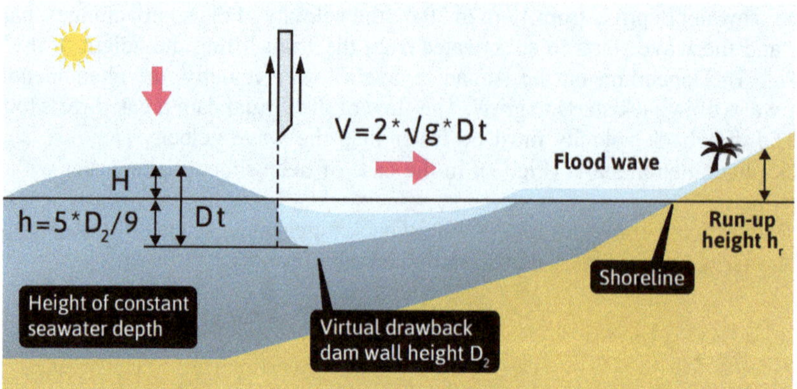

Figure 8: Tsunami flood wave of the shallow shore.

5 SHOALING OF THE DEEP SHORE

Very deep shore – depth over 30 m – does not have drawback and no horizontal flood and only some meters rise of the seawater because of the high velocity of the tsunami wave.

The drawback phenomenon reduces energy of the first tsunami wave and partly second wave. The third tsunami wave is the biggest one. The first wave is weak and can be seen only on a very shallow shore or beach. That is why many times only two tsunami waves will be detected.

In the open ocean, the tsunami wave height is very low; because of the high velocity and long length of the wave, there is no possibility for sucking enough seawater from the wavefront.

6 CONCLUSIONS

Tsunami shoaling theory explains in a simple and understandable way the whole tsunami shoaling process: increasing tsunami wave, drawback, collision of tsunami wave and drawback flow, virtual dam break, massive seawater flow to the shoreline and finally run-up flow onshore.

The results of calculation rules are velocity, depth, volume, kinetic energy, run-up height and the exact times of the shoaling waves. Simulation of different levels of tsunamis is simple by changing the height of seawater drop on the ocean.

With the help of this theory one can innovate and calculate different methods to mitigate or prevent tsunami damages and save human lives. The impact of the kinetic water forces on the coastline buildings and structures [14], [15] can be calculated and simulated according to present shoaling theory.

The shoaling theory presented here does not support the prevailing shoaling theory of the tsunami waves. These tsunami waves are coming to the shallow shore by keeping the shape of the original tsunami waves up to shoreline. In the presented shoaling theory, a tsunami wave will break before the shoreline according to dam break flow theory.

The accuracy of the presented theory depends strongly on the seabed bathymetry and onshore topography. A rough seabed can slow down the coastline flow velocity by 20–30%. The presented theory is based on plain rising seabed because of the corresponding simplicity of the shoaling theory. The most important thing is to find the main influencing general variables that affect the tsunami shoaling. Knowing the most important variables and local conditions can dissipate and prevent damages of the tsunami.

The seawater movement of the tsunami wave is only vertical in the open ocean. When the tsunami wave approaches the shallow shore, the seawater flow will become strongly horizontal. This phenomenon is clearly noticeable in the photos and videos of the tsunamis on the internet.

The tsunami shoaling theory of the very shallow, shallow and deep shore has been described separately.

The seawater flows and kinetic forces on the shore structures and buildings can be simulated by the simple calculation rules with varying tsunami intensities.

The key point presented in the theory is the combination of the tsunami and the dam break flow theory in a creative way.

REFERENCES

[1] Ward, S.N., Tsunami. *Encyclopedia of Solid Earth Geophysics*, Springer Press, 2010.

[2] Green. G., On the motion of waves in a variable canal of small depth and width. *Transactions of the Cambridge Philosophical Society*, **6**, 457–462, 1838.

 © 2019 K. M. Lampela published under CCBY4.0
www.witpress.com, ISSN 1743-3509 (on-line)

[3] Levin, B.W. & Nosov, M.A., *Physics of Tsunamis*, 2nd ed., Springer International Publishing, 2016.

[4] Fritz, H.M. & Borrero, J.C., Somalian field survey after December 2004 Indian Ocean Tsunami. *Earthquake Spectra*, **22**(3), 219–233, 2006.

[5] Chanson, H., *Environmental Hydraulics of Open Channel Flows*, Elsevier Butterworth-Heinemann, 2004.

[6] Chanson, H., Analytical solution of dam break wave with flow resistance. Application to tsunami surges. *Proceedings of the 31st IAHR Biennial Congress*, pp. 3341–3353, Seoul, 2005.

[7] Chanson, H., Analytical solutions of laminar and turbulent dam break wave. *Proceedings of the International Conference on Fluvial Hydraulics*, vol. 1–2, River Flow, 2006.

[8] Chanson, H., A simple solution of the laminar break wave. *Journal of Applied Fluid Mechanics*, **1**(1), pp. 1–10, 2008.

[9] Nunes, M., Ferreira, O. & Luis, J., Tsunami vulnerability zonation in the Algarve coast (Portugal). *Journal of Coastal Research*, **1**(56), pp. 876–880, 2009.

[10] Conde, D.A.S., Babtista, M.A.V., Sousa Oliveira, C. & Ferreira, R.M.L., A shallow-flow model for the propagation of tsunamis over complex geometries and mobile beds. *Natural Hazards and Earth System Sciences*, **13**(10), pp. 2533–2542, 2013.

[11] Aydin, B. & Kanoglu, U., Analytical solution for nonlinear shallow-water wave equations. *Pure and Applied Geophysics*, **174**(8), pp. 3209–3218, 2012.

[12] Lampela, K.M., Theory of tsunami multi drawback and inundation on the coastline. Presented at *15th World Conference on Earthquake Engineering*, Lisboa, 2012.

[13] Lampela, K.M., *Pyörähdyssymmetrisen Rakenteen Epälineaarinen Lämpöjännitysanalyysi Elementtimenetelmällä* (Nonlinear Thermal Stress Analysis of the Axisymmetric Vessel by Finite Element Method), Master's Thesis, University of Oulu, 1976.

[14] Chock, G.Y.K., The new chapter 6 on tsunami loads and effects in ASCE 7-16. *Journal of Structural Engineering*, **142**(11), 2015.

[15] Mase, H., Kimura, Y., Yamakawa, Y., Yasuda, T., Mori, N. & Cox, D., Were coastal defensive structures completely broken by an unexpectedly large tsunami? A field survey. *Earthquake Spectra*, **29**(S1), pp. S145–S160, 2013.

[16] Coastal Carolina University, Oceans in Motion: Waves and Tides, Coastal Carolina University. https://ci.coastal.edu/~sgilman/770Oceansinmotion.htm.

Author index

WIT*PRESS* ...for scientists by scientists

Computational and Experimental Methods in Multiphase and Complex Flow X

Edited by: ***S. HERNÁNDEZ****, University of A Coruña, Spain, and* ***P. VOROBIEFF****, University of New Mexico, USA*

Composed of papers presented at the 10th conference on Multiphase flow this book presents the latest research on the subject. The research included in this volume focuses on using synergies between experimental and computational techniques to gain a better understanding of all classes of multiphase and complex flow.

The presented papers illustrate the close interaction between numerical modellers and researchers working to gradually resolve the many outstanding issues in our understanding of multiphase flow.

Recently multiphase fluid dynamics have generated a great deal of attention, leading to many notable advances in experimental, analytical and numerical studies. Progress in numerical methods has permitted the solution of many practical problems, helping to improve our understanding of the physics involved.

Multiphase flows are found in all areas of technology and the range of related problems of interest is vast, including astrophysics, biology, geophysics, atmospheric process, and many areas of engineering.

The papers in the book cover a number of topics, including: Experimental measurements; Numerical methods; Multiphase flows; Flow in porous media.

ISBN: 978-1-78466-329-2 eISBN: 978-1-78466-330-8
Published 2019 / 250pp

All prices correct at time of going to press but subject to change.
WIT Press books are available through your bookseller or direct from the publisher.

www.ingramcontent.com/pod-product-compliance
Lightning Source LLC
LaVergne TN
LVHW062218200926
848484LV00006B/1422
* 9 7 8 1 7 8 4 6 6 3 3 3 9 *